Mechanisms of Continental Drift and Plate Tectonics

Proceedings of a NATO Advanced Study Institute held at the University of Newcastle upon Tyne, March/April 1979

Mechanisms of Continental Drift and Plate Tectonics

Edited by

P. A. DAVIES and S. K. RUNCORN, *F.R.S.*

School of Physics, The University, Newcastle upon Tyne

1980 · ACADEMIC PRESS

A Subsidiary of Harcourt Brace Jovanovich, Publishers

London · New York · Toronto · Sydney · San Francisco

ACADEMIC PRESS INC. (LONDON) LTD
24/28 Oval Road
London NW1

United States Edition published by
ACADEMIC PRESS INC.
111 Fifth Avenue
New York, New York 10003

British Library Cataloguing in Publication Data

Mechanisms of continental drift and plate tectonics.
1. Plate tectonics—Congresses
I. Davies, P A II. Runcorn, Stanley Keith
551.1'3 QE511.4 80–41582

ISBN 0–12–206160–8

Printed in Great Britain by
John Wright and Sons Ltd. at The Stonebridge Press, Bristol BS4 5NU

Contributors

J. Achace Institut de Physique du Globe, Place Juisseau, 75230 Paris Cedex 05, France

C. A. Anderson Los Alamos Scientific Laboratory, University of California, Los Alamos, New Mexico 87545, USA

E. J. Barron National Center for Atmospheric Research, P.O. Box 3000, Boulder, Colorado 80307, USA

C. Beaumont Department of Oceanography, Dalhousie University, Halifax, Nova Scotia B3H 3J5, Canada

R. J. Bridwell Los Alamos Scientific Laboratory, University of California, Los Alamos, New Mexico 87545, USA

J. A. Canas Universidad C. de Madrid, Facultad de C. Fisicas, Cátedra de Geofisica, Madrid (3), Spain

A. M. Correig Departmento de Fisica de la Tierra y del Cosmos, Facultad de Fisica, Universidad de Barcelona, Barcelona, Spain

V. Courtillot Institut de Physique du Globe, Place Juisseau, 75230 Paris Cedex 05, France

M. Darot Laboratoire de Tectonophysique, 2 Chemin de la Houssinière, 44072 Nantes Cedex, France

P. A. Davies Department of Civil Engineering, The University, Dundee DD1 4HN, Scotland

D. J. Doornbos Vening Meinesz Laboratory, University of Utrecht, Lucasbolwerk 7, Utrecht 2501, The Netherlands

J. Ducruix Institut de Physique du Globe, Place Juisseau, 75230 Paris Cedex 05, France

M. A. Forrest School of Physics, The University, Newcastle upon Tyne NE1 7RU, UK

S. J. Freeth Department of Geology, University College of Swansea, Singleton Park, Swansea SA2 8PP, Wales

R. Freund Institute of Earth Science. The Hebrew University of Jerusalem, Israel

R. W. Girdler School of Physics, The University, Newcastle upon Tyne NE1 7RU, UK

T. Gold Center for Radiophysics and Space Research, Cornell University, Ithaca, New York 14853, USA

Y. Guegen Laboratoire de Tectonophysique, 2 Chemin de la Houssinière, 44072 Nantes Cedex, France

B. H. Hager Seismology Laboratory, California Institute of Technology, Pasadena, California 91109, USA

C. G. A. Harrison Rosenstiel School of Marine and Atmospheric Science, University of Miami, 4600 Rickenbacker Causeway, Miami, Florida 33149, USA

W. R. Jacoby Institut für Meteorologie und Geophysik der Johann Wolfgang Goethe Universität, 6000 Frankfurt, West Germany

U. Kopitzke Institut für Geophysik und Meteorologie, Technische Universität, D 3300 Braunschweig, West Germany

D. Kosloff Institute of Earth Science, The Hebrew University of Jerusalem, Israel

R. Krishnamurti Department of Oceanography and Geophysical Fluid Dynamics Institute, Florida State University, Tallahassee, Florida 32306, USA

F. A. Kulacki Department of Mechanical and Aerospace Engineering, 107 Evans Hall, University of Delaware, Newark, Delaware 19711, USA

J. L. Le Mouël Institut de Physique du Globe, Place Juisseau, 75320 Paris Cedex 05, France

A. Matthews Institute of Earth Science, The Hebrew University of Jerusalem, Israel

R. O. Meissner Institut für Geophysik, Neue Universität, D2300 Kiel, West Germany

B. J. Mitchell Department of Earth and Atmospheric Sciences, St Louis University, St Louis, Missouri 63156, USA

R. J. O'Connell Department of Geological Sciences, Harvard University, Cambridge, Massachusetts 02138, USA

A.-T. Nguyen Department of Mechanical Engineering, The Ohio State University, Columbus, Ohio 43210, USA

G. F. Panza Istituto di Geodesia e Geofisica, Università de Bari, 70122 Bari, Italy

H. N. Pollack Department of Geological Sciences, The University of Michigan, Ann Arbor, Michigan 48109, USA

A. Poma Istituto di Astronomia dell'Universita di Cagliari, Via Ospedale 72, 09100 Cagliari, Italy

E. Proverbio Istituto di Astronomia dell'Università di Cagliari, Via Ospedale 72, 09100 Cagliari, Italy

D. M. Rayburn School of Physics, The University, Newcastle upon Tyne NE1 7RU, UK

C. Relandeau Laboratoire de Tectonophysique, 2 Chemin de la Houssinière, 44072 Nantes Cedex, France

S. K. Runcorn School of Physics, The University, Newcastle upon Tyne NE1 7RU, UK

G. Schubert Department of Earth and Space Sciences, University of California, Los Angeles, California 90024, USA

R. Stephenson Department of Geology, Dalhousie University, Halifax, Nova Scotia B3H 3J5, Canada

J. F. Strehlau Institut für Geophysik der Christian Albrechts Universität, D 23 Kiel, West Germany

D. H. Tarling School of Physics, The University, Newcastle upon Tyne NE1 7RU, UK

D. J. Tritton School of Physics, The University, Newcastle upon Tyne NE1 7RU, UK

D. L. Turcotte Department of Geological Sciences, Cornell University, Ithaca, New York 14853, USA

U. R. Vetter Institut für Geophysik der Christian Albrechts Universität, D 23 Kiel, West Germany

J. Woirgard E.N.S.M.A., rue Guillaume VII, 86034 Poitiers Cedex, France

Preface

The question first considered by Wegener concerning the nature of the processes in the Earth bringing about continental drift and sea-floor spreading (or major displacements of the lithosphere) in the last 100–200 myr, is probably the most important one in geoscience at the present time. The concept of plate tectonics has integrated the evidence for continental drift, sea-floor spreading and global seismicity in a way most stimulating for geology and geophysics. The process has been rightly called by J. T. Wilson the "revolution in the Earth sciences." The global distribution of earthquakes, the world oceanic ridge system, the global pattern of the Tertiary mountain building and the ocean trenches are all generally understood in terms of the parting or collision of plates. Plate tectonics, however, is kinematic and geometrical and is not a theory in the sense of identifying the forces responsible for the plate motions. Thus, while the solution of the geometrical relationships of plate tectonics has absorbed the energies of geophysicists and especially geologists (who have reinterpreted the geological record in the new way, with stimulating effects in the subject), and dynamical questions have been discussed in lively speculative papers, no consensus of opinion has been reached.

We thought it timely to arrange a meeting to examine critically the various theories which have been put forward: gravity sliding of the plates away from the ocean ridges, the pulling of the plates by the sinking of lithospheric slabs in the trenches, mantle plumes, tidal effects, general expansion of the Earth and thermal convection in the mantle. The validity of the physicochemical processes to which these theories appeal need critical examination and the ideas now require formulation in terms of fundamental physics and chemistry. The geophysical and geological evidence which may be used to test them must also be examined thoroughly.

The geotectonics of the geological record prior to Wegenerian continental drift and especially in the Precambrian must be considered. Although the speed and nature of plate movements in recent times are so well documented by the palaeomagnetic record in the ocean floor, no theory of plate motions is satisfactory unless it explains the Earth history record prior to Wegenerian drift – over 20 times as long. Nor can geoscience today ignore the evidence brought back from space missions to the Moon and planets. How does the very different tectonic history of the Moon, Mars and Mercury fit in with ideas developed for the Earth? Does the evidence from Venus reveal continental drift there? Can a theory of continental drift which requires a dynamical model of the mantle provide a basis for understanding?

The Advanced Study Institute on "The mechanism of continental drift and plate tectonics" was held in the School of Physics from Tuesday 27th March until Tuesday 10th April, 1979, and this volume includes a full and representative collection of papers given at the meeting. We are most grateful to NATO who provided generous financial support to the Institute. Such support enabled a large group of Earth scientists (particularly graduate students) from many countries to attend. Their active participation assured the success of the meeting. Not least, we acknowledge the contributions made to the organization of the Institute by Mr W. F. Mavor, Mrs J. Roberts, Mrs D. Orton, Miss Anne Codling and Miss Maureen Hopkinson.

During the preparation of this volume we were all saddened by the death of Dr Raphael Freund – a distinguished Earth scientist who contributed most enthusiastically to the scientific and social activities of this meeting.

Raphael Freund was born in Breslau, Germany and after six months his family emigrated to Israel. After graduating with distinction from high school in Haifa he went on to the Hebrew University in Jerusalem to read geology, obtaining his M.Sc. in 1959 and his Ph.D. in 1963. He was to remain based at the Hebrew University for the rest of his life, becoming successively Instructor (1959), Lecturer (1963), Senior Lecturer (1966), Associate Professor (1972), Full Professor (1977) and first Director of the Institute of Earth Sciences.

As he lived almost on top of the Dead Sea rift, it is not surprising that this became his main object of study. He searched for and found an enormous amount of geological data for elucidating the timing of the shear movements along the Dead Sea rift zone. He was an enthusiastic field geologist and it was fortunate that he was still well enough to be able to show a great deal of the evidence for horizontal shear to the international gathering last year. Further afield, he made contributions to the studies of strike slip faults in New Zealand, and in his last year worked on dynamic models of subduction zones.

November 1980

P. A. Davies
S. K. Runcorn

Contents

The Mechanisms of Continental Drift and Plate Tectonics: Some Boundary Conditions from Surface Phenomena

R. W. GIRDLER

School of Physics, The University, Newcastle upon Tyne, UK

Introduction

The object of this introductory chapter is to present some boundary conditions which may be inferred from surface phenomena for theories of mechanisms of continental drift and plate tectonics.

In 1914, J. Barrell introduced the concept of "lithosphere" and "asthenosphere" for the outermost layers of the Earth. The "lithosphere" was considered to be a shell with high strength overlying the "asthenosphere" with relatively very low strength.

It is now generally accepted that the outer skin of the Earth consists of a number of thin, relatively rigid plates or spherical caps. The margins of these plates are seismically active as a consequence of their moving relative to each other. The seismicity of the Earth, therefore, may be used to give an idea of the size and shape of the plates. Figure 1 shows two faces of the Earth. The regions of deep focus earthquakes (shown as black) are somewhat restricted and mark where the plates are descending into the asthenosphere ("subduction zones") or where the plates are colliding. These regions, of course, also have shallow seismicity. The regions where *only* shallow earthquakes occur locate the rifts (i.e. where the plates are separating) and transform faults (i.e. where the plates are sliding past each other). In the oceans, these regions of shallow earthquakes are very narrow but on the continents they tend to be more diffuse (Fig. 1), presumably reflecting the long history of the continental crust and the presence of many faults and fractures.

Size, thickness and fragility of the plates

As can be seen from Fig. 1, the plates have areas of the order of 10^6–10^7 km^2. By contrast, their thickness is very small. Early workers, studying the gravity anomalies over Fennoscandia and eastern Canada, arrived at estimates of less than 100 km. Tozer (1973), in emphasizing the importance of the temperature dependence of creep processes, obtained even lower estimates. From a consideration of heat transfer processes, he obtained estimates for the thickness of the "quasi-elastic" lithosphere of 7 km for the oceans and 39 km for the continents, noting the continental lithosphere is 5–6 times thicker than the oceanic lithosphere.

If we take a working figure of 50 km for the thickness of the plates, and estimates of 1000–5000 km for their length dimensions, we obtain thickness/length ratios of 20/1 to 100/1. The plates are therefore wafer thin and may be expected to be fragile.

Age of the plates

In addition to the contrast in thickness of the continental and oceanic parts of the plates, there is a tremendous contrast in their relative ages. The continental lithosphere has had a very long history and the oceanic lithosphere a very short history.

The continents are composed of light sialic material which has been at the Earth's surface for thousands of millions of years. This is due to the relatively low density, the material being light enough to escape being recycled into the asthenosphere at the subduction zones.

In contrast, the oceanic lithosphere has all formed within the last 200 Myr and the young lithosphere covers more than two-thirds of the Earth's surface. If we take 4600 Myr for the age of the Earth, we see that the present oceanic lithosphere has formed in less than 5% of the history of the Earth. On the time-scale of Earth history, therefore, we have a process of considerable rapidity.

Creation of the oceanic parts of the plates – development of the idea of sea-floor spreading

The way in which the continents break up and the oceanic lithosphere forms is obviously very important. Considerable progress has been made towards understanding these processes, in particular the evolution of oceanic lithosphere by sea-floor spreading. This concept (and the observational support for it) has probably contributed more than anything towards the understanding of the mechanism of continental drift and plate tectonics.

The idea of sea-floor spreading may be attributed to Arthur Holmes, a Northumbrian who was born at Hebburn-on-Tyne (a few miles from Newcastle) in 1890. The idea is simple and elegant. As the continents drift

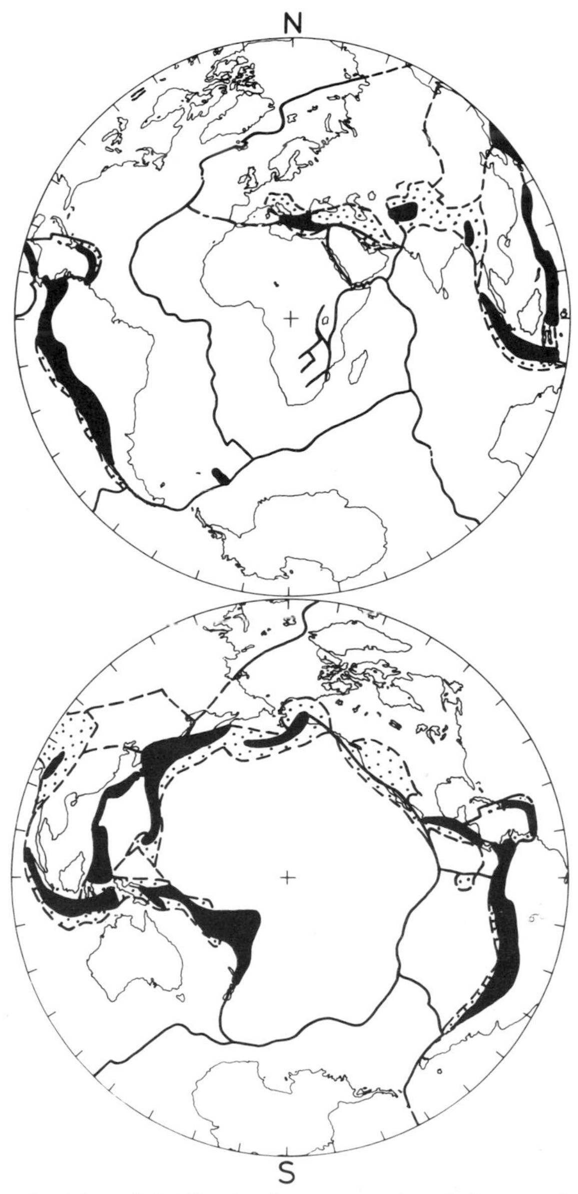

Figure 1. The seismicity of the Earth, shown on azimuthal great circle projections centred on 0°, 20°E and 0°, 160°W and extending for radial distances of 110°. The regions with deep earthquakes (focal depth greater than 100 km) have black shading and the diffuse regions with shallow earthquakes (focal depth less than 100 km) have stippled shading. The narrow regions of shallow seismicity are shown by thick lines.

apart, new ocean is formed by the injection of igneous material at the site of the ever expanding gap. The newest ocean is therefore, found at the middle and the oldest at the margins.

There has been some debate (*J. Geophys. Res.*, 1968) as to whether full credit should be given to Arthur Holmes for what has turned out to be one the most fruitful ideas in the Earth sciences. It is of interest therefore to see how Holmes' ideas evolved. Figure 2 is compounded from two of his publications. Figure 2(a) shows his ideas in the late 1920s; the continents are shown as being carried along by convection currents and an island or "swell" composed of light continental material is shown left behind as an ingenious way of explaining the height of the ridge. The words "new ocean" appear twice between the continents and the island or "swell". Figure 2(b) is taken from Holmes' *Principles of Physical Geology* published fifteen years later in 1944; here the continental relic is replaced by an oceanic island or swell and the words "new ocean" appear only once. By this time studies of rocks dredged from the mid-Atlantic had shown the ridge to be composed of oceanic basalts. The figure caption for the 1944 version describes "ocean floor development on the site of the gap" and leaves little doubt that Holmes fully understood the sea-floor spreading concept. The idea was further developed by Hess and Dietz in the 1960s.

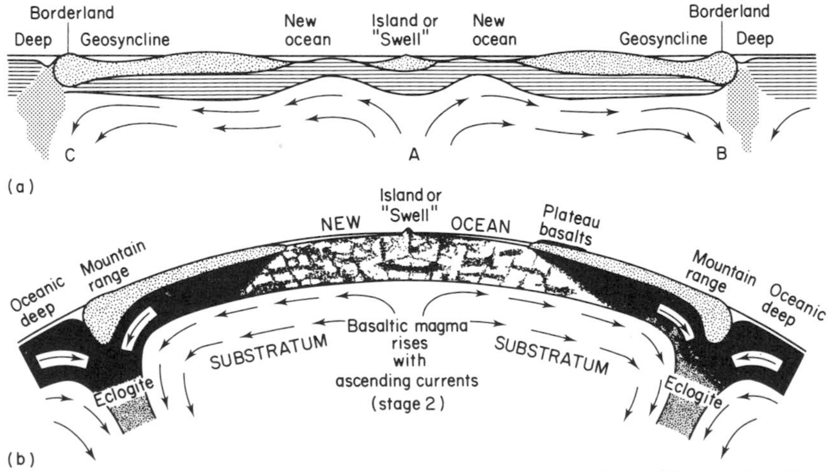

Figure 2. The development of the ideas of sea-floor spreading by (a) Holmes (1929), (b) Holmes (1944).

The remarkably short history of the present day oceanic lithosphere has been inferred from studies of magnetic anomalies following the suggestion by Vine and Matthews (1963) that the sea-floor spreading process should record the reversal history of the Earth's magnetic field. The way in which this happens is depicted in Fig. 3. The early time-scales for reversals were obtained by Vine (1966) and Heirtzler *et al.* (1968) by extrapolation using the radiometric data of Cox, Doell and Dalrymple (1964) for normally and reversely magnetized rocks on land of up to 4 million years old as calibration. The extrapolations, although outrageous, were remarkably successful!

When the observed and computed anomalies are compared, it is possible to deduce the spreading rates as illustrated in Fig. 3. It is found that new oceanic crust is being created at remarkably fast rates of up to 10 cm/yr, the East

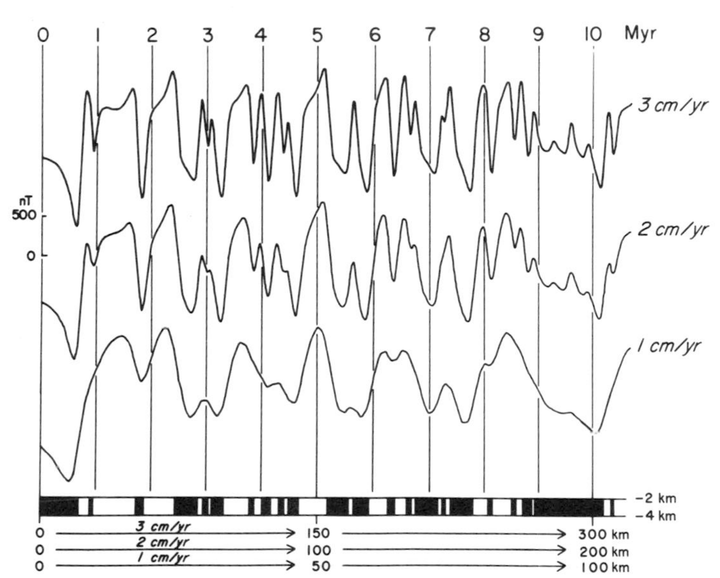

Figure 3. Sea-floor spreading and magnetic anomalies: new sea-floor forms at the left and records the reversal history of the Earth's magnetic field. Once the reversal time-scale has been determined the velocity of the sea-floor spreading can be calculated. The profiles have been stretched to the same lengths and the distance scales varied to demonstrate this. Note that faster spreading rates record more details of the reversal history (as for a tape recorder better fidelity is achieved for higher speeds). The parameters used for these profiles relate to the Gulf of Aden and assume continuous spreading.

Pacific Rise being one of the fastest spreading centres. This sets a further condition on possible mechanisms of continental drift and plate tectonics as presumably the velocities of any driving mechanism must be at least of the order of the magnitudes of the half spreading rates, i.e. greater than about 5 cm/yr.

The reversal time-scale for the Earth's magnetic field has now been constructed back to 160 million years. As mentioned, this is by extrapolation using the radiometric dates of rocks on land over a few million years. It is clearly important to have some check on such extrapolations. This has been done by drilling through the sediments at various locations on the sea-floor and comparing the age of the sediments immediately overlying the magnetic layer with that predicted by the magnetic time-scale. The age of the sediments may be determined independently by examining the fossils contained in them. In nearly all cases the agreement is remarkable, as can be seen from Fig. 4.

Figure 4 shows that the assumption of uniform spreading seems vindicated. In some ways this is not surprising as presumably with the plates having such large areas, the momentum involved in their movement must be considerable. With large momentum, once the plates are set in motion it is likely that they

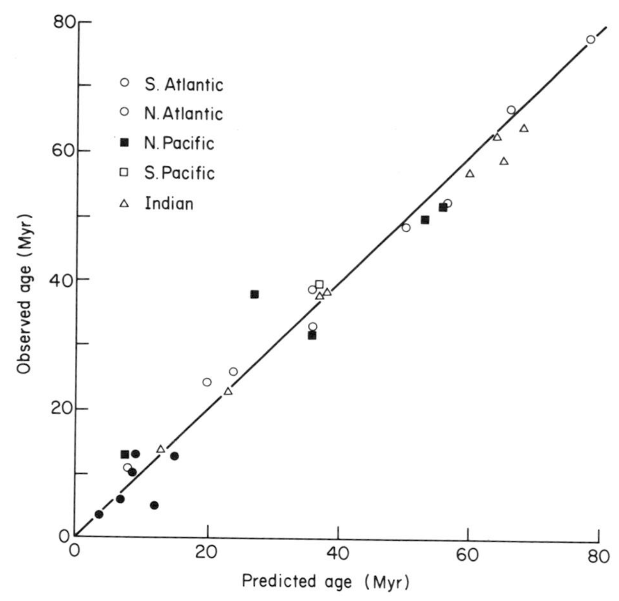

Figure 4. Comparison of the age of oceanic lithosphere as predicted by the magnetic reversal time-scale of LaBrecque *et al.* (1977) and that observed from the age of overlying sediments in Deep Sea Drilling Project (DSDP) boreholes. The 45° line indicates a perfect correlation (Diagram by courtesy of D. J. M. Noy).

will continue to move fairly smoothly. However, evidence is accumulating to suggest that the plate motions may not be so simple.

Possible go–stop–go motion of the plates

A completely different impression is presented by continental geology. Here evidence for the motions of the plates comes mainly from the mountainous regions or collisions zones. Several quite distinct orogenic revolutions have been recognized. To illustrate this, a futher diagram from Arthur Holmes is presented in Fig. 5. It shows graphically how he envisaged the orogenic revolutions to occur with time. Several orogenic phases are seen, some major and some minor, separated by quiet periods of the order of 40 million years. Further, within each major orogenic revolution such as the Alpine, further subdivisions into dynamic and quiet phases can be recognized. This is based

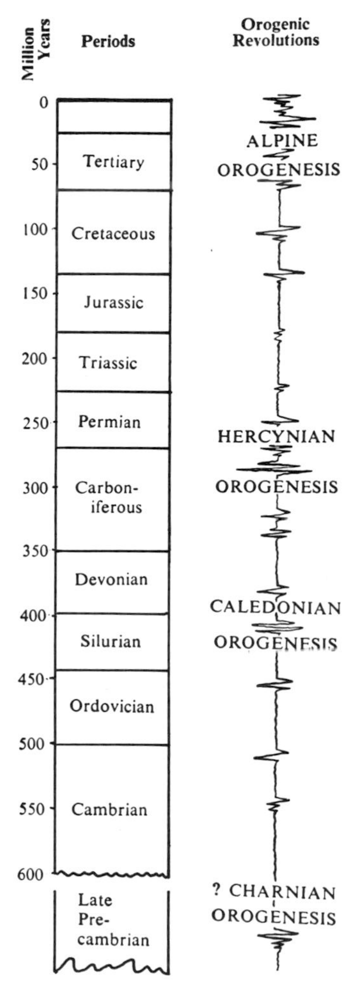

Figure 5. The times of orogenic movements in Europe as envisaged by Holmes (1965).

on the radiometric dating of the igneous activity associated with the mountain building and on studies of the sedimentary record.

Holmes extended this study to include the much larger spate of time represented by the Precambrian and produced a diagram to show the ages of the main radiometrically dated orogenic belts and igneous events through the history of the Earth. The same kind of pattern can be recognized.

The value of this is that it allows an appreciation of thousands of millions of years of Earth history whereas the present oceanic lithosphere is restricted to less than 200 million years. It seems that the plate motions have been going on for most of Earth history and the oceanic lithosphere must have been recycled many times. Any mechanism therefore cannot be of short-lived duration. It

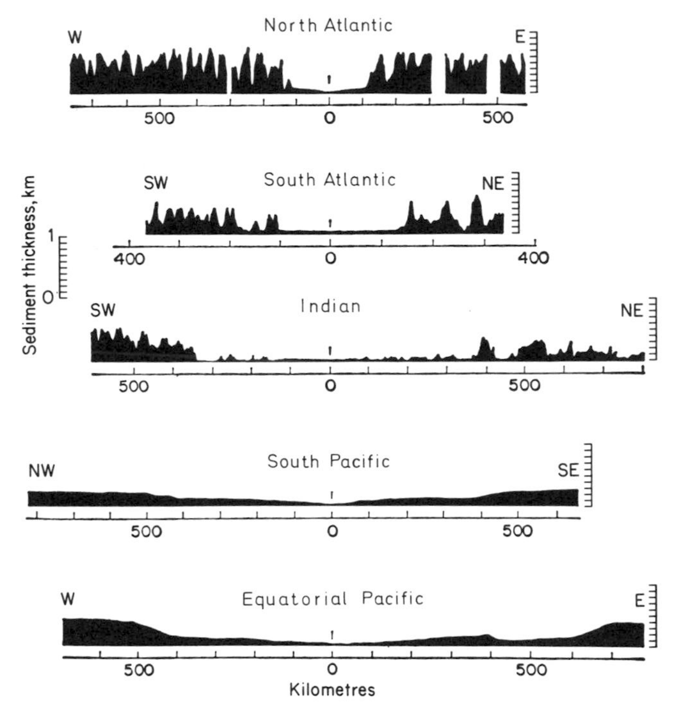

Figure 6. Graphs of sediment thickness versus distance from the ridge crests (after Ewing and Ewing, 1967). The examples include a range of spreading rates, the North Atlantic being about 1 cm/yr and the equatorial Pacific about 5 cm/yr.

also seems we should question the smoothness of the evolution of the oceanic lithosphere and the smoothness of the plate motions.

With regard to the latter, it is interesting to recall some observations by Ewing and Ewing (1967). They mapped the sediment distribution on the mid-ocean ridges and showed there are remarkable discontinuities in the thicknesses of the sediments near the ridge flanks (Fig. 6). They pointed out that this can be due either to discontinuities in the rates of sea-floor spreading or to discontinuities in the rates of sedimentation, the former being more likely. They suspected a period of quiescence during which most of the sediments were deposited followed by comparatively rapid sea-floor spreading. This interesting observation has not been fully explored and is worthy of further study. It is another indicator that the motions of plates may not be uniform.

The relative motions of the African and Arabian plates

Our studies of the Red Sea and Gulf of Aden have presented similar problems (Girdler and Styles, 1974, 1978). In the early stages of continental drift, such as the separation of Africa and Arabia, continuous sea-floor spreading seems unlikely.

Geometrically, the situation is simple. Arabia rotates anticlockwise forming the Toros–Zagros mountains to the north-east and the Red Sea and Gulf of Aden to the west and south (Fig. 7). In time, the situation is more complex. Fortunately, some boundary conditions on the timing have been obtained from studies of the geology along the Aqaba–Dead Sea–Arava transform fault system.

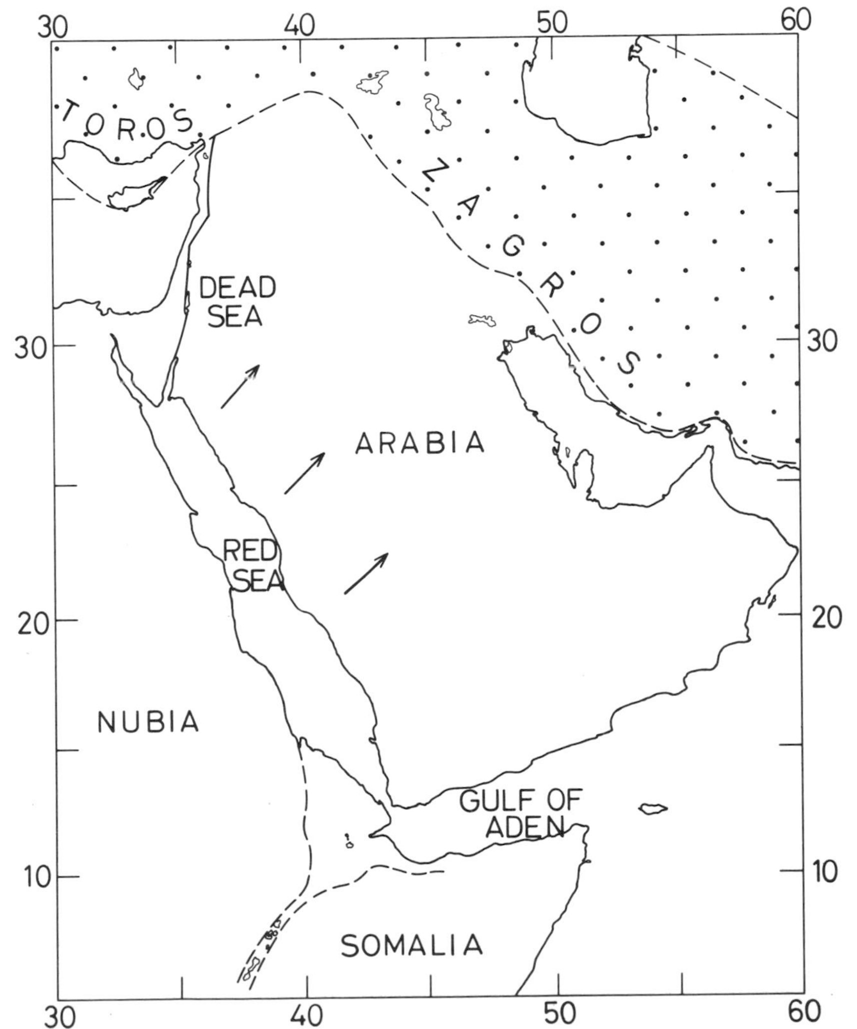

Figure 7. The movement of Arabia and Africa. The Red Sea and Gulf of Aden are new oceans formed as Arabia rotates anticlockwise with respect to Africa; the Gulf of Aqaba–Dead Sea rift is a transform zone with left-lateral shear and the Toros–Zagros mountains are a collision zone to the north of the Arabian plate.

Freund and his collaborators have done much to establish the timing of these movements. In a classic paper, Freund (1965) suggested that the northward anticlockwise movement of Arabia occurred in several phases during upper Cretaceous to Pleistocene times, and in an elegant presentation to the Royal Society in 1969 showed that the shear amounted to 105 km post-Cretaceous with 40–45 km of this offsetting Miocene rocks (Freund *et al.*, 1970). More recently, Freund (personal communication) recognized three

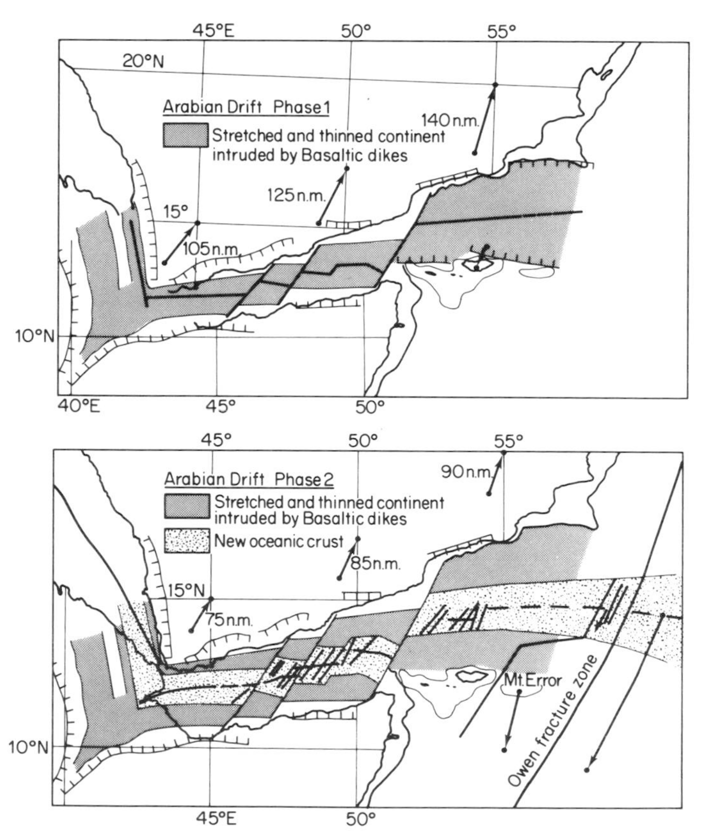

Figure 8. The formation of the Gulf of Aden by the drift of Arabia from Africa (after Laughton, 1966).

phases of shear, i.e. 43 km post-Miocene, 62–64 km in the lowermost Miocene and 10–20 km in the late Cretaceous. These give some valuable boundary conditions for the interpretation of the sea-floor spreading magnetic anomalies over the Red Sea and Gulf of Aden.

In the Gulf of Aden, Laughton (1966) also recognized that the movement of Arabia with respect to Africa must be discontinuous (Fig. 8). One of the major arguments for the two phases here, came from seismic data which showed a discontinuity in the sediment thickness. The main trough has about 0·5–1 km of sediment whereas the central ridge zone has only 0–100 m of sediment. It seems there was an appreciable pause during which time the 0·5–1 km of sediments were deposited (cf. the work of Ewing and Ewing, 1967, referred to above). Further work led Laughton *et al.* (1970) to give up this two phase interpretation and to interpret the magnetic anomalies in terms of continuous sea-floor spreading over the last 10 million years. We shall see that new geophysical data obtained on board the *R.R.S. Shackleton* in 1975 has led us back to the earlier interpretation but in a modified form.

It can be seen from Fig. 7 that if Arabia rotates anticlockwise, the evolution of the Red Sea and that of the Gulf of Aden should be similar. From studies of the magnetic anomalies over the Red Sea, Girdler (1966, 1969), Vine (1966) and Girdler and Styles (1974, 1978) were unable to demonstrate continuous sea-floor spreading beyond about 3–4 million years. Consequenty, Girdler and Styles (1974) were forced to the difficult task of fitting short lengths of profile to

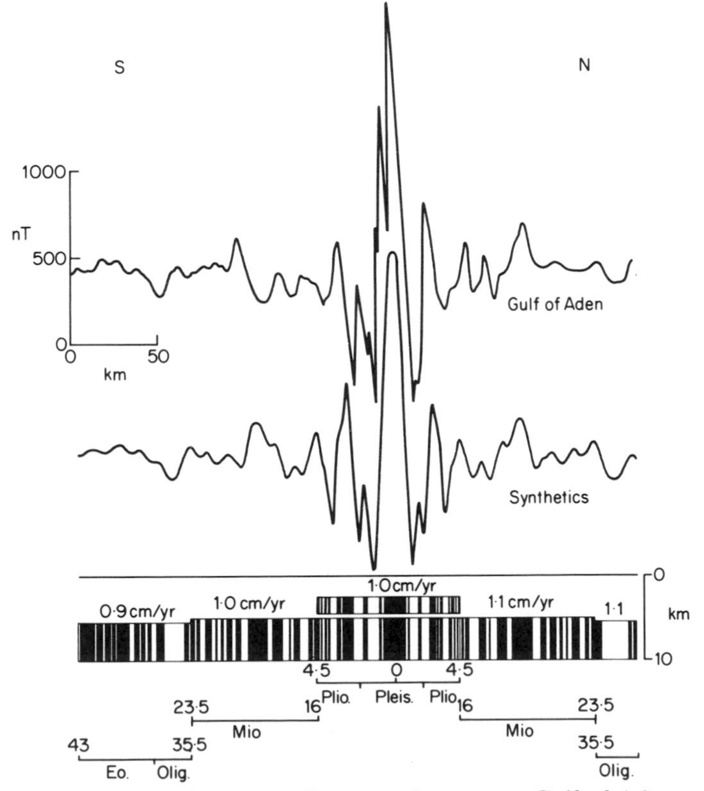

Figure 9. Sea-floor spreading anomalies over the western Gulf of Aden and their possible interpretation. The breaks in spreading correspond to the breaks in the slope of the basement from the seismic reflexion profile (after Girdler *et al.*, 1980). The anomolies are converted to the pole.

the geomagnetic time-scale and concluded there were at least two phases of sea-floor spreading, with an earlier phase in the Oligocene.

The new work in the Gulf of Aden (Girdler *et al.*, 1980) included a seismic reflexion profile over the northern part in the direction of sea-floor spreading. This revealed two distinct changes in the slope of the basement (presumed to be the top of oceanic layer 2) and changes in the characters of the gravity and magnetic anomalies were observed corresponding to the changes in slopes. In accord with recent ideas on the formation and cooling of the oceanic lithosphere (Sclater and Francheteau, 1970), it seems likely there have been three distinct phases of sea-floor spreading. An attempt to fit these to the magnetic time-scale is shown in Fig. 9, the last two phases of spreading corresponding to the times of shear movement along the Dead Sea rift as established by Freund. If this interpretation is correct, there were two pauses in the sea-floor spreading of about 12 million years duration.

State of stress within the plates

So far, most of our deductions have come from studies of the plate margins. The state of stress within the lithospheric plates can give some further constraints concerning the forces acting on the plates. Just occasionally, an earthquake occurs within the plate with sufficient magnitude to enable a fault plane solution to be obtained from first motions on seismic records. These intra-plate earthquakes are little understood. Recent reviews of intra-plate seismicity and the tectonic stresses within the plates have been given by Sykes

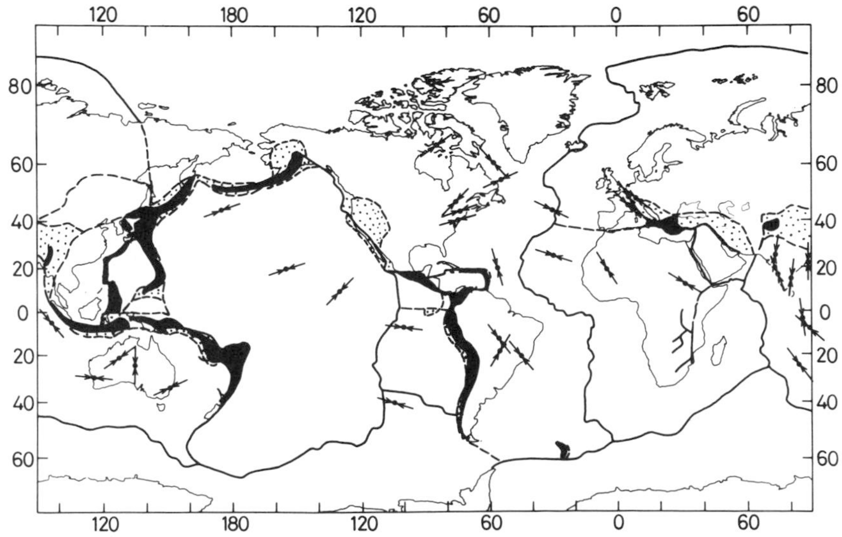

Figure 10. The state of stress within the plates. The plate boundaries are shown as in Fig. 1. The location of the intra-plate earthquakes are shown by dots and the earthquake mechanisms by arrows.

(1978) and Richardson, Solomon and Sleep (1979). The intra-plate earthquake mechanisms are shown in Fig. 10. It is seen that all of the few results available show compressive stresses.

The state of stress may also be obtained from *in situ* measurements and studies of recently deformed geological features. All these studies show that away from the plate margins, compressive stresses dominate.

These horizontal compressive stresses may be transmitted within the lithosphere or may reflect deeper processes in the mantle acting on the plates. They enable certain hypotheses to be eliminated. For example, it is unlikely that the plates are being pulled by gravity acting on the subduction zones, for some of the plates would then be in tension. It seems more likely that the plates are being pushed by injection of material at the ridge zones as emphasized by Mendiguren and Richter (1978). However, this is probably a second order effect as it is hard to envisage how this alone could drive the plates; it seems very likely that deeper processes such as mantle convection are at work. Clearly the compressional state of stress within the plates is another important boundary condition for possible mechanisms.

Conclusions

In this brief review, we have attempted to present some observations and inferences concerning the Earth's surface features which are relevant to any theories for the mechanism of continental drift and plate tectonics. These include:

(1) The plates have length to thickness ratios of about 20/1 to 100/1 giving an impression of wafer thinness and fragility.

(2) The continental parts of the lithospheric plates are more than twenty times the age of the oceanic parts and are presumably about five times as thick. The continental parts are therefore much more difficult to break.

(3) More than two-thirds of the Earth's surface (the oceanic parts of the plates) have formed in less than 5% of the history of the Earth. We have to consider, therefore, a process of considerable rapidity.

(4) From studies of sea floor spreading, any driving mechanism is likely to have a velocity of the order of at least 5 cm/yr.

(5) From studies of the continents, any mechanism cannot be short-lived and is likely to have been operating through most of Earth's history, i.e. for times of the order of thousands of millions of years.

(6) The seismic reflexion data for the oceans show discontinuities in the thickness of the sediments suggesting that the motions of the plates may not be continuous.

(7) Detailed studies of the Arabia–African plate system also suggest discontinuous motions. The go–stop–go sea-floor spreading in the Gulf of Aden and Red Sea is now well established apart from some uncertainty concerning the dates of the earlier movements.

(8) Determinations of the state of stress within the lithospheric plates are, so far, without exception compressional.

It is hoped these observations will give food for thought. The go–stop–go motion need not mean that any driving mechanism has to switch on and off. This could be associated with the jostling of the plates and their getting jammed and releasing themselves. In this case, the analogy of the plates behaving like icebergs is not inappropriate.

References

Barrell, J., 1914. The strength of the Earth's crust. *J. Geol.* **22**, 655–683; 729–741.

Barrell, J. 1915. The strength of the Earth's crust. *J. Geol.* **23**, 27–44; 425–444; 499–515.

Cox, A., Doell, R. R. and Dalrymple, G. B., 1964. Reversals of the Earth's magnetic field. *Science* **144**, 1537–1542.

Dietz, R. S., 1961. Continent and ocean basin evolution by spreading of the sea-floor. *Nature* **190**, 854–857.

Ewing, J. and Ewing, M., 1967. Sediment distribution on the Mid-Ocean Ridges with respect to spreading of the sea floor. *Science* **156**, 1590–1592.

Freund, R., 1965. A model of the structural development of Israel and adjacent areas since upper Cretaceous times. *Geol. Mag.* **102**, 189–205.

Freund, R., Garfunkel, Z., Zak, I., Goldberg, M., Weissbrod, T. and Derin, B., 1970. The shear along the Dead Sea rift. *Phil. Trans. R. Soc.* **A267**, 107–130.

Girdler, R. W., 1966. The role of translational and rotational movements in the formation of the Red Sea and Gulf of Aden. In *The World Rift System*, Rep. Symp. Ottawa, Can., 4–5 Sept. 1965, 65–77. *Geol. Soc. San. Pap.* 66–14, Dept. Mines & Tech. Surveys.

Girdler, R. W., 1969. The Red Sea – a geophysical background. In *Hot Brines and Recent Heavy Metal Deposits in the Red Sea* (Eds Degens, E. T. and Ross, D. A.), 38–58, Springer-Verlag, Berlin.

Girdler, R. W., Brown, C., Noy, D. J. M. and Styles, P., 1980. A geophysical survey of the westernmost Gulf of Aden. *Phil. Trans. R. Soc.* **A298**, 1–43.

Girdler, R. W. and Styles, P., 1974. Two stage Red Sea floor spreading. *Nature* **247**, 7–11.

Girdler, R. W. and Styles, P., 1978. Sea-floor spreading in the Western Gulf of Aden. *Nature* **271**, 615–617.

Heirtzler, J. R., Dickson, G. O., Herron, E. M. Pitman, W. C. and Le Pichon, X., 1968. Marine magnetic anomalies: geomagnetic field reversals and motions of the ocean floor and continents. *J. Geophys. Res.* **73**, 2119–2136.

Hess, H. H., 1962. History of ocean basins. In *Petrologic Studies: A Volume in Honor of A. F. Buddington*, Geol. Soc. Amer., 599–620.

Holmes, A., 1928–9. Radioactivity and Earth movements. *Trans. Geol. Soc. Glasgow* **28**, 559–606.

Holmes, A., 1944. *Principles of Physical Geology* (1st edn), Nelson, London.

Holmes, A., 1965. *Principles of Physical Geology* (new and fully revised), Nelson, London.

LaBrecque, J. L., Kent, D. V. and Cande, S. C., 1977. Revised magnetic polarity time-scale for Late Cretaceous and Cenozoic time. *Geology* **5**, 330–335.

Laughton, A. S., 1966. The Gulf of Aden. *Phil. Trans. R. Soc.*, **A259**, 150–171.

Laughton, A. S., Whitmarsh, R. B. and Jones, M. T., 1970. The evolution of the Gulf of Aden. *Phil. Trans. R. Soc.* **A267**, 227–266.

Mendiguren, J. A. and Richter, F. M., 1978. On the origin of compressional intra-plate stresses in South America. *Phys. Earth Planet. Int.* **16**, 318–326.
Richardson, R. M., Solomon, S. C. and Sleep, N. H., 1979. Tectonic stress in the plates. *Rev. Geophys. Space Phys.* **17**, 981–1019.
Sclater, J. G. and Francheteau, J., 1970. The implications of terrestrial heat flow observations on current tectonic and geochemical models of the crust and upper mantle of the Earth. *Geophys. J. R. Astr. Soc.* **20**, 509–542.
Sykes, L. R., 1978. Intra-plate seismicity, reactivation of pre-existing zones of weakness, alkaline magmatism, and other tectonism postdating continental fragmentation. *Rev. Geophys. Space Phys.* **16**, 621–688.
Tozer, D. C., 1973. The concept of a lithosphere. *Geofis. Int. Mexico* **13**, 363–388.
Vine, F. J., 1966. Spreading of the ocean floor: new evidence, *Science* **154**, 1405–1415.
Vine, F. J. and Matthews, D. H., 1963. Magnetic anomalies over oceanic ridges. *Nature* **199**, 947–949.

A Dynamic Model of Subduction Zones

R. FREUND†, D. KOSLOFF, A. MATTHEWS

Institute of Earth Science, The Hebrew University of Jerusalem, Israel

Introduction

Over a decade ago it was suggested (Elsasser, 1967; Oliver and Isacks, 1967; Isacks, Oliver and Sykes, 1968) that cold slabs of oceanic lithosphere descend obliquely along the Benioff zones, thereby consuming the oceanic lithosphere and producing the deep earthquakes. This model satisfies the geometrical requirements of plate tectonics, and conforms to the compressional nature of mountain building at the leading edges of the plates and to the high-*Q* along the Benioff zones. So far this concept has been accepted without challenge.

Yet despite the great efforts to understand the inclinations of the subduction zones (Ringwood and Green, 1966; Lliboutry, 1969; Luyendyk, 1970; Jacoby, 1973, 1976; Turner, 1973; Forsyth and Uyeda, 1975; Jischke, 1975; Davies, 1977; Stevenson and Turner, 1977; Tovish, Schubert and Luyendyk, 1978), their retrogressive movements (Elsasser, 1971; Garfunkel, 1975), the high heat loss above them (McKenzie and Sclater, 1968; Turcotte and Oxburgh, 1969; Minar and Toksöz, 1970; Griggs, 1972), the formation of marginal seas (Karig, 1971; Matsuda and Uyeda, 1971) and the driving mechanism (Forsyth and Uyeda, 1975; Richter, 1977), no satisfactory physical model of plate tectonics has been proposed so far. This astonishing failure may be due to a false concept rather than to complicated physical processes. In this chapter we propose that the model of obliquely descending lithospheric slabs is not plausible, and that an alternative model of subduction zones composed of separate blocks descending vertically provides answers to the above problems.

The problems of oblique slabs

Very little is known about the nature of the subducting lithosphere, but several constraints are implied:

(1) The consumed lithosphere extends most probably along the oblique Benioff zones.

† The untimely death of Professor Raphael (Rafi) Freund occurred in February 1980.

(2) The slab model requires tight bending of the lithosphere at the trenches and straightening back at a short distance downdip.

(3) If the slabs were not supported by the asthenosphere, the negative buoyancy of the cold lithosphere would create extensive stresses of about 10 kbar at the upper end of the slab (Turcotte and Schubert, 1971). The bending moment would create stresses according to

$$\sigma(0 \pm h) = \pm \tfrac{3}{2} \Delta\rho g \cos\theta_s \, l^2/h, \tag{1}$$

where $\sigma(0 \pm h)$ is the stress at the bending point of the top and of the bottom of a slab with thickness h, length l and density excess $\Delta\rho$ (Stevenson and Turner, 1977). These stresses would be of the order of 100 kbar. Therefore the slabs must be supported to maintain their integrity and inclinations.

(4) The negative gravity anomaly at the trenches can be accounted for only by a downpull of the slab (Uyeda, 1978), and it appears that the negative buoyancy of the subducting lithosphere is the most important force of plate motions (Forsyth and Uyeda, 1975; Richter, 1977). Yet, plates which have no slabs move also, and compression is indicated by focal plane solutions along large portions of the slabs (Isacks, Oliver and Sykes, 1968), as if the slabs are being pushed, certainly along their lower portions.

The models and explanations of slab inclinations ought to be examined according to these constraints. Explanations depending on assumed predetermined geometry (Ringwood and Green, 1966; Jischke, 1975) create equally difficult problems of the origin of this geometry and therefore are not considered. The horizontal momentum of the plate motion is negligibly small and cannot cause the inclination, and the idea that the subduction zones are shaped like a bump in a sphere is contradicted by the various inclinations of the subduction zones, independent of their sizes.

Jacoby's foam rubber (1973) and paraffin (1976) models of plate motions show that oblique subducting slabs occur in plates with great bending resistance (rubber, thick paraffin), but in plates with small bending resistance (thin paraffin) the slabs hang vertically and collapse on the hard bottom. Since the bending stresses of the subducting slabs on the Earth are several orders of magnitude greater than the strength of rocks, it seems that neither the bending resistance of the lithosphere near the surface nor the support on a mesosphere "hard bottom" (Forsyth and Uyeda, 1975) can account for the inclination of the subduction zones. Also the very small radius of curvature at the trenches (about 200 km) seems to exclude elastic plate bending mechanisms because with these curvatures and for quoted values of lithosphere thickness and elastic constants, the fibre stresses would exceed by orders of magnitude the strength of the rocks (Isacks and Barazangi, 1977). If, on the other hand, the bending is explained by an elastic–plastic type model (or by a reduced effective thickness of the lithosphere), it becomes difficult to explain why this weak lithosphere becomes oblique and straight along the Benioff zone (Lliboutry, 1969).

In the viscous models of Turner (1973), Davies (1977) and Hager and O'Connell (1978) the flow is driven by a horizontal velocity boundary condition on the surface from one side, and thereby inclined subduction flows, whose streamlines are approximately parallel to the subduction zones on Earth, are obtained. These appealing results must, however, be regarded with caution, because the horizontal drive of the flow is a physically unexplained boundary condition, and it must be demonstrated that the results are compatible with plausible driving mechanisms of plate tectonics. Moreover, these models do not include subduction slabs, and it is not demonstrated that these flows can support the excess weight of the slabs. This point is elucidated in the following.

In order to examine whether impact from horizontal flow in the mantle can support the inclination of a slab, consider the simple example of 2-D flow in the region shown in Fig. 1. The slab on the left side of the figure has the density

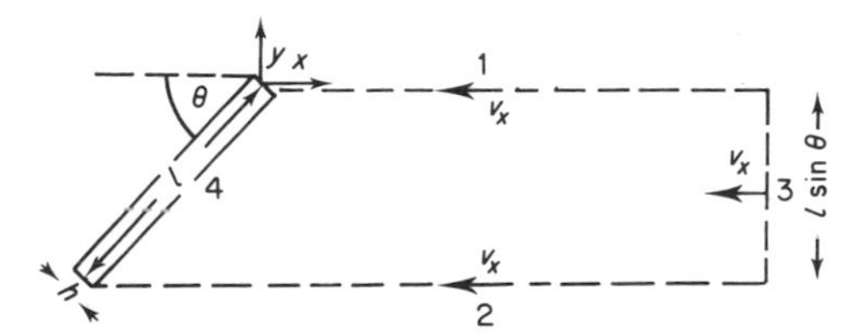

Figure 1. Set up of a flow supporting by impact an inclined slab suspended on a fixed hinge. See calculation in text.

excess $\Delta\rho$ and a thickness h, a length l, dip angle θ, and it is constrained to revolve about a hinge shown in the upper left corner. It is assumed that the velocity is directed everywhere parallel to the x-axis and is equal to v_x along boundaries 1, 2 and 3, and vanishes on the slab. The pressure is assumed to be hydrostatic.

The contribution to the force moment per unit slab depth about the origin from boundaries 1 and 2 vanishes. The contribution from boundary 3 is given by

$$\int_0^{l\sin\theta} \rho y v_x^2 \, dy = \tfrac{1}{2}(l\sin\theta)^2 \rho v_x^2 \qquad (2)$$

where ρ is the density of the fluid. This moment is balanced by the moment of the immersed weight of the slab along boundary 4 given by Eqn (1). Equating (1) and (2) we obtain for v_x

$$v_x = [(\Delta\rho/\rho)\, hg\, (\cos\theta/\sin^2\theta)]^{1/2}.$$

Taking values for the Earth: $\Delta\rho/\rho = 0{\cdot}03$; $h = 6 \times 10^7$ cm; $\theta = 45°$, a velocity of about 4×10^4 cm/s is obtained, which is ridiculously high. Although in reality the flow pattern will be different from that assumed in this example (which does not conserve mass), it is not expected that the order of magnitude of the velocities required to maintain the slab in its inclination will deviate much from the estimate given. It does not seem probable therefore that the flow models mentioned above can support the inclination by impact alone. If, however, viscous drag of the flow is assumed, the inclination may perhaps be maintained, but the flow will increase also the extensile stresses along the slab and contribute to its disintegration.

Stevenson and Turner (1977) and Tovish *et al.* (1978) suggest that the asymmetrical motion of the asthenosphere dragged by the downgoing slab and the plate attached to it (while the opposite plate is stationary), causes a pressure on the side facing the moving plate and suction on the side of the stationary plate. The slab will deviate from vertical due to this pressure difference. Their calculations show, however, that this process can stabilize the slab only at inclinations steeper than 63° with linear viscosity and 55° with non-linear viscosity. According to this cornerflow model inclinations gentler than 55° are impossible, because this process lifts a slab with gentler inclination to a horizontal position. This model fails therefore to account for many gently inclined slabs, In brief, it appears that ten years after the slab model has been proposed, the inclination of the subduction zones, let alone its variations, has not been satisfactorily explained.

The retrogressive movement of subduction zones makes the slab concept most unlikely. Retrogressing slabs should sweep the asthenosphere inward beneath the Pacific Ocean at a rate which leads to countercurrents in the mantle several times larger than the mass transport involved in plate tectonics on the surface of the Earth (Garfunkel, 1975). This enormous horizontal mass movement would leave hardly enough room for vertical, thermally driven mass movements which are required to set the system in motion.

Great effort has gone into the attempts to explain the paradox of excessive heat flow above a cold descending lithospheric slab (McKenzie and Sclater, 1968; Turcotte and Oxburgh, 1969; Minear and Toksöz, 1970), suggesting that it is derived from shear strain heating. Remembering, however, that the entire potential energy of the descending lithosphere (which is the predominant source of plate tectonic energy (Forsyth and Uyeda, 1975)) amounts to less than a quarter of the heat required to warm the same lithosphere to the ambient temperature of the mantle around it, one must provide convincing reasons why this frictional heating rises to the surface instead of heating the adjacent cold slab.

It has been suggested (McKenzie, 1969; Sleep and Toksöz, 1971; Toksöz and Bird, 1977) that the excessive heat rises to the surface by convective motions induced by the downgoing slab. The most carefully worked out model of this kind (Andrew and Sleep, 1973) shows that this can bring the hot material only a few hundred kilometres behind the island arc, and hence the high heat flow on the island arc remains a problem. Moreover, if this process accounts for the heat and the origin of marginal seas, why are they absent above some subduction zones?

A model of subduction zone composed of detached blocks

The idea that the lithosphere may be consumed as detached blocks (Ringwood and Green, 1966) precedes the slab concept. The slabs were suspected to be discontinuous because the earthquake foci are clustered on the Benioff zones between parts devoid of them (Isacks and Molnar, 1969; Isacks and Barazanghi, 1977; Fuchs *et al.*, 1980), and because of different travel times across them under the Andes (Sacks, 1969). Laubscher (1969) speculated that the lithosphere collapses in blocks under the Alps, and Lliboutry (1969) argued that the sharp and localized downbend of the subduction zone at the trenches is more likely due to fragmentation than to bending of the lithosphere. These thoughts have not been followed up so far. We intend to do this in the following in order to show that subduction zones consisting of detached blocks descending vertically provide explanations to the phenomena observed on the subduction zones in nature.

The initiation of the subduction zones may be due to a lithospheric instability created by rapid loading of sediments on the continental rise (Dietz, 1963), or to some other process. There is no distinction between the slab model and the present one with respect to the initiation, and the discussion is therefore limited to well-developed subduction zones.

According to Eqn (1) the bending stresses of a lithospheric slab 80 km thick and about 250 km wide amount to 1 kbar, which is taken as the strength limit of the lithospheric rocks. Thus a block 80 km by 250 km across and several hundred kilometres long breaks off the oceanic lithosphere and begins its vertical journey down (Fig. 2(a)).

The detached block descends vertically under its negative buoyancy until the positive buoyancy of the depression left above balances the downward pull. At this stage the top surface of the block (represented by the deep sea trench) is below the hydrostatic head of the "mantle geoid" (Turcotte *et al.*, 1977) said to be at 3·25 km below sea level. Mantle pressures will therefore induce material to fill up the depression. This may be accomplished by two competing mechanisms; by inward movement of the adjacent plates (hereafter termed case I) or by rising mantle material into the depression (case II). The mechanism of case I dominates when the resistance of the adjacent plates to horizontal motions is small, and case II is more important when the resistance is large compared to the viscous retardation of the mantle material. The nature of the process depends also on the type of the plates involved – whether oceanic or continental – and whether it occurs on the margin of a plate or in its middle, as demonstrated in the following.

In case I with oceanic and continental plates which approach one another to cover the gap above the descending block (Fig. 2(b)) the respective velocities are inversely related to the resistance holding each plate back. The space between the descending block and the overriding plates is narrow (otherwise the light mantle material would leak to the surface as in case II), and thus the descending block exerts a downward pull on the edges of the two plates above it. The buoyancy of the light continental plate keeps it afloat, but the heavy oceanic plate is bent down until a second block is broken off and starts its own journey down (Figs. 2(b), (c)).

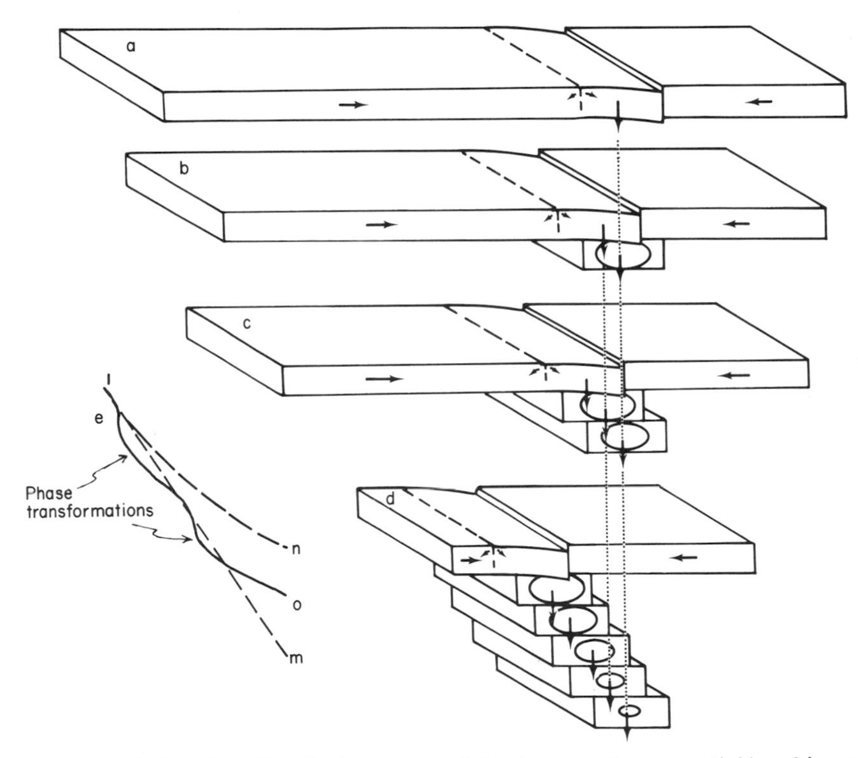

Figure 2(a–d). Creation of an Andean-type subduction zone by sequential breaking-off and vertical descent of oceanic lithospheric blocks and the convergence of both plates above them. The subduction zone "moves" retrogressively at the velocity of the continental plate, and mantle material moves across it. (e) The inclination is decreased (l–m to l–n) by reduction in effective block size due to heating, and increased locally (l–n to l–o) due to phase transformations.

The second block is offset oceanward from the first one by the amount of movement of the continental plate, and so is the third relative to the second. The repetition of detachment of blocks creates therefore an inclined zone of heavy lithospheric blocks. Each of the blocks descends vertically, yet the whole zone has an apparent, retrogressive motion whose "velocity" is equal to the velocity of the continental upper plate (Fig. 2(d)). On their way down the blocks are heated by friction and by the hot mantle around them, thereby their effective size is reduced. Mantle material can therefore rise around the deeper blocks, and the velocity of the diminishing blocks decreases downwards so that the outline of the zone becomes concave (line 1–n in Fig. 2(e)). Phase transformations from basalt to eclogite (Ringwood and Green, 1966) and from olivine to spinel (Anderson, 1967) cause local acceleration of the downward movement, and consequently the outline of the subduction zone should be wavy (line l–o, Fig. 2 (e)). Like the Tonga (Mitronovas and Isacks, 1969) and the Aleutian (Engdahl, 1977) Benioff zones.

In general, the blocks are decelerating on their way down, so that usually the higher of any two blocks descends faster than the deeper one. On the occasions that they touch one another the subvertical shear between them may create earthquakes whose focal plane solution appears as compression along the subduction zone. Where the blocks are accelerating due to phase transformation (or to other reasons such as rotation of the narrow side of the block downwards) the relative shear between adjacent blocks may cause earthquakes whose focal plane solution appears as tension along the subduction zone. The relative motion between the two plates and the highest block is a shear on inclined surfaces in both the subduction direction and the obduction direction. The bending of the lithosphere may create normal fault type earthquakes (Stauder, 1968), and a large earthquake which has probably detached the entire lithosphere (Kanamori, 1971) may perhaps be attributed to the detachment of a descending block.

When case I takes place between two oceanic plates, the two sides are able to subduct, and blocks may be detached from both sides. This can lead to either two oppositely inclined subduction zones or to one vertical subduction zone. However, sediments skimmed off the descending lithosphere accumulate rapidly above the descending block. This light material becomes attached to the edge of one side or the other, protecting it from further subsidence. This side becomes thus the equivalent of the continental plate, and the process proceeds as shown in Fig. 2, with one exception. The protecting edge is a narrow feature which may occasionally switch sides, forcing the subduction to switch sides in the opposite direction.

Case II concerns plates which cannot move to close the depression above the descending block. Such a situation may perhaps occur when the subduction starts as a local feature (e.g. a large volcanic pile) in the middle of an oceanic plate. In this case a single block will subside slowly, being replaced by mantle material rising around it and freezing into a new piece of oceanic lithosphere. This process terminates, because the subsidence of the first block does not induce the creation of others.

If, however, the process occurs on a moving plate (whose motion may be caused by another subduction zone) the descending block is overrun by the moving plate (Fig. 3(b)), and this can pull down and detach another block. The repetition of this process creates a subduction zone in the direction of the movement (Figs 3(c), (d)). The gap in front of the subduction zone is filled up by hot mantle material, forming an inter-arc marginal basin, whose spreading velocity is equal to the velocity of the moving plate, and with high heat flow.

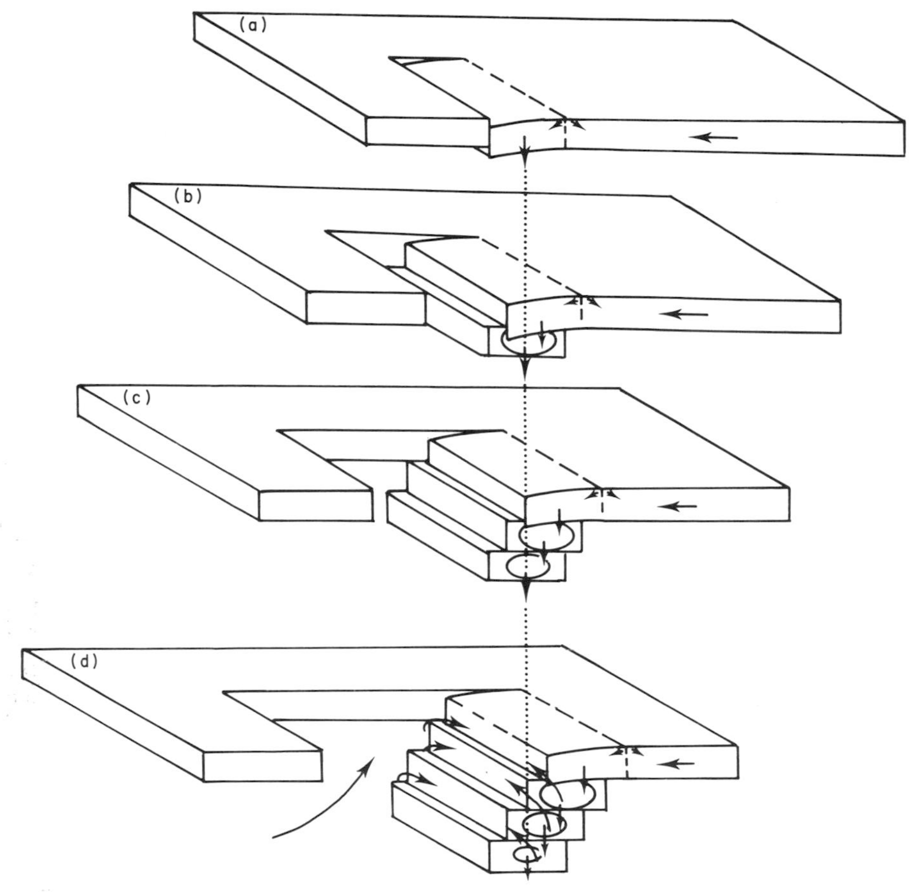

Figure 3. Creation of a Mariana-type subduction zone in the middle of a moving oceanic plate. Note the upward traction at the ends of the blocks; this may lead to the arcuate shape of the subduction zones and island arcs.

The viscous drag is larger at the edges of the subsiding blocks, where it acts on three sides, than in their middle where it acts only on two sides. Therefore the viscous drag may bend the ends of the long blocks upwards relative to their median portion (Fig. 3(d)). Consequently the centre of the blocks is detached more easily and sooner at the surface than the ends, leading to the arcuate shape of the island arcs.

Case II may occur also if the plates on the two sides of the subduction zone are retarded (as for example by the large friction under large plates) so that the gap above the subsiding blocks is only partly covered by the plates (Fig. 4), the remainder becoming marginal seas. In both examples of case II (Figs 3 and 4), the island arcs, which consist of light material skimmed off the subducting blocks, are driven by the rising mantle material towards the following subsiding block, i.e. towards the trench.

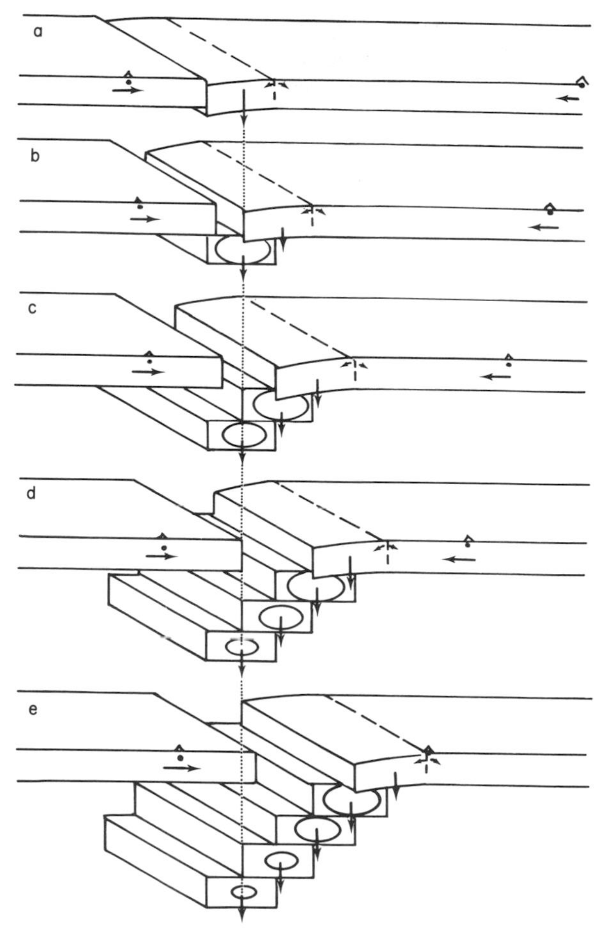

Figure 4. The retardation of very large plates by the great drag resistance on their bottom leaves vacant space above the descending blocks. This is filled by rising asthenospheric material, and appears as marginal seas with high heat flow.

The case I model shown in Fig. 2 is applicable to the Andes, which occur along a boundary between two rather small plates (Fig. 5(a)); that of case II shown in Fig. 4 is applicable to the western (Aleutian to New Zealand) side of the Pacific Ocean which is between two large plates (Fig. 5(c)), and that of Fig. 3 applies to the Marianas and the Caribbean. The Indian–Pacific subduction zone extending from Samoa to the Philippines is explained by the ocean–ocean variant of case I.

The hitherto described block model may seem unlikely because it requires the descent of the top of the block to the level of the bottom of the lithosphere before the two plates can move inwards, whereas the depth of the trenches is always much shallower than such a depression. Clearly, the two motions must

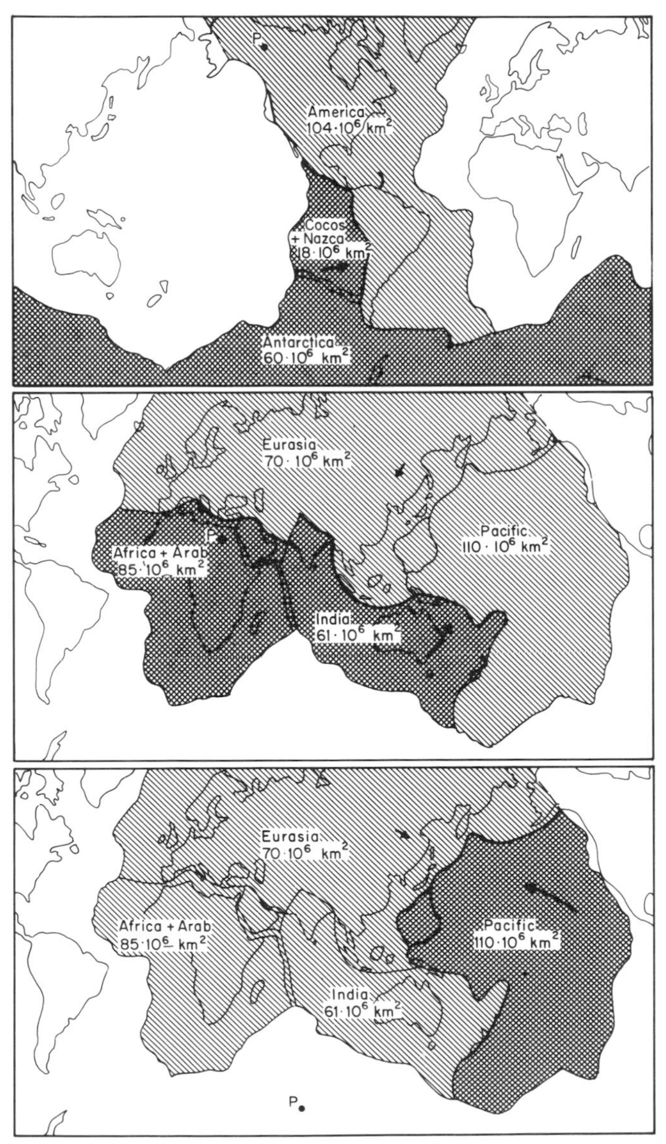

Figure 5. The three subduction zones and the plates pulled by them. In each case the smaller plates move faster.

occur concurrently. Figure 6 illustrates that by the means of reverse faults in both the subduction and the obduction orientations, which are confirmed by focal plane solutions of earthquakes and by geological observations, the simultaneous movement of the descending block and the inward moving plates can indeed by accomplished.

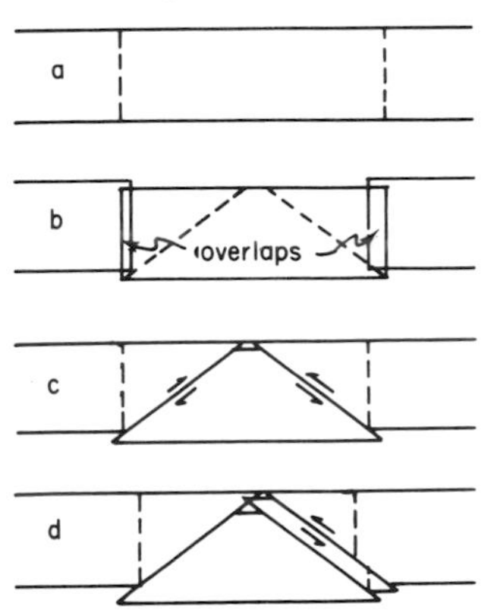

Figure 6. Diagram demonstrating a way in which the downward motion of the lithospheric block may occur simultaneously with the inward motion of the two plates by repeated thrusting. The overlaps which would appear at the first increment of the simultaneous motion (b) are eliminated by reverse faults on both sides (c), e.g. in both obductive and subductive senses, and the surface depression is eliminated as well. At each further increment (d) more reverse faults are created on the ocean side.

Velocities and stresses

To be viable the block model must provide reasonable mechanical parameters of velocity and stress. The velocity of descent and the stresses of the blocks are therefore examined here, as well as the stresses involved in driving the plates, all this being limited by our poor knowledge of the properties of the asthenosphere.

For the calculation of the velocity of descent of the blocks the asthenosphere is modelled as an infinite uniform Newtonian viscous fluid and the blocks are modelled as rigid prolate ellipsoids of revolution falling individually by the force of gravity. The terminal velocity of such bodies is given by (Appendix 1):

$$v = \frac{2ab\Delta\rho g}{9R_{\text{eff}}\eta}, \tag{3}$$

where η denotes the fluid viscosity, $\Delta\rho$ is the density contrast between the ellipsoid and its surroundings, a and b respectively denote the length of the major and minor axes of the ellipsoid, g is the gravitational acceleration and R_{eff} is the effective drag radius of the ellipsoid (Appendix 1). For an ellipsoid which falls with the major axis horizontal and with dimensions $a = 500$ km, and $b = 50$ km, and with the values $\eta = 10^{21}$ poise, $\Delta\rho = 0{\cdot}09$ g/cm^3, a terminal velocity of 4 cm/yr is obtained.

Since this calculation approximates a triaxial ellipsoid by an ellipsoid of revolution, and it ignores the decelerating effect of the free surface and the accelerating effect of the interaction between the blocks, it is only a rough estimate of the possible velocities. However, if the blocks are about three times wider than their thickness, the descent velocity of 4 cm/yr conforms to the observed plate convergence velocities of 10 cm/yr.

The magnitude of stress involved can be estimated by dividing the buoyant weight of block by its surface area when the block is equally supported all around, or by its upper surface when its motion is retarded from above. For the ellipsoid described above one obtains values of 300 bars in the first case and 1 kbar in the second. These values give an estimate of the magnitude of the tensile stress which the block exerts on the lithosphere above it during the initial stage when they are mutually juxtaposed. These results conform to the stress required to pull the deep sea trenches down and to maintain their excessive negative gravity anomaly (Uyeda, 1978). Indeed, the discontinuous nature of the heavy lithosphere implied by the block model accounts much better for the negative gravity anomaly of the trenches than does the supposed downward pull which is required by the slab model.

Although the block model does not necessarily imply that the gravitational pull of the blocks is the single driving force of plate tectonics, let us examine this possibility. Lachenbruch (1976) has demonstrated that the lithospheric plate can be pulled despite the small tensile strength of the lithospheric rocks. If a plate 80 km thick and 8000 km long is pulled by 300–1000 bars at its edge, force equilibrium requires that the shear resistance at its base does not exceed 3–10 bars. If in addition this plate (e.g. the Pacific) moves at 10 cm/yr on a mantle with a viscosity of 10^{21} poise, the thickness of the mantle dragged along with the plate is 300–1000 km. The thickness would be smaller if the viscosity is smaller (Artyushkov, 1974).

The symmetry of the block model requires that the resisting force acting on the plates located on one side of a subduction zone should balance with those acting on the other side. Assuming a viscous Newtonian rheology in the mantle these forces should be the product of the "absolute" velocity of the plates (Morgan, 1972; Minster *et al.*, 1974) multiplied by their area. It has been shown that there is no direct or inverse correlation between the sizes of the plates and their velocities (Forsyth and Uyeda, 1975), but in every subduction zone the plates with the smaller area on one side move faster than the larger ones on the other side (Fig. 5). If, however, this relation is examined quantitatively it is found that in the two semi-independent subduction–plate systems (one consisting of America, Nazca, Cocos and Antarctica, and the other of Pacific, India, Eurasia and Africa) the mutual pull of the opposing plates does not balance out if equal drag resistance is assumed at the base of all plates. However, if the drag resistance of continental areas is about ten times that of oceans, and the drag of mountain ranges is three times larger than that of continents, the mutual pull of the plates across the subduction zones balances out. This agrees with results by Solomon, Sleep and Richardson (1975) and with the suggestions of a thicker continental lithosphere or "tectonosphere" (Jordan, 1978).

Inclination and age

The inclination (θ_p) and the velocity (v_p) of every point on a subducting slab can be determined (Fig. 7) according to the assumption that the slab maintains its inclination (θ_s) and length. v_a is the velocity of the subducting plate and v_b is the velocity of the upper plate and the retrogressive movement of the slab. Point a on Fig. 7 moves to a' in order to let point b reach b' at velocity v_b. At the

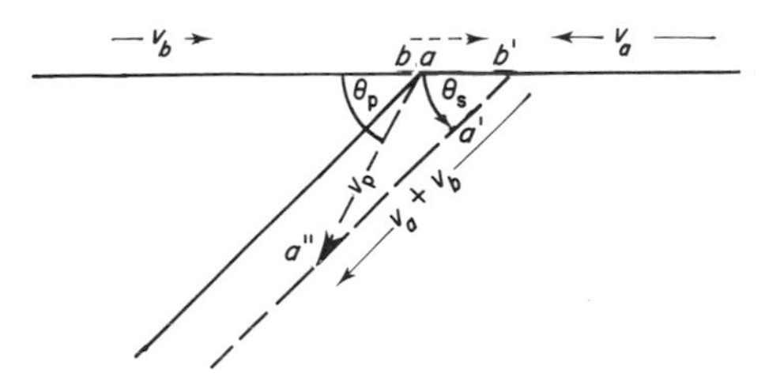

Figure 7. The geometry of the slab model showing the velocity and the inclination (θ_p) of the movement of any point on the slab.

same time point a moves with the subducting plate at velocity v_a from a' to a''. All the points on the slab move parallel to a–a'' whose inclination is

$$\cot \theta_p = \frac{v_b}{(v_a + v_b)\sin \theta_s} - \cot \theta_s$$

so that the points are descending straight down only when

$$\cos \theta_s = \frac{v_b}{v_a + v_b}$$

which is either when θ_s is very large (the subduction zone very steep) or when $v_b > v_a$, which never happens (Morgan, 1972). In general each point on the subduction slab moves obliquely down at a steeper angle than the inclination of the slab.

The velocity, v_p of every point on the slab is

$$v_p = (v_a + v_b)\sin \theta_s / \sin \theta_p$$

which is always larger than v_a and smaller than $v_a + v_b$. So that every point on the slab moves faster relative to the mantle than every point on the horizontal plate attached to it.

The age, t_p of a point on the slab (i.e. the time that has passed since this point has left the surface of the earth) is

$$t_p = L/(v_a + v_b),$$

where L is the length of the slab to the said point. As L does not exceed 1000 km and $(v_a + v_b)$ is about 10 cm/yr, t_p does not exceed 10 Myr.

In the block model θ_p is vertical and the inclination of the subduction zone, θ_s, is directly related to the ratio of the vertical velocity v_d to the horizontal velocity v_b of the upper plate (Fig. 8)

$$\tan\theta_{s'} = v_d/v_b$$

for blocks with a given aspect ratio h/l (h is the thickness of the lithosphere and l is the width of the block). $\theta_{s'}$ is likewise related to h/l for blocks with a given v_b

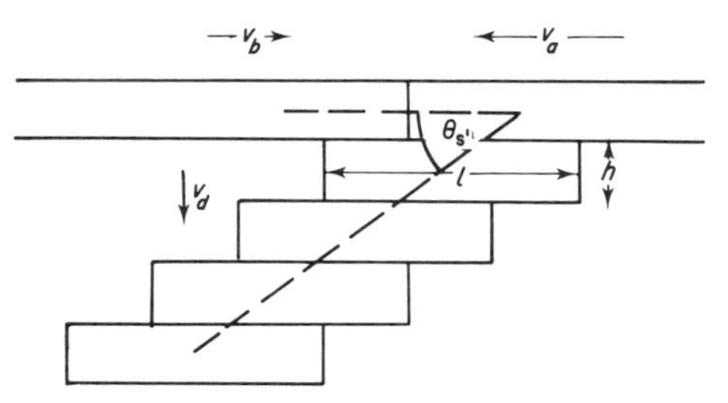

Figure 8. The geometry of the block model. The subduction inclination is determined by the aspect ratio, h/l, of the blocks and by the ratio, v_d/v_b, of descent velocity to upper plate velocity.

and a given ratio of the velocity of the two converging plates, as (Fig. 8)

$$h/l = v_d/(v_a + v_b).$$

The term l^2/h which appeared in Eqn (1) shows that if the lithosphere is assumed to be of uniform strength, the aspect ratio h/l of the blocks increases with the thickness of the lithosphere. Blocks broken off thicker lithosphere will therefore descend faster, as their mass increases by the power of 3/2 while their width increases only by the power of 1/2.

The gentle inclination of the subduction zone under the Andes (Isacks and Barazangi, 1977) may be explained in this way by the relatively young lithosphere involved and by the relatively high velocity of South America (3·3 cm/yr according to Minster and Jordan, 1978). The gentle inclination of the subduction zone under Japan can be accounted for by the block model only if the spreading rate of the Sea of Japan amounts to about 2 cm/yr.

The inclination of subduction zones discussed under case II increases with the velocity of the subducted plate if all other conditions are equal. This is confirmed by correlation between the velocity and inclination along several subduction zones shown by Luyendyk (1970).

The age, $t_{p'}$, of a point on the block model subduction zone is $t_{p'} = D/\bar{v}_d$, where D is its depth and $\bar{v}_d$ is the average velocity of its descent to the position. As $\bar{v}_d$ of the distal point of the subduction zone can be estimated as $v_d/2$ and $D = L\sin\theta_{s'}$

$$t_{p'} = 2Ll\sin\theta_{s'}/(v_a + v_b)\,h.$$

$t_{p'}$ is about three times longer than t_p as

$$t_p/t_{p'} = (h/2l)\,\mathrm{cosec}\,\theta_{s'} = (v_b/4(v_a+v_b)+h^2/4l^2)^{1/2} \approx \tfrac{1}{3}$$

because $h/l \approx \frac{1}{3}$ and $(v_a+v_b)/v_b \approx 3$. The age of the terminal part of a 1000 km long subduction zone is approximately 10 Myr according to the slab model, and more than 30 Myr according to the block model.

Indeed this is also evident from the geometry of the block model (Figs 2–4), which shows that the subducting lithosphere is shortened by about three times in the block model whereas its length does not change according to the slab model.

This result may be compared to the time in which the cold lithosphere is expected to disappear due to heating. For the calculation of the heating of the blocks during their descent, the example of the heating of a sphere with a unit initial temperature difference to its surroundings is employed. It is assumed that the sphere and its surroundings have the same thermal properties. The solution to this problem is given by

$$T(r,t) = \sqrt{\left(\frac{\chi t}{\pi}\right)}\frac{1}{r}\left\{\exp\left[-\frac{(r+a)}{4\chi t}\right]-\exp\left[-\frac{(r-a)}{4\chi t}\right]\right\}$$
$$+\frac{1}{2}\left\{\mathrm{erf}\left[\frac{r+a}{2\sqrt{(\chi t)}}\right]+\mathrm{erf}\left[\frac{a-r}{2\sqrt{(\chi t)}}\right]\right\},$$

where T is the temperature, r is the distance from the centre of the sphere, a is the radius of the sphere and χ is the thermal conductivity. Taking a as 50 km and $\chi = 4\times10^{-5}$ km/yr, the relaxation time for the sphere (the time when the temperature difference between the sphere and the surrounding medium is approximately half its initial value) is about 10 Myr, so that the sphere practically ceases to exist after about 50 Myr. This result is in agreement with the calculations made previously (McKenzie, 1969) for the heating of the descending lithospheric slab.

Although it is not known at what thermal stage the lithosphere disappears seismically, it may be suggested that the thermal considerations indicate that the subduction zones exist during a much longer time span than 10 Myr, in agreement with the block model and contrary to the slab model.

Moreover, if the plates are driven by the subducting lithosphere (Forsyth and Uyeda, 1975), the fact that the Indian plate is still moving 40 Myr after its collision with Eurasia (Molnar and Tapponier, 1977) shows that the subduction zones survive much longer than supposed by the slab model. The same is implied by occurrence of heavy lithospheric mass under the transverse range of California, where the subduction of the Pacific has ceased 30 Myr ago (Hadley and Kanamori, 1977).

Heat sources of island arcs and mountain ranges

It has been suggested that the rise of hot asthenospheric material provides the high heat flow of marginal seas. It is still necessary to explain the high heat flow on the mountain ranges and island arcs, which are situated above shallow blocks which are closely packed and avoid the rise of hot asthenospheric matter. We suggest two possible heat sources: the exothermic hydration of basalt at shallower levels, and an upward heat transport by water derived from eclogitization of amphibolites at deeper levels.

Basalt may become amphibolitized by reaction with saline water which is brought with it to the depth of 12–16 km and temperatures of 300–500 °C. The hydration reaction is exothermic (Bowen and Tuttle, 1949). The heat evolved can be calculated using standard thermodynamic methods. We approximate the process of amphibolitization of basalt by the reaction

$$2\ \text{diopside} + 3\ \text{enstatite} + \text{quartz} + H_2O \longleftrightarrow \text{tremolite.}$$

From the calculation in Appendix 2 it appears that about 37 kcal of heat energy are released for every mole of tremolite (i.e. amphibole) formed, which amounts to 45 cal/g. For an amphibole with a specific heat of 0·25 cal/g/°C, this is sufficient to heat the amphibole by 180 °C. This may provide a significant heat source for low pressure–high temperature types of metamorphism and melting of rocks with granitic composition.

The extent to which this heat will be available is also dependent on the degree of amphibolitization which occurred prior to the oceanic material reaching the subduction zone. Studies of oceanic ridges have shown that extensive hydrothermal alteration is occurring. However, we are assuming that a significant quantity of basalt remains unaltered until tectonic processes bring it down to the depth where the amphibolitization is resumed.

The hydration reaction depletes the basaltic rocks of free water, concentrating pore fluid with saline salts, and thus lowering the pressure and the activity of water. These conditions may lead to the formation of eclogite at greater depths by direct reaction from unaltered basalt and by destabilization of the amphibolite. The role of the liquid phase is critical. Eclogite formation from basalt occurs in the presence of small amounts of water which provides an important kinematic function in catalysing basalt/eclogite reactions (Fry and Fyfe, 1969). Under completely dry conditions this would occur too slowly to have any geological validity (Ahrens and Schubert, 1975), provided that $P_{H_2O} < P_{total}$ or otherwise amphibolites will form. Experimental work (Essene, Hensen and Green, 1970) suggests that for water-saturated quartz tholeite composition, eclogite will be stable at $P_{H_2O} = 20$–25 kbar for the temperature range 600–1000 °C, corresponding perhaps to a depth of 80–100 km. If the water saturation conditions are not maintained, this process probably occurs gradually with descent.

The critical feature of the eclogitization of amphiboles is that it must be endothermic and involve dehydration. The heat is provided by the surrounding mantle material, and the hot water rises, heating the higher level by this heat and by repeated hydration.

Summary and conclusions

A model of subduction zones composed of separate blocks descending vertically is proposed in this chapter. This model accounts for the phenomena explained by the slab model in a somewhat different way, and in addition explains almost all the problems unaccounted for by the slab model.

(1) The inclination of subduction zones, which has not been satisfactorily explained by the slab model, is the result of the consecutive detachment of blocks from the oceanic lithosphere (Figs. 2–4). The variation of the inclinations of the subduction zones is related to the thickness of the lithosphere and to the velocity of the upper plate (Fig. 8). The mechanical problem of tight bending of the lithospheric slab at the trench and its straightening downdip disappears.

(2) Though the blocks are descending vertically, the subduction zone appears as if it is moving retrogressively at the velocity of the upper plate. As the subduction zone is not coherent, the asthenospheric material can flow through it and thus the problem of the large countercurrents driven by the retrogressing slabs disappears.

(3) Marginal seas with their high heat flow occur where the plates cannot move fast enough to close the gap formed above the descending block, and they are absent where the plates can move fast enough (Figs 2–5).

(4) The depression of the deep sea trenches and the associated negative gravity anomaly are due to the lag in the inward movement of the plates and the upward movement of the asthenosphere behind the downgoing block. The seaward positive anomaly (Watts and Talwani, 1974) is due to the downbending of the plate prior to the break-off of the next block.

(5) The quasi-symmetrical inward motion of the plates on both sides of the descending block creates reverse faults of both subduction and obduction directions simultaneously.

(6) The top convergence velocity of *c.* 10 cm/yr irrespective of the sizes of the plates involved, is the result of the calculated velocity of descent of the blocks amounting to *c.* 4 cm/yr. The division of the convergence velocity between the plates on the two sides of the subduction zones is determined by the resistive viscous drag on the bottom of the plates, which depends on the size of the plates assuming that the resistive drag of continents is ten times greater than that of the oceans.

The six points listed hitherto were not satisfactorily explained by the slab model. The following three are explained by the block model in a different way from that of the slab model.

(7) The deep earthquakes of the Benioff zone explained by vertical shear between the vertically descending blocks, not by compression and extension along the slab.

(8) The block model subduction zones are strongly shortened and contain 2–4 times more material than the slab. The age of the terminal point is likewise 2–4 times older. This explains the relics of heavy lithosphere under the Himalayas and the Californian Transverse Ranges, where the lithosphere should have disappeared long ago according to the slab model.

(9) The fragmented nature of the block model subduction zone deteriorates its high-Q property, but the increased quantity of cold lithospheric material improves it. The local variations in Q observed under the Andes (Sacks, 1969) are better explained by the block model.

Neither the slab model nor the block model explain the high heat flow (magmatism, metamorphism) of the island arcs and the mountain ranges. This phenomenon is attributed to the exothermic reaction of amphibolitization, and to the heat transfer by water released by eclogitization. Together with this suggestion it seems that the block model is capable of providing a coherent physical mechanism to plate tectonics.

Acknowledgements

The critical comments and encouragement of Drs A. Lachenbruch, C. L. Drake, D. J. Stevenson, L. Knopoff, E. Okal, Z. Reches and Y. Kolodny are gratefully acknowledged.

Appendix 1

The drag force on an ellipsoid moving through a viscous medium is given by Stokes' formula (Lamb, 1932)

$$F_{\mathrm{d}} = 6\pi\eta R_{\mathrm{eff}}\, U, \tag{4}$$

where η is the viscosity of the fluid, R_{eff} is the effective radius of the ellipsoid defined below and U is the velocity of the ellipsoid. Equation (4) is applicable for flows for which $\rho U R_{\mathrm{eff}}/\eta \ll 1$, where ρ is the fluid density. This condition holds in the Earth for R_{eff} of the order of plate thicknesses and U of the order of plate velocities. For a free falling ellipsoid the drag force is balanced by the gravity force,

$$F_{\mathrm{g}} = \tfrac{4}{3}\pi abc\Delta\rho g, \tag{5}$$

where a, b, c respectively denote the lengths of the axes of the ellipsoid, $\Delta\rho$ is the difference in density of the ellipsoid with respect to the surrounding fluid and g denotes the gravitational acceleration. By equating (4) and (5) one obtains for the terminal velocity of the ellipsoid,

$$U = \frac{2}{9}\frac{abc\Delta\rho g}{R_{\text{eff}}\,\eta}. \tag{6}$$

For ellipsoids of revolution R_{eff} can be given in closed form.

For a sphere $a = b = c = R$, $R_{\text{eff}} = R$ and (6) gives

$$U = \frac{2\Delta\rho g R^2}{9\eta}$$

which is the result obtained by Stokes.

For a prolate ellipsoid of revolution $a > b = c$ moving parallel to its major axis the effective radius is given by

$$R_{\text{eff}} = \frac{8ab^2}{3(\chi_0 + a^2\alpha_0)},$$

where

$$\alpha_0 = 2(\xi^2 - 1)(\xi \coth^{-1}\xi - 1)$$

and

$$\chi_0 = \frac{2a^2(\xi^2 - 1)}{\xi}\coth^{-1}\xi,$$

and

$$\xi = \frac{a}{(a^2 - b^2)^{1/2}}$$

is a measure of the eccentricity.

When the prolate ellipsoid moves sideways the effective radius is given by

$$R_{\text{eff}} = \frac{8ab^2}{3(\chi_0 + b^2\gamma_0)},$$

where

$$\gamma_0 = \xi^2 - (\xi^2 - 1)\xi\coth^{-1}\xi.$$

Appendix 2

The exothermic heat of amphibolitization

$$2CaMgSi_2O_6 + 3MgSiO_3 + SiO_2 + H_2O \longrightarrow Ca_2Mg_5Si_8O_{22}(OH)_2,$$

$$\Delta H^P_{r,T} = \Delta(\Delta H^0_{f,T}) + \int_1^P \Delta V(1 - \Delta\alpha T)\,dP.$$

[$\Delta H^P_{r,T}$ is the enthalpy of reaction at pressure P and temperature T; $\Delta H^0_{f,T}$ is the standard enthalpy of formation per mole at T and 1 bar, ΔV is the molar volume change and α the coefficient of thermal expansion.]

Separating solid terms from H_2O terms

$$\Delta H^P_{r,T} = \Delta(\Delta H^0_{f,T})\,\text{solids} + \int_{1\,\Delta V(\text{solids})}^{P} (1 - T\Delta\alpha_{\text{solids}})\,dP$$

$$-\Delta H^0_{f,T}(H_2O) - \int_{1\,V_{H_2O}}^{P} (1 - T\alpha_{H_2O})\,dP$$

using $\Delta(\Delta H^0_{f,T})$ solids and $\Delta H^0_{f,T}$ (H_2O) data (Robie and Waldbaum, 1968) and estimating the pressure integral for water (Burnbaum, Holloway and Davis, 1969). The solid pressure term can be approximated

$$\int_1^P \Delta V\,\text{solids}\,(1 - T\Delta\alpha_{\text{solids}})\,dP \cong P\Delta V_{\text{solids}}(1 - T\Delta\alpha_{\text{solids}})$$

by data for ΔV_{solids} and $\Delta\alpha_{\text{solids}}$ (Robie and Waldbaum, 1968; Skinner, 1966). The results can be expressed for two widely differing conditions:

T (°C)	P (kb)	$H^P_{r,T}$ (kcal)
423	4	−37·4
923	10	−37·1*

* At 923 °C, the standard enthalpy of formation of tremolite was estimated by extrapolation (Robie and Waldbaum, 1968).

References

Ahrens, T. J. and Schubert, G., 1975. Gabbro eclogite reaction rate and its geological significance. *Rev. Geophys. Space Phys.* **13**, 383–400.

Anderson, D. L., 1967. Phase change in the upper mantle. *Science* **157**, 1165–1173.

Andrew, D. J. and Sleep, N. H., 1973. Numerical modelling of tectonic flow behind island arcs. *Geophys. J. R. astr. Soc.* **38**, 237–251.

Artyushkov, E. V., 1974. Can the Earth's crust be in a state of isostasy? *J. Geophys. Res.* **79**, 741–752.

Bowen, N. L. and Tuttle, O. F., 1949. The system MgO–SiO_2–H_2O. *Bull. Geol. Soc. Am.* **60**, 439–460.

Burnbaum, C. W., Holloway, J. R. and Davis, N. F., 1969. Thermodynamic properties of water up to 1000 °C and 10·000 bars. *Geol. Soc. Am. Sp. Paper* **132**.

Davies, G. F., 1977. Viscous mantle flow under moving lithospheric plates and under subduction zones. *Geophys. J. R. Astr. Soc.* **49**, 557–563.

Dietz, R. S., 1963. Collapsing continental rises: an actualistic concept of geosynclines and mountain building. *J. Geol.* **71**, 314–333.

Elsasser, W. M., 1967. Convection and stress propagation in the upper mantle. In *The Application of Modern Physics to Earth and Planetary Interiors* (Ed. Runcorn, S. K.), 223–246, Wiley Interscience, New York.

Elsasser, W. M., 1971. Sea-floor spreading as thermal convection. *J. Geophys. Res.* **76**, 1101–1112.

Engdahl, E. R., 1977. Seismicity and plate subduction in central Aleutians. In *Island Arcs, Deep Sea Trenches and Back Arc Basins* (Eds Talwani, M. and Pitman, W. C. III) Amer. Geophys Union, Maurice Ewing Series **1**, 259–271.

Essene, E. J., Hensen, B. J. and Green, D. H., 1970. Experimental study of amphibole and eclogite stability. *Phys. Earth Planet. Int.* **3**, 387–384.

Forsyth, D. and Uyeda, S., 1975. On the relative importance of the driving forces of plate motion. *Geophys. J. R. Astr. Soc.* **43**, 163–200.

Fry, N. and Fyfe, S. W., 1969. Eclogite and water pressure. *Contr. Miner. Petrol.* **24**, 1–6.

Fuchs, K., Bonjer, K. P., Bock, G., Cornea, I., Radu, C., Enescu, D., Jianu, D., Nourescu, A., Merkler, G., Moldaveanu, T. and Tudirache, G. 1980. The Rumanian earthquake of March 4, 1977. II: Aftershocks and migration of seismic activity. *Tectonophysics* (in press).

Garfunkel, Z., 1975. Growth, shrinking and long term evolution of plates and their implication for the flow pattern in the mantle. *J. Geophys. Res.* **80**, 4425–4432.

Griggs, D. T., 1972. The sinking lithosphere and the focal mechanism of deep earthquakes. In *The Nature of the Solid Earth* (Ed. Robertson, E. C.), 361–384, McGraw-Hill, New York.

Hadley, D. and Kanamori, H., 1977. Seismic structure of the Transverse Ranges, California. *Geol. Soc. Am. Bull.* **88**, 1469–1478.

Hager, B. H. and O'Connell, R. J., 1978. Subduction zone dip angles and flow driven by plate motion. *Tectonophysics* **50**, 111–134.

Isacks, B. and Barazangi, M., 1977. Geometry of Benioff zones: lateral segmentation and downward bending of the subducted lithosphere. In *Island Arcs, Deep Sea Trenches and Back Arc Basins* (Eds Talwani, M., and Pitman, W. C., III). Amer. Geophys. Union, Maurice Ewing Series **1**, 99–114.

Isacks, B. and Molnar, P., 1969. Mantle earthquake mechanism and the sinking of the lithosphere. *Nature* **223**, 1121–1124.

Isacks, B., Oliver, J. and Sykes, L., 1968. Seismology and new global tectonics. *J. Geophys. Res.* **73**, 5855–5899.

Jacoby, W. R., 1973. Model experiments in plate movements. *Nature* **242**, 130–134.

Jacoby, W. R., 1976. Paraffin model experiments of plate tectonics. *Tectonophysics* **35**, 103–113.

Jischke, M. C., 1975. On the dynamics of descending lithospheric plates and slip zones. *J. Geophys. Res.* **80**, 4809–4813.

Jordan, T. H., 1978. Composition and development of the continental tectonosphere. *Nature* **274**, 544–548.

Kanamori, H., 1971. Great earthquakes at island arcs and the lithosphere. *Tectonophysics* **12**, 187–198.

Karig, D. E., 1971. Origin and development of marginal basins in the western Pacific. *J. Geophys. Res.* **76**, 2542–2560.

Lachenbruch, A. H., 1976. Dynamics of a passive spreading centre. *J. Geophys. Res.* **81**, 1883–1902.

Lamb, H., 1932. *Hydrodynamics*, Dover, New York.

Laubscher, H. 1969. Mountain building. *Tectonophysics* **7**, 551–563.

Lliboutry, L., 1969. Sea floor spreading, continental drift, and lithosphere sinking with asthenosphere at melting point. *J. Geophys. Res.* **74**, 6525–6540.

Luyendyk, B. P., 1970. Dip of downgoing lithospheric plate beneath island arcs. *Bull. Geol. Soc. Am.* **81**, 3411–3416.

McKenzie, D. P., 1969. Speculations about the consequences and causes of plate motions. *Geophys. J. R. Astr. Soc.* **18**, 1–32.

McKenzie, D. P. and Sclater, J. G., 1968. Heat flow inside the island arcs in the north-west Pacific. *J. Geophys. Res.* **73**, 3173–3179.

Matsuda, T. and Uyeda, S., 1971. On the Pacific-type orogeny and its model: extension of the paired belt concept and possible origin of marginal seas. *Tectonophysics* **11**, 5–27.

Minear, J. W. and Toksöz, M. N., 1970. Thermal regime of a downgoing slab and new global tectonics. *J. Geophys. Res.* **75**, 1397–1419.

Minster, J. B. and Jordan, T. H., 1978. Present day plate motions. *J. Geophys. Res.* **83**, 5331–5354.

Minster, J. B., Jordan, T. H., Molnar, P. and Haines, E., 1974. Numerical modelling of instantaneous plate tectonics. *Geophys. J. R. Astr. Soc.* **36**, 541–576.

Mitronovas, W. and Isacks, B., 1971. Seismic velocity anomalies in the upper mantle beneath the Tonga–Kremadec island arc. *J. Geophys. Res.* **76**, 7154–7180.

Molnar, P. and Tapponier, P., 1977. The collision between India and Eurasia. *Scientific American* **236**, 30–41.

Morgan, W. J., 1972. Deep mantle convection plumes and plate motions. *Am. Ass. Petrol. Geol. Bull.* **56**, 203–213.

Oliver, J. and Isacks, B., 1967. Deep earthquakes zones, anomalous structure in the upper mantle and the lithosphere. *J. Geophys. Res.* **72**, 4259–4275.

Richter, F. M., 1977. On the driving mechanism of plate tectonics. *Tectonophysics* **38**, 61–88.

Ringwood, A. E. and Green, D. H., 1966. An experimental investigation of the gabbro eclogite transformation and some geophysical implications. *Tectonophysics* **3**, 383–427.

Robie, R. A. and Waldbaum, D. R., 1968. Thermodynamic properties of minerals and related substances at 298·15 K (25·0 °C) and at one atmosphere (1·013 bars) pressure and at higher temperatures. *Bull. U.S. Geol. Surv.* **1259**.

Sacks, I. S., 1969. Distribution of absorption of shear waves in South America and its tectonic significance. *Carnegie Institution Yearbook* **67**, 339–344.

Skinner, J. B., 1966. Thermal expansion. In *Handbook of Physical Constants* (Ed. Clarke, S. P. Jr), *Mem. Geol. Soc. Am.* **97**, 78–96.

Sleep, N. H. and Toksöz, M. N., 1971. Evolution of marginal basins. *Nature* **33**, 548–550.

Solomon, S. C., Sleep, N. H. and Richardson, R. M., 1975. On forces driving plate tectonics inferred from absolute plate velocities and interplate stress. *Geophys. J. R. Astr. Soc.* **42**, 769–801.

Stauder, W., 1968. Tensional character of earthquake foci beneath the Aleutian trench with relations to sea floor spreading. *J. Geophys. Res.* **73**, 7693–7701.

Stevenson, D. J. and Turner, J. S., 1977. Angle of subduction. *Nature* **270**, 334–336.

Toksöz, M. N. and Bird, P., 1977. Formation and evolution of marginal basins and continental plateaus. In *Island Arcs, Deep Sea Trenches and Back Arc Basins* (Eds. Talwani, M. and Pitman, W. C. III), Am. Geophys. Union, Maurice Ewing Series **1**, 379–394.

Tovish, A., Schubert, G. and Luyendyk, B. P., 1978. Mantle flow pressure and angle of subduction; non-Newtonian corner flow. *J. Geophys. Res.* **83**, 5892–5898.

Turcotte, D. L., Haxby, W. F. and Ockendon, J. R., 1977. Lithosphere instabilities. In *Island Arcs, Deep Sea Trenches and Back Arc Basins* (Eds Talwani, M. and Pitman, W. C. III), Amer. Geophys. Union, Maurice Ewing Series **1**, 63–70.

Turcotte, D. L. and Oxburgh, E. R., 1969. Convection in a mantle with variable physical parameters. *J. Geophys. Res.* **74**, 1458–1474.

Turcotte, D. L. and Schubert, G., 1971. Structure of the olivine–spinel phase boundary in the descending lithosphere. *J. Geophys. Res.* **76**, 7980–7987.

Turner, J. S., 1973. Convection in the mantle; a laboratory model with temperature dependent viscosity. *Earth Planet. Sci. Lett.* **17**, 369–374.
Uyeda, S., 1978. *The New View of the Earth*, W. H. Freeman, San Francisco.
Watts, A. B. and Talwani, M., 1974. Gravity anomalies seaward of the deep sea trenches and their tectonic implications. *Geophys. J. R. Astr. Soc.* **36**, 57–90.

Thermomechanical Models of the Rio Grande Rift

R. J. BRIDWELL, C. A. ANDERSON

Los Alamos Scientific Laboratory, University of California, Los Alamos, New Mexico, USA

Introduction

Genesis of a continental rift is closely tied to understanding its thermal history. Interactions of convective heat and mass with materials of the lithosphere cause thinning, magma genesis, uplift, an extensional stress state, normal faulting and, ultimately, extrusion of surficial volcanics.

Various static models have been proposed along the Rio Grande rift to discuss the geometry of the continental lithosphere. Decker and Smithson (1975) utilized Bouguer gravity data to suggest a low density sill at depths of 31 km beneath the southern rift. Surface heat flows $>2{\cdot}5$ HFU, (Reiter *et al.*, 1975) (1 HFU = μcal/cm²/s = 41·8 mW/m²) provided a geotherm which intersects typical mantle solidi at depths of 25–31 km in a conductive lithosphere (Decker and Smithson, 1975). Ramberg, Cook and Smithson (1978) models suggest crustal attenuation may be accounted for by extension and partial melting in the southern rift (Cook *et al.*, 1979). In the northern rift, a mass deficiency of $\sim 0{\cdot}1$ g/cm³ for the upper mantle is caused by a mantle diapir with high temperatures beneath the rift (Bridwell, 1976); a second paper, using steady state conductive models and estimates of the geotherm from surface heat flow and a conductive lithosphere predicted a viscosity minima of 10^{20} poise beneath the rift with an associated increase in effective stress gradients in the crust and upper mantle beneath the rift (Bridwell, 1978a). The physical state of the continental lithosphere has been characterized as having high temperatures, low density, low viscosity, and high effective stress gradients beneath the rift at depths of 40–80 km (Bridwell, 1978b).

In this chapter, we treat the dynamic aspects of lithospheric thinning by providing coupled thermomechanical models, using the finite element technique, of a continental rift and platform. A model is presented to evaluate the dynamic process of uplift and continental rifting. This model uses non-linear flow laws, thermally dependent material properties, thermal gradients varying from continental rift to platform, and the Boussinesq approximation of temperature dependent buoyancy to drive the uplift. Constraints for new dynamic models are based on geological and geophysical data such as uplift, uplift rates, Cenozoic volcanism in New Mexico, crust and mantle geotherms from rift and platform, and present surface heat flow. Uplift rates and magnitudes for northern New Mexico are calculated from regionally averaged topography, palaeobotanical data and stratigraphic relations. Crust and mantle geotherms are determined petrologically using natural xenoliths which provide P, T data. A suite of temperature conditions is analysed to show the spectrum of thermomechanical behaviour during development of a continental rift. This dynamic study is the first to consider the deformable lithosphere and convective flow in the underlying asthenosphere for the Rio Grande rift.

Regional uplift

Modelling regional uplift requires knowledge of an initial undeformed structural datum, the amount of uplift, and its duration. A number of geologic events, such as formation of the late Eocene erosional surface of the southwestern United States, deposition of widespread Miocene basins, temporal

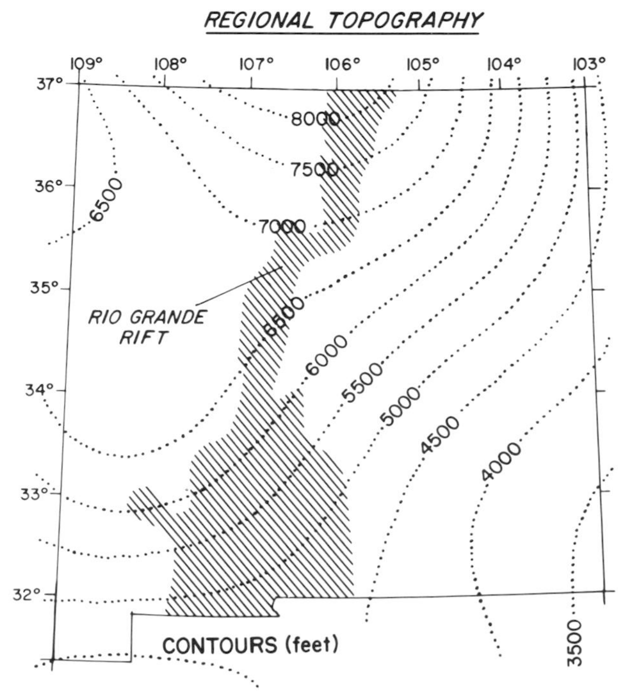

Figure 1. Regional topography of New Mexico. The Rio Grande rift is outlined by the hatched region. Contour interval in feet. Modified from Aiken *et al.* (1978).

nature of magmatism and volcanism including a mid-Miocene lull in volcanism, duration of uplift, and present elevations, bear directly on the mechanism of continental rifting.

The post-Laramide, late Eocene erosion surface of the southern Rocky Mountains and Datil–Mogollon volcanic field of west-central New Mexico is a fundamental regional structural datum (Epis and Chapin, 1975). The late Eocene erosion surface had initially low relief of ~0·5–0·9 km ~35 Myr ago. Regional extension began ~29 Myr ago (Chapin and Seager, 1975), with little uplift and development of broad basins in Miocene time along the length of the rift from 26 to 10 Myr ago (Chapin, personal communication, 1979). Present uplift occurs over a broad region in New Mexico. Figure 1 shows regionally averaged topography ranging in elevation from ~1·2 to 2·4 km (Aiken *et al.*, 1978). The majority of the Rio Grande rift occurs in regions showing topography of 1·5–2·1 km. If the Eocene erosion surface was at elevations of 0·5–0·7 km (Epis and Chapin, 1975; Axelrod and Bailey, 1976), then relative regional motion would be 1–1·4 km. Since geologic relations in latest Miocene time ~10 Myr ago date the onset of motion, regional uplift also occurred with a relative velocity of 1–1·4 km/10 Myr.

Cenozoic volcanism and thermal evolution of rifting

Thermal evolution of the Rio Grande rift is indicated by the spatial, temporal and chemical nature of volcanic and plutonic rocks. Chapin and Seager (1975) documented 161 K/Ar and fission track dates of late Eocene to Holocene age in New Mexico as shown in Fig. 2(a). There are clearly two temporal maxima at 40–20 Myr and 15–1 Myr ago separated by the Miocene lull. Taken at face value, this would suggest two thermal pulses within the lithosphere since formation of the late Eocene erosion surface.

Application of a spatial filter reduces the complexity of the thermal pulses. Most of the rocks shown in Fig. 2(a) in the time span 40–20 Myr ago are from the Datil–Mogollon volcanic pile in south-western New Mexico although most of the stocks and dikes are associated with the Rio Grande rift (Chapin, personal communication, 1979). Hence, although some volcanism is associated with the rift prior to the Miocene lull, it is small in volume compared to the basaltic volcanism of late Miocene to recent time (e.g. Baldridge *et al.*, 1980). Note in Fig. 2(b) the regional distribution of volcanism from Colorado to central New Mexico.

The presence and age of basaltic volcanism of <10 Myr age indicates melting and magma formation in the lowermost crust and uppermost mantle. Although basaltic volcanism (basaltic andesites of Chapin and Seager, 1975) occurred prior to the Miocene lull, only small volumes of rocks occur in the Espanola, Albuquerque, southern San Luis and Arkansas basins. The relatively voluminous fields of the Taos Plateau, Jemez Mountains, Cerros del Rio, Mt Taylor, Cat Hills, Wind Mesa and others followed the Miocene lull. It

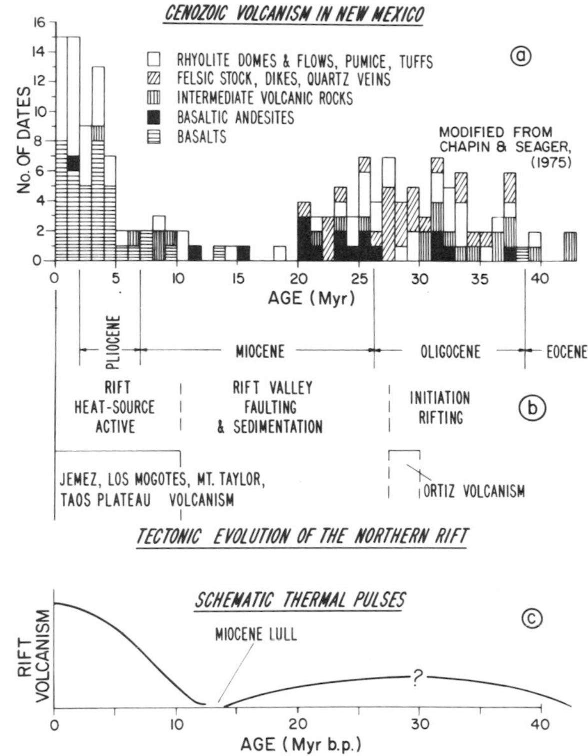

Figure 2. Cenozoic volcanism, tectonic evolution and schematic thermal history of Rio Grande rift. (a) 161 K/Ar and fission track dates from Chapin and Seager (1975). (b) Tectonic evolution of Rio Grande rift north of Socorro, New Mexico. (c) Postulated thermal pulse associated with rifting in New Mexico. The pre-Miocene-lull thermal pulse is schematic because most of the data set represent age dates from the Datil–Mogollon volcanic field.

is tempting to postulate thermal changes such as mass flux in the lithosphere and increased surface heat flow based on the simple function shown in Fig. 2(c).

Thermal gradients of a continental rift

A continental platform geotherm has been determined for the Colorado Plateau to depths of several hundred kilometres by McGetchin and Silver (1972), shown in Fig. 3. The continental rift geotherm, shown in Fig. 3, is a composite geotherm utilizing geobarometry from crust and mantle xenoliths principally at Kilbourne Hole Maar, New Mexico. Geobarometry data are discussed by Bridwell and Anderson (1980). The crustal segment has a slope of 32 °C/km and agrees well with an average surface heat flow of 2·5 HFU determined by Decker and Smithson (1975) as shown by Padovani and Carter (1977). The asthenosphere is modelled as a subsolidus regime beneath the

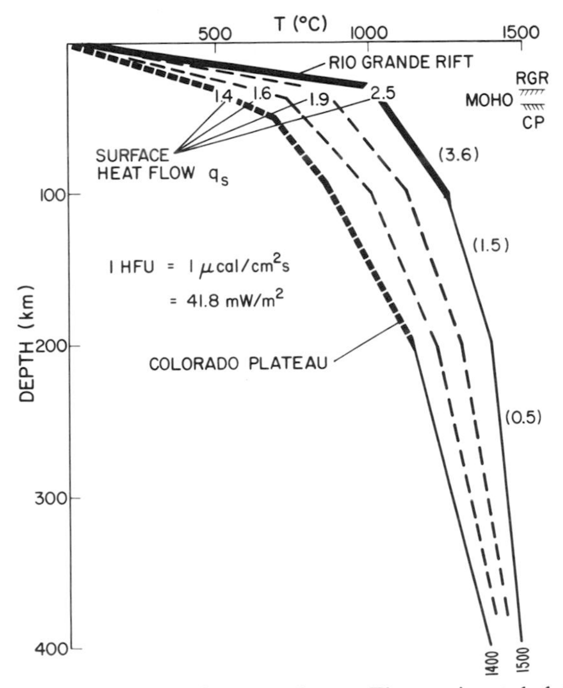

Figure 3. Continental rift and platform geotherms. The continental platform geotherm is shown as a heavy dashed line (McGetchin and Silver, 1972). The continental rift geotherm is shown as a heavy solid line. The remaining solid lines are extrapolated to temperatures of the olivine–spinel phase transition. Numbers in parentheses above the rift geotherm represent gradients in °C/km. Numbers at Moho depths represent the surface heat flow in heat flow units (HFU).

continent. Since temperatures beneath the continental rift are excessive at relatively shallow depths in the presence of Mercier's (1977) 4 °C/km gradient to 100 km, we assume a superadiabatic gradient to 200 km and an adiabatic gradient (0·5 °C/km) for the rift geotherm below 200 km. This results in a temperature increase of ~100 °C relative to the continental geotherm at the 400 km olivine–spinel phase transition.

Thermomechanical model

Recent work has produced one-dimensional models of the coupled thermo-mechanical structure of oceanic and continental lithospheres (Schubert and Turcotte, 1972; Froidevaux and Schubert, 1975; Schubert, Froidevaux and Yuen, 1976). A quasi-two-dimensional model of oceanic mantle circulation with partial shallow return flow considers plate stresses (Schubert *et al.*, 1978). In this paper, we consider the fully two-dimensional, coupled thermo-mechanical behaviour of a continental rift and platform. The thermal and mechanical structure of a continental lithosphere and asthenosphere are coupled through several effects; the dependence of viscosity on temperature, temperature-dependent material properties and the Boussinesq approximation of density change due to temperature.

A conceptual two-dimensional model of initial conditions for a continental rift and platform in plane strain, symmetric about the ordinate, is shown in Fig. 4. The relatively cool continental platform on the right has negative buoyancy whereas the less-dense relatively hot continental rift on the left has positive buoyancy. The relative positive buoyancy causes asthenospheric flow toward the rift at depth, hot buoyant upward diapiric motion beneath the rift and lateral shear flow outward beneath the crust associated with uplift of the shoulder of the rift. In Fig. 4, we list thermal and mechanical quantities required by our models.

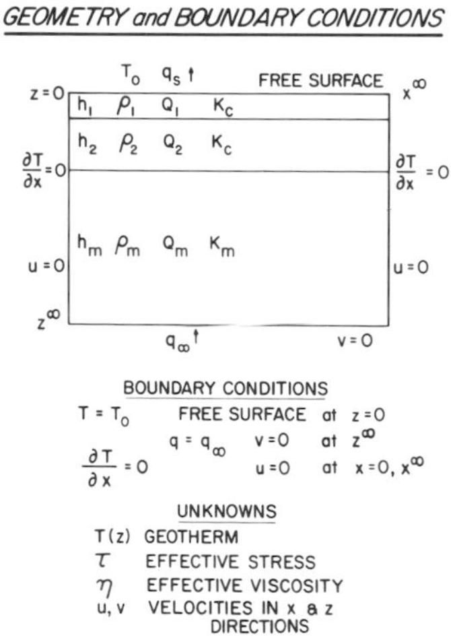

Figure 4. Schematic diagram of thermomechanical model, constraints and unknowns for numerical calculations of rifting.

At the surface, we specify the temperature $T = T_0$. At great depth (400 km), we specify the mantle flux, q_{∞}. Adiabatic conditions are specified on the vertical boundaries. The heat generation, H, and the thermal conductivity, K, vary within each layer. The densities are shown for each layer. Mechanical boundary conditions consist of no flow across either the 400 km olivine phase transition or the vertical boundaries. The upper two layers are lithosphere with crust and upper mantle whereas the lower layer is asthenosphere. The unknowns of temperature, effective viscosity, heat flow and topography are provided by model calculations.

Our model includes the following features:

(1) linear or non-linear flow law relating shear stress, shear stress rate and strain rate, Eqn (1);

(2) temperature- and pressure-dependent viscosity;

(3) temperature-dependent thermal conductivity;

(4) temperature-dependent bulk coefficient of thermal expansion;
(5) temperature-dependent density through Boussinesq approximation for buoyancy, Eqn (4);
(6) material convection using a frame invariant stress rate tensor;
(7) a free surface; and
(8) an elastically deformable crust.

Mathematical formulation

The equations governing the thermomechanical behaviour of a continental rift and platform are the constitutive law and the two-dimensional equations of energy conservation and stress equilibrium (Anderson and Bridwell, 1980).

The constitutive law used for describing the mechanical behaviour of the model is of the form

$$\dot{\varepsilon}_{ij} = \frac{(1+\nu)}{E}\sigma_{ij}^{\nabla} - \frac{\nu}{E}\delta_{ij}\sigma_{kk}^{\nabla} + \frac{1}{\eta}s_{ij}, \tag{1}$$

where $\dot{\varepsilon}_{ij}$ is the strain rate tensor, σ_{ij}^{∇} is a frame invariant stress rate tensor (Prager, 1961), s_{ij} is the stress deviator tensor, ν and E are Poisson's ratio and Young's modulus respectively, and η is the material viscosity. In Eqn (1) the indicial notation is applied $(i,j = 1, 2, 3)$ with repeated index implying a summation and with $\delta_{ij} = 0$, $i \neq j$ or 1, $i = j$ being the Kronecker delta. The viscosity is determined by the state of stress and the temperature through an equation of the Weertman form,

$$\eta = \frac{\exp(E^* + pV^*)/RT}{A\tau^{n-1}}, \tag{2}$$

where A and n are experimentally determined constants, E^* and V^* are activation energy and volume respectively, p is the mean stress, R is the universal gas constant, T is the absolute temperature and τ is the effective stress given by

$$\tau = (\tfrac{1}{2}s_{ij}s_{ij})^{1/2}. \tag{3}$$

The Boussinesq approximation assumes that density variations due to changes in temperature occur in the gravitational or buoyancy term of the equations of motion. All other variations in density are neglected. The approximation is justified if variations are small compared with the mean density of the rock. Dynamically important variations in density, ignoring the effect of phase changes, can be written as a function of temperature alone

$$\rho = \rho_0[1 + \alpha(T_r - T)], \tag{4}$$

where ρ_0 is the mean density at a reference temperature T_r, T is the temperature of an element and α the bulk or volume coefficient of thermal expansion.

The energy equation is

$$\rho C_P \frac{\partial T}{\partial t} = H + K \frac{\partial^2 T}{\partial x_i \partial x_i} + \sigma_{ij} \dot{\varepsilon}_{ij}, \tag{5}$$

where C_P is heat capacity, H is heat generation, K is thermal conductivity, and σ_{ij} and $\dot{\varepsilon}_{ij}$ are the stress and strain-rate tensors. Again, repeated index implies summation over that index. The first term represents material convection, the second heat generation, the third heat conduction and the last term is shear strain heating.

Equations (1)–(4) together with the equilibrium equations, the strain rate–velocity relations and the definition of the frame invariant stress rate (Prager, 1961) constitute the physical equations of the thermomechanical model. The equations are made discrete using the finite element method and a numerically stable time-stepping algorithm is used to advance the physical quantities stress, temperature and velocity from an initially specified equilibrium stress and temperature state. Details are given in Anderson and Bridwell (1980).

Thermomechanical material properties

The behaviour of the equations of motion depends on such material parameters as thermal conductivity, heat capacity, density, bulk coefficient of thermal expansion, elastic modulus and non-Newtonian rheological coefficients. Each will be discussed in detail below.

Thermal conductivities have been measured in surficial granitic rocks. Of 100 values, the average is 2·3 W/m °C (Blackwell, 1971). No temperature variation is known. For upper mantle rocks, such as pyrolite or spinel lherzolite, Schatz and Simmons (1972) found a temperature dependence in the range 325–1625 °C for polycrystalline forsterite-rich olivines. In our model, conductivities of the crust are from measured values cited by Blackwell (1971) and conductivities of the mantle are from Schatz and Simmons (1972). The heat capacity is assumed constant at ~1250 J/kg °C.

Densities are based on seismic profiles using the well-known relation between P-wave velocities and density. Densities are 2·7 g/cm^3 for the upper crust, 2·85 g/cm^3 for the lower crust, 3·3 g/cm^3 for the cold upper mantle and 3·2 g/cm^3 for the low velocity zone in the mantle (Olsen, Keller and Stewart, 1979).

The bulk or volume coefficient of thermal expansion varies linearly with temperature. Several minerals are shown to increase their thermal expansion linearly to 800 °C (Skinner, 1966). We use crust and mantle thermal expansions from Skinner extrapolated to 1400 °C because of the clear separation in α's for feldspars and olivine/pyroxenes. For feldspars of Ab_{56}–An_{44}, α ranges from 12 to 24×10^{-6} °C whereas for olivines and pyroxenes α ranges from 26 to 40×10^{-6} °C.

Elastic moduli are chosen as a maximum of 1 Mbar for crustal materials. Because of the stability criterion on time steps for the finite element method (Anderson and Bridwell, 1980) the elastic moduli are decreased as a function of temperature to values ~100 bars in regions at temperatures of 1400 °C.

We assume that the creep law describes the Newtonian rheology of the crust and the non-Newtonian rheology of the mantle, using for data those of high-temperature dislocation creep for dunite. For the Newtonian behaviour, we choose a crustal viscosity which provides uplift rates of the free surface consistent with geologic data. The mantle rheology is non-Newtonian and we use an exponent of 3 on the effective stress. A set of models was evaluated to predict values of E^*, V^* and A for the rift. We eventually ran models described with $E^* = 100$ kcal/mol, $V^* = 17\,cm^3/mol$ and $A = 3{\cdot}4\,kbar^{-n}s^{-1}$. These values are consistent with experimental determinations of a "wet" dunite rheology as defined by Carter (1976). A recent determination of $V^* = 14 \pm 3\,cm^3/mol$ has been provided by Ross *et al.* (1979).

Lateral distribution of heat generation is assumed to give smooth temperature profiles ranging from low temperatures of a continental platform (600 °C) to a successively increasing suite of high temperatures for a continental rift (>900 °C) at the crust–mantle interface. The rift heat generation values from Table I are arbitrarily imposed 60 km from the left margin of the model, which, by symmetry, is the centre of the rift.

TABLE I *Heat generation to produce thermal gradients*

	Continental rift HGU		Continental platform HGU	
Depth (km)	$T = 750\,°C^a$ $q_s = 71\,mW/m^2$ = 1·6 HFU	$T = 875\,°C$ $q_s = 88\,mW/m^2$ = 1·9 HFU	$T = 1000\,°C$ $q_s = 105\,mW/m^2$ = 2·5 HFU	$T = 600\,°C$ $q_s = 59\,mW/m^2$ = 1·4 HFU
0–10	2·1	2·1	2·1	2·1
10–40	1·18	1·37	1·65	0·88
40–100	0·23	0·20	0·186	0·025
100–200	0·095	0·073	0·069	0·025
2000–400	0·026	0·021	0·013	0·025

(1) A basal flux of $4{\cdot}18\,mW/m^2$ is applied at a depth of 400 km.
(2) 1 HGU = $10^{-13}\,cal/cm^3\,s = 0{\cdot}418\,\mu W/m^3$.
(3) q_s is surface heat flow where 1 HFU = $10^{-6}\,cal/cm^2\,s = 41{\cdot}8\,mW/m^2$.
[a] Temperatures at base of crust (40 km).

Discussion of results

The fully two-dimensional thermomechanical problem is formidable! The mathematical formulation provides temperature T, velocities u, v and effective viscosity η as unknowns. We reduce some of the non-uniqueness by requiring vertical temperature profiles on the left- and right-hand boundaries of the model to concur with Fig. 3. We require the mantle viscosities to be determined by experimental measurements of dunite and regional geotherms although crustal viscosities are truly unknown. Since there is a range of measurements for flow coefficients of Eqn (2) as well as a range of temperatures from Fig. 3 which determine the viscosity and hence the velocity fields, we have conducted a number of numerical experiments to predict the viscosity. We have estimated the crustal viscosity by varying coefficients to obtain free surface velocities in quasi-steady-state flow which are consistent with geologic calculations of surface motions mentioned in the section on regional uplift.

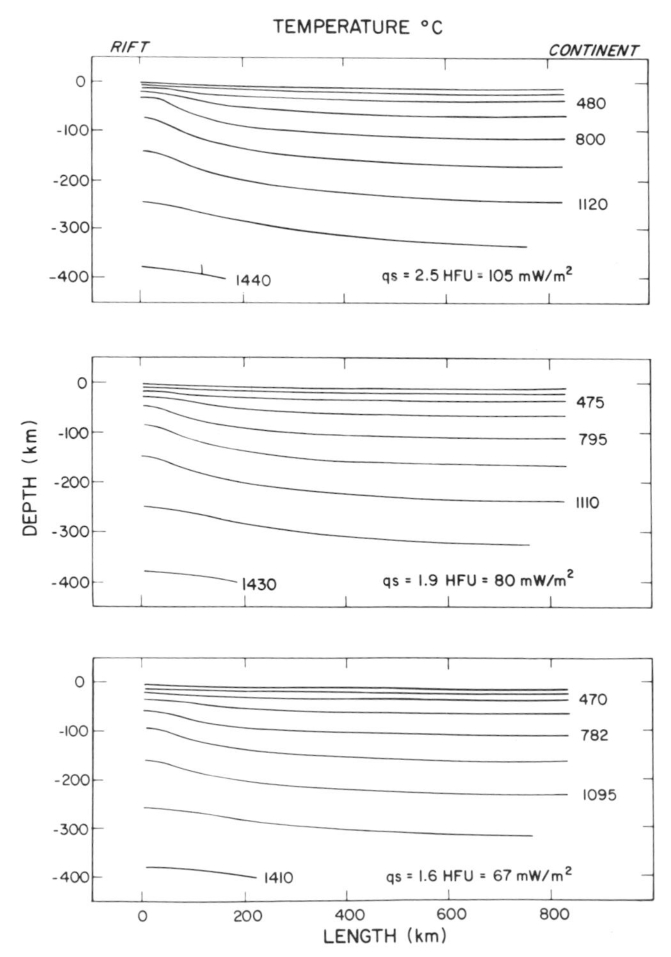

Figure 5. Temperature profiles beneath a continental rift and platform. The notation of increase in surface heat flow from 1·6 to 2·5 HFU is shown on the top left boundary of the model. Note increasingly higher temperatures at shallow depths. These cases are shown in profile from Fig. 3.

From these numerical experiments we present calculations of the viscosity, temperature and velocity fields associated with formation of a continental rift.

A succession of thermal solutions was investigated to evaluate thermo-mechanical behaviour of continental rifts. Equation (5) is solved at t_{∞} with H specified in Table I. These solutions produce temperature at the base of the crust of 750, 875 and 1000 °C which are equivalent to heat flows of 1·6, 1·9 and 2·5 HFU respectively. Figure 5 shows isotherm contours of a rift model symmetric about the left ordinate. These thermal fields produce sufficient temperature differences to generate density changes (cf. Eqns (4) and (5)) which drive buoyancy and create uplift. Density changes of less than 3% of regional values are sufficient to produce regional uplift.

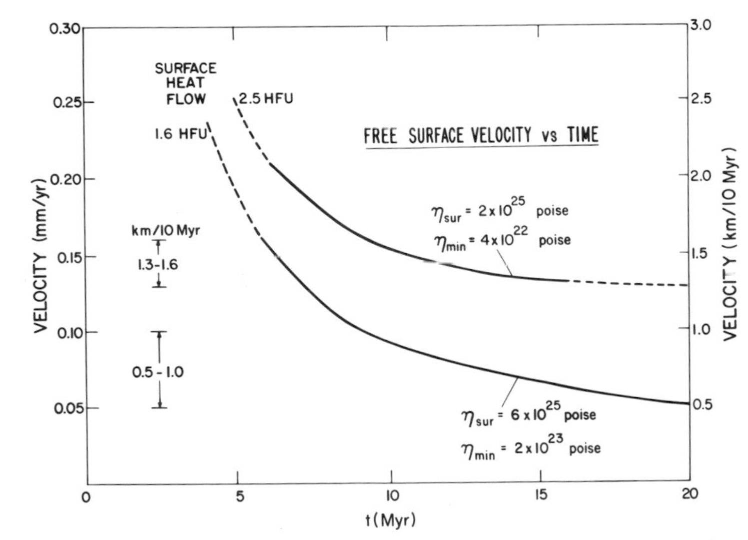

Figure 6. Plot of free surface velocities as a function of time for surface heat flow $1{\cdot}6 < q_s < 2{\cdot}5$ HFU. Velocity profiles occur on symmetric centreline of model.

Positive buoyancy associated with the temperature increase causes sub-solidus flow of crust and mantle materials from regions of relatively cool continent to hot rift. Free surface velocities are one of the unknowns of the numerical calculation. The vertical free surface motion at the centre of the rift is shown in Fig. 6 for surface heat flows having the range $1{\cdot}6 < q_s < 2{\cdot}5$ HFU. With time, velocity shows an initial rapid decay for ~8 Myr followed by a 15–20 Myr period where velocities are quasi-steady-state. The initial decay represents relaxation from the estimated elastic pre-stress state to the steady-state-creep flow behaviour. We shall use flow velocities from the model for $8 < t < 25$ Myr for further discussions of steady-state creep and formation of continental rifts.

Although temperatures at the base of the crust vary considerably, vertical crustal velocities from Fig. 6 for $t > 8$ Myr fall in a relatively narrow band of

0·5–1·3 km/10 Myr. Figure 7 is a plot of vertical velocity versus depth for the symmetric centreline of the rift model. Velocities of the hot buoyant mantle diapir are ~1·2 km/Myr. An explanation for the variations in relative velocities lies in the differing behaviour of the lithosphere and asthenosphere.

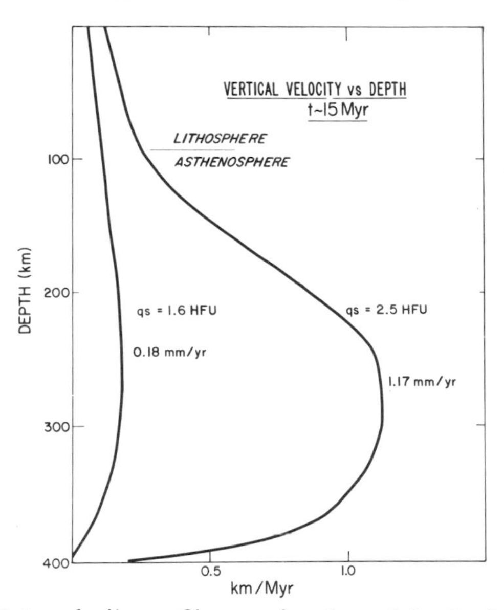

Figure 7. Steady-state velocity profile as a function of depth, the lithosphere and asthenosphere for continental rift. This figure is a snapshot of the instantaneous velocity field for $t \sim 15$ Myr as q_s ranges from 1·6 to 2·5 HFU. *A priori*, the base of the lithosphere is that change in slope where the velocity increases markedly. Note that there is little change where $q_s \sim 1·6$ HFU whereas the increase in heat transport thins the lithosphere as $q_s \sim 2·5$ HFU.

The lithosphere is capped by an elastic crust with large temperature changes, yet which has only limited ability to creep. The upper mantle and asthenosphere are composed of mantle materials whose creep behaviour is quite sensitive to increased temperature although these temperature changes are smaller. The lithosphere moves rather slowly, regardless of the surface heat flow, only showing a small increase in velocity with increasing temperature. In effect, it tends to damp out motions of the hot buoyant asthenosphere. In contrast, the buoyant asthenosphere beneath the rift moves 10 times faster, convecting heat and mass upward.

Horizontal motion occurs concomitantly with uplift. Horizontal velocities at the free surface are zero at the symmetric centreline and increase to ~1 km/10 Myr at distances of 90–150 km from the rift. The horizontal velocities at the crust–mantle interface at 5–10% higher than surface velocities, indicating crustal shear. The assumptions of a continuum crust, high crustal viscosities and a zero horizontal velocity on the left boundary provide spreading rates approximately equal to uplift rates.

Two-dimensional cross-sections of the steady-state velocity field as a function of increasing surface heat flow are shown in Fig. 8. The instantaneous velocity vectors of the lithosphere ($Z < 100$ km) are small ($V < 0{\cdot}05$ mm/yr) whereas the velocities in the asthenosphere beneath the rift ($V \sim 1$ mm/yr) are

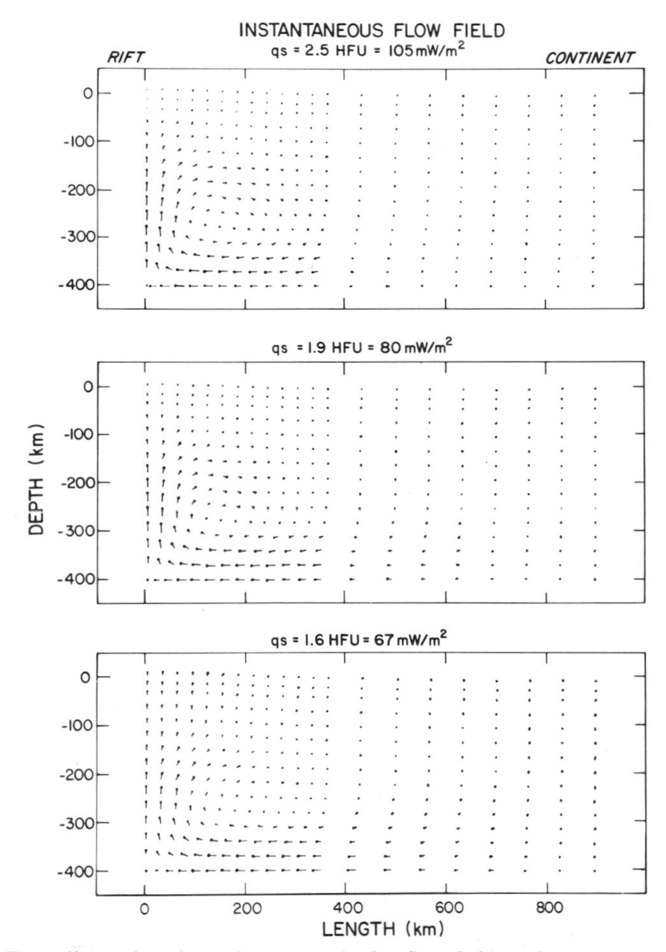

Figure 8. Two-dimensional steady-state velocity flow fields as heat transport increases beneath Rio Grande rift. Successive snapshots of the instantaneous flow field at $t \sim 15$ Myr show the growth of the hot, buoyant mantle diapir beneath a continental rift.

large. These cross-sections illustrate the shape of the velocity flow field where hot, buoyant asthenosphere flows laterally towards the centre of the rift, ascends vertically and spreads outward beneath the lithosphere. As heat transport and surface heat flow increase, the velocity function of the hot buoyant mantle diapir increases by a factor of 10 as shown in Fig. 7, the half flow field narrows from >175 km to ~ 100 km as shown in Fig. 8, a marked temperature increase occurs in the lithosphere beneath the rift, and transverse shear increases between lithosphere and asthenosphere. The actual hot, buoyant mantle diapir is ~200 km wide at depths of 200–300 km, increases its width to 300–400 km at the lithosphere–asthenosphere boundary, thins the lithosphere and creates broad crustal upwarp associated with major continental rifts. In effect, the buoyant process narrows and concentrates heat transport beneath the rift, increases upward velocity, arches the lithosphere and generates broad surface swells, a well-known characteristic of continental rifts since Gregory (1921) defined the structure and surface features of African rifts.

The *a priori* choice of a 400 km deep model provides a local rather than a global solution to mantle convection. This local solution does predict uplift rates and mantle viscosities which are in accord with geological data and rebound studies of viscosity (Cathles, 1975; Hager and O'Connell, 1979). The viscosity structure is discussed in a following section. Recent studies indicate convection may be confined to a series of superimposed layers (Richter, 1979). If convection occurs at all depths but is confined to discrete layers such as upper mantle ($Z < 670$ km) and lower mantle ($670 < Z < 2900$ km), our local model is consistent with convection theory.

Creep coefficients A, E^*, V^* and n from Eqn (2) have a range of behaviour depending on $\dot{\varepsilon}$, T and material (Carter, 1976). In general, as E^* and V^* increase, the viscosity increases and velocities decrease (both in a non-linear fashion). The parameter A linearly scales the viscosity (Eqn (2)). The free surface velocities are determined by the viscous creep of Eqn (2). For the quasi-elastic crust, n is assumed to be 1, $V^* = 0$, whereas A and E^* are scaled to produce a viscosity. This viscosity is actually unknown – its value controls surface uplift and hence if we can predict free surface velocities, a constraint has been placed on crustal viscosities associated with rifting. A wide range of material parameters for A, E^*, V^* and n was studied for upper mantle materials. An $A = 3{\cdot}4\,\text{kbar}^{-n}\,\text{s}^{-1}$, $n = 3$, $E^* = 110$ kcal/mol and $V^* = 17\,\text{cm}^3/\text{mol}$ is used in this study.

The viscosity of the lithosphere and asthenosphere is shown in Fig. 9. Figure 9(a) shows the general character of the two-dimensional viscosity field for continental rift and platform. The increase of temperature at the base of the crust and subsequent increase of surface heat flow from 1·6 to 2·5 HFU produced a decrease in crustal viscosity by a factor of 2 from 7×10^{25} to $3{\cdot}5 \times 10^{25}$ poise (see Fig. 9) and an increase in free surface velocity by a factor of 3 from 0·5 to 1·3 km/10 Myr (see Fig. 6). The numerical simulations suggest a crustal viscosity of $> 10^{25}$ poise is appropriate. A broad, relatively low viscosity region occurs below the rift in the asthenosphere. A relatively small lateral viscosity variation occurs in the lithosphere, with equivalent variation occurring in the crust. As one progresses away from the rift, viscosity increases and the lithosphere thickens. Figure 9(b) shows viscosity minima of $\sim 10^{22}$ poise beneath the rift and $\sim 10^{23}$ poise beneath the platform.

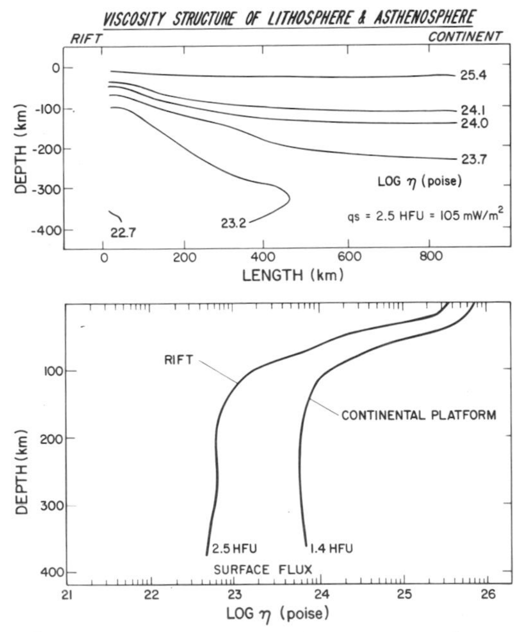

Figure 9. Viscosity of a continental rift and platform. (a) Two-dimensional variation of viscosity for present surface heat flow of Rio Grande rift.
(b) Range of viscosity and minima as temperature increases at base of crust beneath rift.

The viscosity structure of the Earth has been studied in the past 45 years using post-glacial rebound data, non-tidal, secular angular acceleration of the earth, polar wandering and microscopic mechanisms of flow in solids by a host of researchers. The notable studies of Cathles (1975), Peltier (1976) and Hager and O'Connell (1979) review the viscous nature of the solid Earth. Present evidence is consistent with a mantle with viscosity on the order of 10^{22} poise overlain by a low viscosity layer of undetermined thickness and viscosity (Hager and O'Connell, 1979). It is likely that mantle viscosities are 1–2 orders of magnitude lower under oceanic and tectonically active continents than under shields. The global kinematic study of large-scale mantle flow of Hager and O'Connell (1979) found that the predicted dip of flow for subduction zones agreed best with seismic dip using the following model. A Newtonian viscosity of 10^{25} poise, extending to depths of 64 km, overlies a thin low viscosity layer of 4×10^{20} poise. The lower mantle is modelled as a constant viscosity 10^{22} poise. This model is similar to that of Cathles (1975). Note that the viscosity models of Fig. 9 are similar, e.g. a lithosphere of $\sim 10^{25}$ poise overlying an asthenosphere of $\sim 10^{23}$ poise with a tectonically active continental rift having a viscosity of 10^{22} poise.

A tenuous trail is available to explain initial convection of mass and heat into the lithosphere. The rate of rift volcanism, a surface expression of deep seated convection, is small during Oligocene to late Miocene time (40–20 My ago). A sharp increase in rift volcanism occurred following the Miocene lull 17–13 Myr ago, indicating a major change in crustal magmatism and implying convective sources of heat in the lower lithosphere and rapidly increasing temperatures at the base of the crust (see Fig. 2(c)). If the proto-rift region had a nearly continental geotherm and average surface heat flow ($\sim$1·4 HFU), prior to the Miocene lull, its viscosity would have been greater than shown in Fig. 9 resulting in a smaller free surface velocity. Hence uplift might have been small. Convective heat transfer following the lull could have increased crustal temperature, decreased mantle and crustal viscosities, and increased surface velocities to values of 1 km/10 Myr. Free surface velocities of 1–1·5 km/10 Myr for present surface heat flow of 2·5 HFU and crustal temperatures of $\sim$1000 °C satisfy geologic evidence for uplift. Regional uplift of 1–1·5 km produced present topography of $\sim$2 km (see Fig. 1); uplift began $\sim$10–15 Myr ago based on palaeobotanical evidence and block faulting. Recall though, that this is a trade-off because crustal viscosities are unknown. If crustal viscosities are larger (e.g. 2×10^{26} poise) free surface velocities are smaller (e.g. 0·1–0·5 km/10 Myr) and regional uplift would take 30 Myr for present temperature and heat flow. Figure 10 shows a comparison of measured, regionally averaged, and calculated topography for the Rio Grande rift.

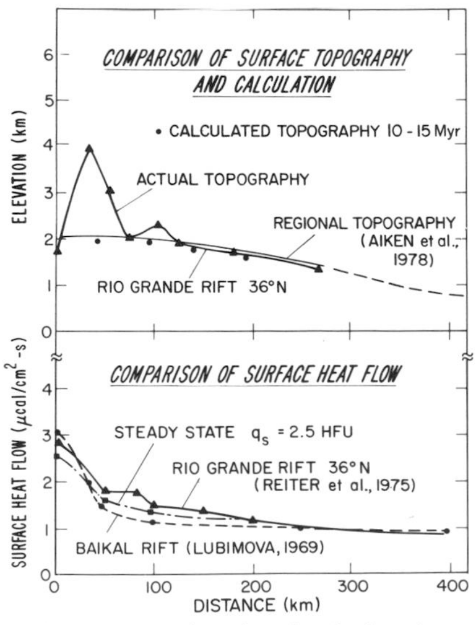

Figure 10. Comparison of surface, regional and calculated topography. Surface heat flow for Baikal and Rio Grande rifts are also shown.

Conclusions

Present geologic features of the Rio Grande rift include regional topography >2 km, surface heat flow of >2·5 HFU, a geotherm with gradient

$>30\,°C/km$, temperatures at the base of the crust of $\sim 1000\,°C$ and an increase in rift volcanism in late Miocene time. Geologic constraints on rate of uplift, based on inception of block faulting and palaeobotanical evidence, suggest this uplift was initiated 10–15 Myr ago on a broad regional scale. A concurrent outpouring of volcanic matter occurred over a large portion of the rift. These geologic data are used to provide constraints for the many unknowns of the thermomechanical problem. A suite of thermal profiles is evaluated to describe the viscosity structure, velocity profiles, and uplift of the rift.

Temperatures at the base of the crust have the range $600 < T < 1000\,°C$ for continent and rift. The crustal viscosity, which provides final uplift velocities of 1·3 km/10 Myr at a surface heat flow of 2·5 HFU, is 6×10^{25} poise. The coefficients of the Weertman creep law, Eqn (2), which provide reasonable flow velocities of upper mantle and viscosities of 10^{22}–10^{23} poise, are $A = 3{\cdot}4\,\text{k bar}^{-n}\,\text{s}^{-1}$, $E^* = 110$ kcal/mol, $V^* = 17\,\text{cm}^3$/mol and $n = 3$. The steady-state creep field has a zone of flow toward the rift at asthenospheric depths of >325 km, a region of hot, buoyant uplift ~ 150–200 km wide between 200 and 300 km depth, and an outward spreading lobe ~ 300–400 km wide in the upper asthenosphere beneath the rift. Free surface velocities range from 0·5 to 1·5 km/10 Myr as heat flow increases from 1·6 to 2·5 HFU for the relatively stiff, viscous lithosphere. Vertical velocity profiles through the mantle diapir increase from 1·8 km/10 Myr beneath the cold continent to 1·2 km/Myr in the buoyant mantle diapir beneath the rift, a factor of ~ 10. Horizontal spreading velocities are ~ 1 km/10 Myr to provide spreading rates approximately the same as uplift rates. The rift process narrows and concentrates heat transport beneath the rift, increases upward velocity and broadly arches the lithosphere. Calculated topography reaches a maximum value of 2 km at the continental rift and diminishes in accord with measured topography towards the continental platform.

The crust, acting as an elastic/highly viscous body, and the relatively cold upper mantle, which together comprise the lithosphere, exhibit a profound effect on thermomechanical processes of continental rifting. This “cap” damps out convective motion, aids in broadly spreading uplift because of shear tractions at its base, absorbs vast quantities of heat and resists brittle deformation. Flow in the underlying asthenosphere becomes decoupled through the increase in velocity and sharp reduction in viscosity.

Acknowledgements

The early interaction and support of Tom McGetchin, and his incessant good cheer, started the initial chain of events leading to this chapter. Many discussions with K. Bailey, S. Baldridge, J. Callender, C. Chapin, R. Girdler, R. Keller, P. Mohr, G. Palmason, I. Ramberg, B. Seager and T. Shankland have benefitted the evolving ideas of rift processes. W. S. Baldridge and T. J. Shankland provided critical reviews.

References

Aiken, C. L. V., Laughlin, A. W. and West, F. G., 1978. Residual Bouguer gravity map of New Mexico. Los Alamos Scientific Laboratory Report, LA–7466–MAP.

Anderson, C. A. and Bridwell, R. J., 1980. A finite element creep method for studying the transient nonlinear thermal creep of geological structures. *Int. J. Num. Anal. Meth. in Geomechanics* **4**, 255–276.

Axelrod, D. I. and Bailey, H. P., 1976. Tertiary vegetation, climate, and altitude of the Rio Grande depression, New Mexico–Colorado. *Palaeobiology* **2**, 235–254.

Baldridge, W. S., Damon, P. E., Shafiquallah, M. and Bridwell, R. J., 1980. Evolution of the central Rio Grande rift, New Mexico: new potassium–argon ages. *Earth Planet. Sci. Lett.* (in press).

Blackwell, D. D., 1971. The thermal structure of the continental crust. In *Am. Geophys. Union Monograph Ser.*, Vol. 14 (Ed. Hancock, J. G.), 169–184.

Bridwell, R. J., 1976. Lithospheric thinning and the late Cenozoic thermal and tectonic regime of the northern Rio Grande rift. *New Mexico Geol. Soc. Guidebook*, 27th Field Conf., 283–292.

Bridwell, R. J., 1978a. The Rio Grande rift and a diapiric mechanism for continental rifting. In *Tectonics and Geophysics of Continental Rifts* (Eds Ramberg, I. B. and Neumann, E. R.), Proc. NATO Adv. Study Inst. on Palaeorift Systems with Emphasis on Permian Oslo Rift **2**, 73–80, Reidel, Boston.

Bridwell, R. J., 1978b. Physical behavior of upper mantle beneath northern Rio Grande rift. In *Guidebook to Rio Grande Rift in New Mexico and Colorado* (Ed. Hawley, J. W.), NM Bur. Mines and Min. Res., Circ. 163, 228–230.

Bridwell, R. J. and Anderson, C. A., 1980. Dynamic thermomechanical structure of continental rifts (in preparation).

Carter, N. L., 1976. Steady-state flow of rocks. *Res. Geophys. Space Phys.* **14**, 301–360.

Cathles, L. M., III, 1975. *The Viscosity of the Earth's Mantle*, Princeton University Press, Princeton.

Chapin, C. E. and Seager, W., 1975. Evolution of the Rio Grande rift in the Socorro and Las Cruces Areas, *New Mexico Geol. Soc. Guidebook*, 26th Field Conf., 297–322.

Cook, F. A., McCullar, D. B., Decker, E. R. and Smithson, S. B., 1979. Crustal structure and evolution of the southern Rio Grande rift. In *Rio Grande Rift: Tectonics and Magmatism* (Ed. Riecker, R. E.), Am. Geophys. Union Sp. Pub., 195–208.

Decker, E. R. and Smithson, S. B., 1975. Heat flow and gravity interpretation across the Rio Grande Rift in southern New Mexico and west Texas. *J. Geophys. Res.* **80**, 2542–2552.

Epis, R. C. and Chapin, C. E., 1975. Geomorphic and tectonic implications of the post-Laramide, late Eocene erosion surface in the southern Rocky Mountains. *Geol. Soc. Mem.* **148**, 45–74.

Froidevaux, C. and Schubert, G., 1975. Plate motion and structure of the continental asthenosphere: a realistic model of the upper mantle. *J. Geophys. Res.* **80**, 2553–2564.

Gregory, J. W., 1921. *Rift Valleys and Geology of East Africa*, Seclay Service, London.

Hager, B. R. and O'Connell, R. J., 1979. Kinematic models of large-scale flow in the Earth's mantle. *J. Geophys. Res.* **84**, 1031–1048.

Lubimova, E. A., 1969. Heat flow patterns in Baikal and other rift zones. *Tectonophysics* **8**, 457–468.

McGetchin, T. R. and Silver, L. T., 1972. Compositional relations in minerals from kimerlite and related rocks in the Moses Rock Dike, San Juan County, Utah. *Am. Mineral.* **55**, 1738–1771.

Mercier, J.-C. C., 1977. Natural peridotites: chemical and rheological heterogeneity of the upper mantle. Ph.D. Thesis, State University of New York, Stonybrook.
Olsen, K. H., Keller, G. R. and Stewart, J. N., 1979. Crustal structure along the Rio Grande Rift from seismic refraction profiles. In *Rio Grande Rift: Tectonics and Magmatism* (Ed. Riecker, R. E.), Am. Geophys. Union Sp. Pub., 127–144.
Padovani, E. R. and Carter, J. L., 1977. Aspects of the deep crustal evolution beneath south central New Mexico. In *The Earth's Crust, Am. Geophys. Union Monograph Ser.*, Vol. 20 (Ed. Hancock, J. G.), Am. Geophys. Union, Washington DC.
Peltier, W. R., 1976. Glacial isostatic adjustment. II The inverse problem. *Geophys. J. R. Astr. Soc.* **46**, 669–706.
Prager, W., 1961. *Introduction to Mechanics of Continua*, 154–157, Ginn and Co., Aylesbury.
Ramberg, I. B., Cook, F. A. and Smithson, S. B., 1978. Structure of the Rio Grande Rift in southern New Mexico and west Texas based on gravity interpretation. *Bull. Geol. Soc. Am.* **89**, 107–123.
Reiter, M., Edwards, C. L., Hartman, H. and Weidman, C., 1975. Terrestrial heat flow along the Rio Grande rift. New Mexico and southern Colorado *Bull. Geol. Soc. Am.* **86**, 811–818.
Richter, F. M., 1979. Focal mechanisms and seismic energy release of deep and intermediate earthquakes in the Tonga–Kermadec region and their bearing on the depth extent of mantle flow. *J. Geophys. Res.* **84**, 6783–6795.
Ross, J. V., Ave' Lallemant, H. G. and Carter, N. L., 1979. Activation volume for creep in the upper mantle. *Science* **203**, 261–163.
Schatz, J. F. and Simmons, M. G., 1972. Thermal conductivity of earth material at high temperatures. *J. Geophys. Res.* **77**, 6966–6983.
Schubert, G., Froidevaux, C. and Yuen, D. A., 1976. Oceanic lithosphere and asthenosphere: thermal and mechanical structure. *J. Geophys. Res.* **81**, 3525–3540.
Schubert, G. and Turcotte, D. L., 1972. One-dimensional model of shallow mantle convection. *J. Geophys. Res.* **77**, 945–951.
Schubert, G., Yuen, D. A., Froidevaux, C., Fleitout, L. and Sourian, M., 1978. Mantle circulation with partial shallow return flow: effects on stresses in oceanic plates and topography of the sea floor. *J. Geophys. Res.* **83**, 745–758.
Skinner, B. J., 1966. Thermal expansion, handbook of physical constants. *Mem. Geol. Soc. Am.* **97**, 75–96.

The Tectonic Evolution of the Earth's Surface and Changing Lithospheric Properties

D. H. TARLING

School of Physics, The University, Newcastle upon Tyne, UK

Introduction

Any model that attempts to explain the main tectonic features of the Earth's evolution must obviously take into account the geological evidence for the nature of those features in the past, the nature of the tectonic forces presently operating on the Earth's surface, and the probable physical and chemical conditions that existed at the time of the Earth's formation and the way in which these are likely to have evolved to the present situation. As there is little agreement concerning the interpretation of specific geological features in the past, even less of the present mechanisms to tectonics operating today, and models for the original conditions of the formation of the Earth are changing rapidly, it is evident that any model can only be, at best, provisional. Nevertheless, it is important that such paradigms be erected in order to evaluate which are the critical features to distinguish between different models. However, at this stage, there are remarkably few models that even attempt to explain the gross features of the evolution of the Earth's surface. In general, most views (Dewey and Spall, 1975) merely restrict themselves to assertions that plate tectonic activity, as witnessed by tectonic developments during the last 200 million years, has either (a) only been a recent feature of tectonics and did not operate during the Precambrian, or (b) operated throughout Earth history but that the evidence has been largely obscured by later events or has not yet been recognized. A model has been suggested (Tarling, 1978, 1980) which proposes that certain aspects of "modern" plate tectonics have been continuously active, particularly sea-floor spreading, but other features have been discontinuous, such as continental splitting and growth. On this basis, it is thought that apparently opposing views can be reconciled and that the model offers possible explanations for other features in the tectonic development of the Earth. However, it is evident that the model depends fundamentally on an evaluation of the nature of the tectonic forces operating today and it is hence necessary to discuss the nature of these forces prior to assessing their operation in the past.

The present driving force of plate tectonics

A fundamental difficulty in any attempts to examine the forces operating in the Earth today is that there is still no unambiguous physical definition of a tectonic plate (Tarling, 1978). None the less the concept of a rigid upper surface of the Earth, comprising both crustal rocks and some mantle components, that is broken into discrete "plates" of large lateral but shallow vertical dimensions, is widely accepted. These lithospheric plates are regarded as having a high viscosity – well in excess of 10^{23} poise – and to be overlying "soft" mantle rocks with viscosities of 10^{21} poise or less. The boundary between these two viscosity conditions is widely considered to be sharp and to occur at a level where the geothermal gradient from the surface intersects with the mantle rock solidus. This intersection results in a 0·1% partial melt which accounts for the seismic low velocities observed in the mantle at depths ranging systematically from close to zero at an active oceanic ridge crests to some 120 km beneath old oceanic crust. On most conventional models (Isacks, Oliver and Sykes, 1968), therefore, the seismic low velocity zone is interpreted as a zone of partial melt and this is equated directly with the soft asthenosphere. The conventional model therefore incorporates primary mantle differentiation at the world's oceanic ridges where new oceanic crust is generated from the mantle rocks. The new crustal rocks, with an increasing thickness of mantle attached to them, move perpendicularly away from the ridges (as indicated by the magnetic striping – Vine and Mathews, 1963), before eventually descending back into the mantle at subduction zones. The subduction is accompanied by magmatic differentiation of mantle rocks, but from much greater depths than at the ridges, and forms the calc-alkaline volcanics of island arcs (Ringwood, 1975; Taylor, 1979).

As in this model the base of the lithosphere is considered to be a zone of partial melt, it appears reasonable to construct models in which the asthenosphere and lithosphere are decoupled from each other. The source mechanisms for intra-plate earthquake activity indicate that the lithosphere is either tightly coupled or completely decoupled from the asthenosphere (Richardson, Solomon and Sleep, 1979) and the general view appears to be that the plates must therefore move by the action of forces that lie within the plates themselves. Two main locations have been suggested for such intra-plate forces. As the ridges stand 2–3 km higher than their surrounding oceanic basins, there is clearly a component of gravity that is operating away from the ridge crests, in the plane of the oceanic plate (Hales, 1969). The forces operating at the ridge crests are thought to be of the order of 8×10^{18} J/yr (Forsyth and Uyeda, 1975) and are therefore only barely adequate to account for the energy released by seismic activity. It seems unlikely, therefore, that such a force is the main driving force for plate tectonics as a whole. The other intra-plate driving force is that caused by phase changes within the descending oceanic lithosphere (Schubert and Turcotte, 1971; Harper, 1975; Neugebauer and Breitmayer, 1975; Ringwood, 1975). As the oceanic lithosphere gets older and cooler, its buoyancy must gradually reduce. When it descends at a subduction zone, it is much cooler than the surrounding mantle and hence phase changes will occur in its lithospheric minerals at higher levels than in the surrounding mantle. These phase changes are, of course, associated with significant increases in density and will clearly give rise to a major downward force on the descending oceanic lithosphere. If this descending lithosphere remains coupled to oceanic lithosphere still at the surface, then such lithosphere could readily be pulled over the decoupled asthenosphere away from the ridges. Estimates for the magnitude of this force (McKenzie, 1969; Richardson *et al.*, 1979) indicate that it is adequate to account for the observed tectonic activity of the world, although such calculations usually have to discount any possible interference by the energy requirements of the process involved in the phase changes.

There can be little doubt that these intra-plate forces exist. Undoubtedly many of the features of ridge crests and subduction zone tectonics are directly attributable to their operation. None the less, there are numerous reasons for considering that these are not the driving force for modern plate tectonic activity. The most fundamental observation is that both of these forces need to be initiated. In the case of the ridge crest forces, the ridge must first be elevated, while in subduction zones, the main phase changes occur at depths of 300 and 600 km (Press, 1971; Ringwood, 1979), so that another mechanism must exist to carry some 400 km or more of oceanic lithosphere downwards at an angle of about 45° before the uppermost major phase change can begin to operate. It has been suggested that this may occur by means of an eclogitic sinker providing a strong negative buoyancy when basaltic layers cool sufficiently that basalt converts to eclogite (Ringwood, 1975). An oceanic lithosphere of several 100 km extent must already have formed before even this mechanism can begin to operate and, in any case, there are examples where younger oceanic lithosphere is being subducted in preference to older thicker oceanic lithosphere – as in the New Hebrides, Solomons, New Britain, etc. The presence of discrete levels at which such phase changes operate would also imply that the attached oceanic lithosphere would be subjected to accelerations when such depths were reached in a newly forming subduction zone, yet the oceanic anomaly patterns do not appear to indicate such accelerations. However, in terms of the subductive "pull" mechanism, this force is clearly not available to form the Siberian Arctic, the North and South Atlantic, or West Indian Oceans – none of these have associated subduction zones. The seismic evidence in subduction zones also seems to indicate that, in many areas, the descending lithosphere breaks into discrete slabs, in which case these would become decoupled from the rest of the oceanic lithosphere (Baranzangi *et al.*, 1973; Liu, 1975).

One of the fundamental features of the intra-plate mechanism arguments is that the seismic low velocity zone corresponds to an area of partial melting and thus corresponds with the asthenosphere. There are strong grounds for suspecting the existence of these associations. The upper boundary of the seismic low velocity zone is almost certainly an isotherm as there is a clear correspondence between its depth, the heat flow and the topography away from the oceanic ridges (Froidevaux, Schubert and Yuen, 1977). However, if this really were the level at which the geothermal gradient and mantle solidus intersect, then it cannot simultaneously be the boundary between the lithosphere and the asthenosphere. The viscosity of the mantle rocks of the lithosphere must drop drastically some 200 °C below their melting temperature and they would therefore deform plastically long before reaching their melting temperature. If the top of the low velocity zone is a zone of partial melt, the lithosphere/asthenosphere boundary must lie much higher. However, convective motion of the asthenosphere would also prevent significant melting as such motions are self-regulating and maintain a constant viscosity with adiabatic temperature gradients when operating on scales in excess of a few 100 km^3 (Tozer, 1972, 1978). On this basis localized melting can occur, for example by frictional heating or localized pressure release, but a volume of the Earth of the scale of the seismic low velocity zone could not persist. It seems probable that the seismic low velocity zone is simply due to the presence of water as a free phase within the mantle. As an intergranular film, such water would account for the seismic properties observed, and could have formed, for example, by the dehydration of amphibolites at temperatures above some 800 °C.

If the seismic low velocity zone is not a zone of partial melt, then the intraplate earthquake source mechanisms are equally consistent with a strong asthenosphere–lithosphere coupling (Richardson *et al.*, 1979), as would arise if the viscosity changed from that of the asthenosphere, 10^{21}–10^{22} poise, to that of the lithosphere over a wider range of mantle than in the "decoupled model". It thus becomes realistic to visualize radiogenically heat-driven mantle convection as operating and driving the lithospheric plates in a broadly similar way to that proposed by Holmes (1927, 1944) and Daly (1940). The problem then becomes one of establishing the probable scale of such convection.

In the "intra-plate" driving force models, the motion of the lithosphere plates was assumed to induce a return flow within the seismic low velocity zone. The restriction of mantle motion within the seismic low velocity zone is, however, impossible even on the conventional partial melt model. There is clear seismic evidence that the low velocity zone is absent beneath the Eastern Canadian (Jordan and Frazer, 1975) and the West Australian shield (Gonz and Cleary, 1976), and a similar situation is suspected beneath most cratonic shields (Poupinet, 1979). Older areas of the continents are also characterized by low heat flow values, yet are characteristically enriched in radiogenic heat-producing elements. In order to resolve this paradox, and to account for other observations, the underlying rocks must be depleted in heat-producing elements and the amount of heat conducted from the mantle must also be low and correspond with depletion in radiogenic heat-producing elements within the underlying mantle, possibly down to depths of the order of 300–400 km (Pollack and Chapman, 1977; Allis, 1979). The restriction of almost all kimberlites (Dawson, 1977; Meyer, 1979) to cratonic blocks, irrespective of the age of the intrusion, similarly indicates that at least 180–200 km of the upper mantle must travel with the cratonic blocks. These figures for the thickness of mantle carried with the continent (Osmaston, 1977) therefore correspond with the thickness of the continental lithosphere and so return flow, within the mantle, must take place at even greater depths. As the travel-time delays on seismic waves indicate an essentially uniform chemical composition for the mantle below some 200–250 km (Birch, 1952) it seems reasonable to assume that such convective motions extend throughout the mantle, although not excluding smaller scale motions. The heat flow from the Earth's interior is of the order of 10^{21} J/yr and the energy required for seismic and volcanic activity is only 0·5% of this, suggesting that plenty of energy must be available from this source for plate motions.

A model for crustal evolution

Pre-Archaean

It is clear from studies of extinct radionuclei that planets of the solar system must have formed shortly after the creation of elements heavier than Fe by a supernova explosion (Reeves, 1975; McGill, 1977). The Moon, Mars, Venus and Mercury established rigid lithospheres very quickly as indicated by the preservation, on these bodies, of the evidence of the intense meteoritic bombardment characteristic of the early phase of the solar system. In the case of the Moon, a rigid lithosphere must have formed by 4·3 Gyr ago which, by 3·9 Gyr ago was some 200 km thick in order to support the mare basalts that are still characterized by major gravitational anomalies – mascons (Taylor, 1975). As the Earth is much larger than the Moon, the gravitational energy released during differentiation of its core would be much greater than for the Moon so that it seems probable that the formation of the core took place much more rapidly in the Earth. Shortly after its formation, therefore, the Earth's interior would have been characterized by very rapid convective motions in order to remove the gravitational energy released, thus preventing depletion of the Earth in noble elements. After core formation, this convective motion would then become more ordered as it adjusted to the removal of radiogenic heat alone. The exact conditions during the pre-Archaean are, of course, difficult to ascertain. On the Earth, there is, so far, no evidence preserved for a lithosphere before 3·9 Gyr (Goldich and Hedge, 1974; Moorbath, O'Nions and Pankhurst, 1975) and oldest rocks were not derived from pre-existing crustal rocks (Gunn, 1976; Lambert, 1976; Moorbath, 1976). At the start of the Archaean, the surface temperatures must have been similar to those of today, as there is evidence for the presence of running water. It is not easy to determine the total radiogenic heat being generated within the Earth at the start of the Archaean, but most estimates (Lambert, 1976; Bickle, 1978) suggest that this was 5–6 times greater than today. On this basis, it seems that in pre-Archaean times, the Earth must also have had a lithosphere as surface temperatures seem unlikely to have been radically different at least by late pre-Archaean times. The higher heat productivity within the mantle would mean that convective rates would have been just over twice as fast as today – as their rate depends, fundamentally, on the square of the heat production (Tozer, 1972). The need to remove 5–6 times as much heat yet retain constant surface temperatures, means that there would be much steeper geothermal gradients at that time, but the surface of the Earth would have had to comprise solid lithosphere. Oceanic ridges would therefore form as today, generating oceanic lithosphere. The upper surface of this lithosphere would be essentially similar to oceanic crust today as it formed from similar materials in similar pressure/temperature conditions. There would, however, be some differences reflecting the generally "wetter" nature of the mantle (which then contained all continental differentiates) and the higher geothermal gradient. The main implication is that the basaltic layer would be much thicker. The oceanic lithosphere would also be thin as the rise in viscosity with depth must mirror the geothermal gradient. When oceanic lithosphere was subducted, therefore, it would be heated and dehydrated at shallow depths. Water, contained within it or within its constituent minerals, would not reach the depths from which calc-alkaline volcanics could be generated. In other words, oceanic crust and lithosphere was being produced along active oceanic ridges, but not continental crust, and hence there were no continental lithospheres. There were, therefore, no light continental differentiates to accumulate over subduction zones and hence no material to be preserved for examination today.

Archaean

As the Earth cooled, because radiogenic heat production would necessarily decrease, the average geothermal gradient would become less so that the decrease in viscosity from lithospheric to asthenospheric conditions would take place at increasingly deeper average depths. The first preservation of continental rocks could only take place after the oldest oceanic lithosphere reached sufficient thickness that conductive heating and dehydration was delayed until the subducting slabs had reached depths of the order of some

80–100 km, from which level calc-alkaline differentiation occurs today. The model therefore predicts that there would be a gradual increase in the potassic content of the continental differentiates, lanthanides, rare earth elements, etc., as the Archaean progressed. This would result from the gradual increase in the depth of the level at which water and other volatiles were expelled from the descending plate and would thus mirror the compositional changes with depth along traverses perpendicular to oceanic trenches today (Kuno, 1959; Jakes and White, 1972; Dickinson, 1975; Miyashiro, 1975; Ringwood, 1975). Although the observations are controversial, there does seem to be some evidence for such variations (Engel *et al.*, 1974; Veizer, 1976; Veizer and Compston, 1976).

If the descending oceanic lithosphere was warmed to ambient mantle temperatures before reaching a depth of 300 km or so, then the phase changes within it would not provide a significant downward force and it seems probable that, under such conditions, the descent of Archaean oceanic lithosphere would be at shallower angles than normal today. The thinness of the lithosphere would also mean that its negative buoyancy would be small. However, the thin lithosphere estimated for this time would have other major consequences. In particular, the depth at which isostatic adjustments would take place, the isopeistic level, would also be much shallower (Hargraves, 1976). In the model suggested here, the maximum topographic difference that could be sustained is estimated to be only some 3–4 km. This would prevent the accumulation of thick sedimentary sequences – no Archaean sediments are known thicker than 0·5 km – and any isostatic adjustments to tectonic forces or erosion would take place on a rapid, local scale, probably giving rise to coarsely sorted sediments. Similarly, tectonics would occur on a small scale, i.e. a single gneissic belt could act as a microcontinent, but it is unlikely that larger units would behave uniformly.

Although the Archaean lithosphere is considered to be thin, this would not drastically alter the maximum thickness of continental crust that could accumulate, although the thickness obviously could not exceed that of the maximum thickness of continental lithosphere for that period. Changing convective motions would, however, readily break through either continental or oceanic lithosphere, although the weaker strength of the continental lithosphere may have made it slightly more susceptible to splitting by convective changes. The Archaean greenstones could thus be the equivalent of the modern-day flood basalt associated with continental splitting. In some areas, these appear to have flowed over pre-existing gneissic terrains, but it is also probable that some greenstones were developed by back-arc spreading, although it is not clear from modern day systems whether the proposed shallower subduction would retard or enhance back-arc spreading. It seems probable, however, that komatiites, with melting temperatures of 1650 °C, would have been incorporated within the submarine greenstones by pressure-release melting in an area of much higher geothermal gradient than today.

Proterozoic

The change from Archaean tectonic conditions to those of the Proterozoic has previously been more puzzling than the absence of crustal rocks in the pre-Archaean. Proterozoic conditions appear to have developed over only 100–200 million years on a global scale, possibly following a world-wide magmatic event (Moorbath, 1976). The main features appear to be very large scale tectonic stability, with the possibility for forming carbonatites within the continents, and accumulating considerable thicknesses of sediments on their continental shelves. Palaeomagnetic evidence is somewhat ambiguous as to whether the world's continents were united as one or two major blocks, but the palaeomagnetic and palaeoclimatic evidence is quite clear in indicating that continental drift was occurring in at least Proterozoic times. (Palaeomagnetic methods are ineffective for earlier times as the magnetization of Archaean rocks has almost invariably been re-set by at least greenschist metamorphism at the end of the Archaean, or younger.)

The model proposes that most Archaean rocks were, at the very end of the Archaean, in amphibolitic grade of metamorphism. This is based on analogies with the metamorphism associated with modern island arcs (Miyashiro, 1961; Ernst, 1975), probably reflecting the observation that the major process involved in both instances is identical, i.e. it involves the addition of water and volatiles to mantle rocks in subduction zone environments. Some granulites could, of course, have formed and would have been stable, but the predominant grade would be amphibolitic. However, as mantle rocks cooled and were "underplating" the lithosphere, the conditions at the base of the lithosphere were changing as its depth increased. The Archaean continental lithosphere would have been slightly thicker than that of the oceanic as the differentiation of the continental margins somewhat reduced the volatile content of the associated mantle rocks, hence increasing their melting temperature and thereby decreasing their viscosity relative to other mantle rocks at the same depth. Under the very late Archaean continents, therefore, the lower lithosphere extended into higher pressure regions where granulites became more stable than amphibolites and only a small trigger was required to initiation dehydration of the amphibolites in the mantle and lower crust. Once initiated, dehydration would occur throughout a thick section of the upper mantle, probably involving at least the upper 200 km as the stability field of granulites almost exactly matched the predicted continental geothermal gradient at the end of Archaean times. It is proposed that this dehydration of the subcontinental mantle was the major factor accounting for the onset of Proterozoic tectonic conditions, and that the differentiation of the continental crust also took place at this time as the volatiles migrated upwards, scavenging out most lithophile elements and concentrating them in the upper crust. There certainly seems to have been a major change in the rare earth element patterns in sediments of Proterozoic as compared with Archaean age (Taylor, 1979).

As the formation of a very thick continental lithosphere took place quickly and on a continent-wide scale, it had drastic effects on the Early Proterozoic continental conditions. Clearly there could be major mountain belts within the continents and they could also support great thicknesses of sediments. However, the changes in the oceans were equally dramatic, but essentially the converse of those occurring in the continental regions. As the Earth was still generating radiogenic heat 2–3 times as fast as today, this heat still had to be dissipated, but this would now be very slow and inefficient through the continental lithosphere as it could only gradually remove heat by conduction. This meant that the cooling of mantle convective cells was restricted to the oceanic areas. This would obviously result in an increase in the oceanic geothermal gradient and hence a thinning of the oceanic lithosphere. Indeed it seems probable that the oceanic lithosphere was thinned to be comparable with that of pre-Archaean times. This implies that the creation of new continental crust was also inhibited, although it is difficult to evaluate the situation where such thin lithosphere was subducted contiguous with very thick continents. The thin oceanic lithosphere would also mean that oceanic topographies were restricted and, more importantly, would break readily. Changes in convective patterns would therefore always break through pre existing oceanic lithosphere rather than the thinned oceanic lithosphere.

The last 10^9 years

As the Earth was still cooling, the average thickness of the lithosphere must have been increasing. During the Proterozoic, the radiogenic heat was being largely lost through the oceanic lithosphere and it seems likely that the thickness of the continental lithosphere remained, essentially, the same. The oceanic lithosphere was therefore thickening at a much faster rate than the continental lithosphere. As the Proterozoic progressed, therefore, the strength of the oceanic lithosphere was increasing, while that of the continents remained little changed. As the compositions of the two lithospheres differ, they have different strengths per unit thickness. It is proposed that some 10^9 years ago, the weaker zones of the continental lithosphere became comparable with the strength of some of the older, thicker parts of the oceanic lithosphere. When convective changes occurred, therefore, the newly developing convective patterns would find it as easy to break through these weak areas of continental lithosphere as through pre-existing oceanic lithosphere. Hence continental splitting which was strongly inhibited at the start of the Proterozoic, became increasingly easy during the Late Proterozoic until it was then probably easier for new convective systems to break through continental lithosphere than through thick oceanic lithosphere.

Comments

It must be emphasized that, throughout this chapter, the meaning of an "average" lithosphere thickness must be interpreted cautiously. Clearly an oceanic lithosphere, at any time, will be more variable in thickness than a continental lithosphere as its ridges must have essentially zero thickness. The lower average thickness of oceanic lithosphere will therefore probably arise by the presence of a greater total length of oceanic ridges – which must, of course, mean shorter distances between either spreading centres or between a spreading centre and a subduction zone. In contrast, the continental lithosphere is likely to be more uniform in thickness, with most variation occurring between cratonic blocks and active orogenic zones.

Generalizations about the tectonic situation at any one time are, quite clearly, dangerous. It is certainly clear that the onset of continental splitting during the Proterozoic will have occurred at different places at different times. Similarly, the onset of Proterozoic conditions was likely to have been at a somewhat different time in different cratonic blocks – although many people have commented on the contemporaneity of this change. However, one possible explanation of the so-called "Pan-African" (Clifford, 1970) orogeny (which also occurred in many other parts of the world) could be in a delayed effect of the amphibolite/granulite conversion. If the cratonic blocks were stabilized at the end of the Archaean, as approximately circular units, each unit could still be surrounded by continental rocks that were still in their "Archaean" amphibolite grade. Such marginal areas would be metastable at depth, but protected by the interconnected mantle lithosphere beneath them. It was only when continental collisions began to occur in the Late Proterozoic that sufficient impetus was given to trigger conversion, with consequential rise of volatiles, granites, etc. as a thermal event around the cratonic margins. Such behaviour would not require movement between the individual cratons. A distant continental collision would perhaps not be necessary to initiate such changes as these areas would have become increasingly less stable with time as the continental lithosphere continued to cool. However, this change in Late Proterozoic times is likely to have been small, and it seems more likely that the Pan-African, and similar orogenies elsewhere, were in response to some particular event, such as the first major inter-continental collision since the start of the Proterozoic.

The proposed model therefore offers scope for the explanation of many features of the Earth's history. On first consideration, it seems odd to consider that lithosphere conditions could evolve and cool from stable amphibolitic grade to stable granulitic grade. Nonetheless, it is clear that the lithosphere must have changed drastically during the evolution of the Earth. As it cooled, it would also have thickened, so that the pressure at the base of the lithosphere would have increased, yet its temperature must have remained constant, unless the actual composition of the mantle has changed drastically since pre-Archaean times. The constant temperature for the base of the lithosphere arises because this base can only be defined by its viscous properties, and these are determined by a specific temperature for a specific composition, i.e. some 200 °C less than the melting point of its constituent materials. The base of the lithosphere can therefore be regarded as a zone in which the pressure is increasing with time, yet the temperature remains constant. Such a model therefore incorporates changes in both the geochemical and geophysical properties of the Earth's surface. As such, it can only provide a framework for further evaluation and testing, but it is important that the Earth be treated as a complex interacting physico-chemical unit in which slight changes in any one property may have drastic effects on another.

Acknowledgements

The author wishes to thank D. C. Tozer, in particular, for many stimulating thoughts even though we may still not agree in many respects.

References

Allis, R. G., 1979. A heat production model for stable continental crust. *Tectonophysics* **57**, 151–165.

Baranzangi, M., Isacks, B. L., Oliver, J., DuBois, J. and Pascal, G., 1973. Descent of lithosphere beneath New Hebrides, Tonga, Fiji and New Zealand: evidence for detached slabs. *Nature* **242**, 98–101.

Bickle, M. J., 1978. Heat loss from the earth: a constraint on Archaean tectonics from the relation between geothermal gradients and the rate of plate production. *Earth Planet. Sci. Lett.* **40**, 301–315.

Birch, F., 1952. Elasticity and constitution of the Earth's interior. *J. Geophys. Res.* **57**, 227–286.

Clifford, J. N., 1970. The structural framework of Africa. In *African Magmatism and Tectonics* (Eds Clifford, T. N. and Gass, I. G.), 1–26, Oliver and Boyd, Edinburgh.

Daly, R. A., 1940. *Strength and Structure of the Earth*, 434, Prentice-Hall, Englewood Cliffs.

Dawson, J. B., 1977. Sub-cratonic crust and upper mantle models based on xenolith suites in kimberlite and nephelenitic diatremes. *J. Geol. Soc.* **134**, 173–184.

Dewey, J. and Spall, H., 1975. Pre-Mesozoic plate tectonics: how far back in Earth history can the Wilson cycle be extended? *Geology* **3**, 422–424.

Dickenson, W. R., 1975. Potash-depth (K–*h*) relations in continental margin and intra-oceanic magmatic arcs. *Geology* **3**, 53–56.

Engel, A. E. J., Itson, S. P., Engel, C. G., Stickney, D. M. and Gray, Jr, E. J., 1974. Crustal evolution and global tectonics: a petrogenic view. *Bull. Geol. Soc. Am.* **85**, 843–858.

Ernst, W. F. (Ed.), 1975. *Subduction Zone Metamorphism*, 445, Dowden, Hutchinson and Ross, Pennsylvania.

Forsyth, D. and Uyeda, S., 1975. On the relative importance of the driving forces of plate motion. *Geophys. J. R. Astr. Soc.* **43**, 162–200.

Froidevaux, C., Schubert, G. and Yuen, D. A., 1977. Thermal and mechanical structure of the upper mantle: a comparison between continental and oceanic models. *Tectonophysics* **37**, 233–246.

Goldich, S. S. and Hedge, C. E., 1974. 3800-Myr granitic gneiss in south-western Minnesota. *Nature* **252**, 467–468.

Gonz, J. H. and Cleary, J. R., 1976. Variations in the structure of the upper mantle beneath Australia from Rayleigh Wave observations. *Geophys. J. R. Astr. Soc.* **44**, 507–516.

Gunn, B. M., 1976. A comparison of modern and Archaean oceanic crust and island-arc petrochemistry. In *The Early History of the Earth* (Ed. Windley, B. F.), 389–403, Wiley, London.

Hales, A. L., 1969. Gravitational sliding and continental drift. *Earth Planet. Sci. Lett.* **6**, 31–34.

Hargraves, R. B., 1976. Precambrian geologic history. *Science* **193**, 363–371.

Harper, J. F., 1975. On the driving forces of plate tectonics. *Geophys. J. R. Astr. Soc.* **40**, 465–474.

Holmes, A, 1927. Some problems of physical geology and the Earth's thermal history. *Geol. Mag.* **65**, 263–278.

Holmes, A., 1944. *Principles of Physical Geology*, 532, Nelson and Sons, London.

Isacks, B., Oliver, J. and Sykes, L. R., 1968. Seismology and the new global tectonics. *J. Geophys. Res.* **73**, 5855–5899.

Jakes, P. and White, A. J. R., 1972. Major and trace elements abundances in volcanic rocks of orogenic areas. *Geol. Soc. Am. Bull.* **83**, 29–40.

Jordan, T. H. and Frazer, L. B., 1975. Crustal and upper mantle structure from Sp phases. *J. Geophys. Res.* **80**, 1504–1518.

Kuno, H., 1959. Origin of Cenozoic petrographic provinces of Japan and surrounding areas. *Bull. Volc.* **20**, 37–76.

Lambert, R. St J., 1976. Archaean thermal regimes, crustal and upper mantle temperatures, and a progressive evolutionary model for the earth. In *The Early History of the Earth* (Ed. Windley, B. F.), 363–373, Wiley, London.

Liu, H. S., 1975. Dynamical model for the detachment of descending lithosphere. *Geophys. J. R. Astr. Soc.* **42**, 607–619.

McGill, G. E. 1977. Craters as "fossils": the remote dating of planetary surface materials. *Bull. Geol. Soc. Am.* **88**, 1102–1110.

McKenzie, D. P., 1969. Speculations on the consequences and causes of plate motions. *Geophys. J.* **18**, 1–32.

Meyer, H. O. A., 1979. Kimberlites and the mantle. *Rev. Geophys. Space Phys.* **17**, 776–788.

Miyashiro, A., 1961. Evolution of metamorphic belts. *J. Petrol.* **2**, 277–311.

Miyashiro, A., 1975. Petrology and plate tectonics. *Rev. Geophys. Space Phys.* **13**, 94–98.

Moorbath, S., 1976. Age and isotope constraints for the evolution of Archaean crust. In *The Early History of the Earth* (Ed. Windley, B. F.), 351–360, Wiley, London.

Moorbath, S., O'Nions, R. K. and Pankhurst, R. J., 1975. The evolution of early Precambrian crustal rocks at Isua, West Greenland – Geochemical and isotopic evidence. *Earth Planet. Sci. Lett.* **27**, 229–239.

Neugebauer, H. J. and Breitmayer, G., 1975. Dominant creep mechanism and the descending lithosphere. *Geophys. J. R. Astr. Soc.* **43**, 873–895.

Osmaston, M. F., 1977. Some fundamental aspects of plate tectonics bearing on hydrocarbon location. In *Developments in Petroleum Geology* (Ed. Hobson, G. D.), **1**, 1–52, Applied Science, London.

Pollack, H. N. and Chapman, D. S., 1977. Mantle heat flow. *Earth Planet. Sci. Lett.* **34**, 174–184.
Poupinet, G., 1979. On the relation between P-wave travel time residuals and the age of continental plates. *Earth Planet. Sci. Lett.* **43**, 139–161.
Press, F., 1971. The Earth and the Moon, *Q. J. R. Astr. Soc.* **12**, 232–243.
Reeves, H., 1975. L'origine du systeme solair. *La Recherche* **6**, 808–817.
Richardson, R. M., Solomon, R. C. and Sleep, N. H., 1979. Tectonic stress in the plates. *Rev. Geophys. Space Phys.* **17**, 981–1019.
Ringwood, A. E., 1975. *Composition and Petrology of the Earth's Mantle*, McGraw-Hill, New York.
Ringwood, A. E., 1979. *Origin of the Earth and Moon*, 295, Springer-Verlag, New York.
Schubert, G. and Turcotte, D. L., 1971. Phase changes and mantle convection. *J. Geophys. Res.* **76**, 1424–1432.
Tarling, D. H. (Ed.), 1978. *Evolution of the Earth's Crust*, Academic Press, London.
Tarling, D. H., 1980. Lithosphere evolution and changing tectonic regimes. *Q. J. Geol. Soc.* **137**, 459–466.
Taylor, S. R., 1975. *Lunar Science: A Post-Apollo View*, Pergamon, New York.
Taylor, S. R., 1979. Chemical composition and evolution of the continental crust: the rare earth element evidence. In *The Earth: Its Origin, Structure and Evolution* (Ed. McElhinny, M. W.), 353–376, Academic Press, London.
Tozer, D. C., 1972. The concept of a lithosphere. *Geofis. Internac.* **13**, 363–388.
Tozer, D. C., 1978. Terrestrial planetary evolution and the observational consequences of their accumulation. In *The Origin of the Solar System* (Ed. Dermott, S. F.), 433–462, Wiley, London.
Veizer, J., 1976. $^{87}Sr/^{86}Sr$ evolution of seawater during geologic history and its significance as a index of crustal evolution. In *The Early History of the Earth* (Ed. Windley, B. F.), 569–578, Wiley, London.
Veizer, J. and Compston, W., 1976. $^{87}Sr/^{86}Sr$ in Precambrian carbonates as an index of crustal evolution. *Geochim. Cosmochim. Acta.* **40**, 905–914.
Vine, F. J. and Matthews, D. H., 1963. Magnetic anomalies over oceanic ridges. *Nature* **199**, 947–949.

Evolution of the Earth's Lithosphere*

G. F. PANZA

Instituto di Geodesia e Geofisica, Università di Bari, Italy

Introduction

The properties of the lithosphere at present are the result of the interaction between the different plates which form the outermost part of the Earth. There are remarkable differences in origin, thickness and age of the lithosphere under the oceans and in the continents. It is now generally accepted that the present distribution of continental and oceanic plates is the result of the break-up and joining together of different continental masses. Through the concept of sea-floor spreading and the theory of plate tectonics it is easy to understand most of the main features of the evolution of the lithosphere during the last 200 million years.

Attempts have been made to apply the plate tectonic model to Palaeozoic and late Precambrian orogenic terrains, such as the Uralides, and to the Appalachian–Caledonian orogeny (Bird and Dewey, 1970; Hamilton, 1970; McElhinny and Briden, 1971). Furthermore, it has been hypothesised that ancient mobile belts, and even the Archaean greenstone belts, represent the impact scars of protocontinental fragments colliding with each other, with island arcs or with oceanic regions Gibb, 1971; Gibb and Walcott, 1971; White, Jakes and Christie, 1971; Condie, 1972; Talbot, 1973). To confirm the hypothesis that sea-floor spreading and continental drift have operated throughout geological time, the best place to look for evidence is on the continents, where the lithosphere is much older than the present oceanic lithosphere.

When the thickness of continental lithosphere decreases, isostatic subsidence will occur aided by rifting. If the thinning process continues the low velocity material from the upper mantle reaches the surface and contributes to the formation of new oceanic lithosphere. The ageing oceanic lithosphere thickens as it moves away from the ridge crest to the subduction zones where it

* Publication no. 223, P. F. Geodinamica, CNR Rome, Italy.

is consumed. Part of it returns to the surface, contributing to the formation of continental lithosphere in volcanic island arcs, through the recycling of part of the subducted oceanic lithosphere and surrounding asthenosphere. The continental lithosphere gets thicker and cooler after its formation unless it undergoes tectonic processes, as for example rifting or continent–continent collision. This evolution scheme is perfectly in agreement with the classical model of plate tectonics and explains nicely the "orogenetic cycles" of Andean type. When the oceanic lithosphere lying between two continental blocks is completely consumed a continent–continent collision takes place, and a mountain range of the Alpine type is developed. According to the classical scheme of plate tectonics, when the continent–continent collision occurs, subduction terminates. This model is inadequate, as indicated by evidence from palinspastic reconstructions (Laubscher, 1971) and seismological investigation (Panza and Müller, 1978) of the Alpine area, which support the idea of subduction of continental lithosphere, along the lines of the "verschluckungs" hypothesis formerly proposed by Ampferer (1906).

Thus in the continent–continent collision a significant amount of continental lithosphere is destroyed and recycled in the asthenosphere. At this point the cycle of lithospheric evolution is complete. This concept, less schematic than the classical plate tectonics theory, provides the explanation for many geophysical and geological observations, such as the pattern of seismicity, the distribution and properties of volcanic belts, heat flow observations and the characteristics of mountain ranges both of Alpine and Andean type.

Oceanic lithosphere

Most of the knowledge we have about the oceanic lithosphere is based on dispersion measurements of surface waves carried over the Pacific Ocean. Different approaches have been followed to regionalize the long dispersion profiles available in the area (Leeds, Knopoff and Kausel, 1974; Forsyth, 1975; Yoshida, 1978). Owing to the non-uniqueness of the inversion, the conclusions of the different authors differ in some aspects. However, independently of the approach used, there is noticeable agreement on the fact that the oceanic lithosphere is getting thicker with increasing age, through a thickening of both crust and lid, as shown in Fig. 1. In fact the lid is very thin or absent near the ridge crest, reaches a thickness of about 60 km after 50 million years and about 85 km after 100 million years. After this age the lid continues to grow and probably reaches a thickness of about 135 km after 150 million years. An increase of the velocity of S_n, the S-wave velocity in the lid, with age is also consistent with the available data. For the Pacific Ocean, Forsyth (1975) gives an increase from about 4·3 km/s to 4·5 km/s from 0 to about 40 Myr, while, for the Atlantic Ocean, Hart and Press (1973) give 4·58 km/s and 4·7 km/s as the velocity of S_n respectively for the regions younger and older than 50 million years. In the Atlantic, surface wave dispersion measurements have recently been made in the zone of plate contact between the Azores and Gibralter by Müller, Marillier and Panza (1979), who detected a pronounced difference between the structure of the lithosphere–asthenosphere system on both sides of the plate contact. According to the time–evolution model described above,

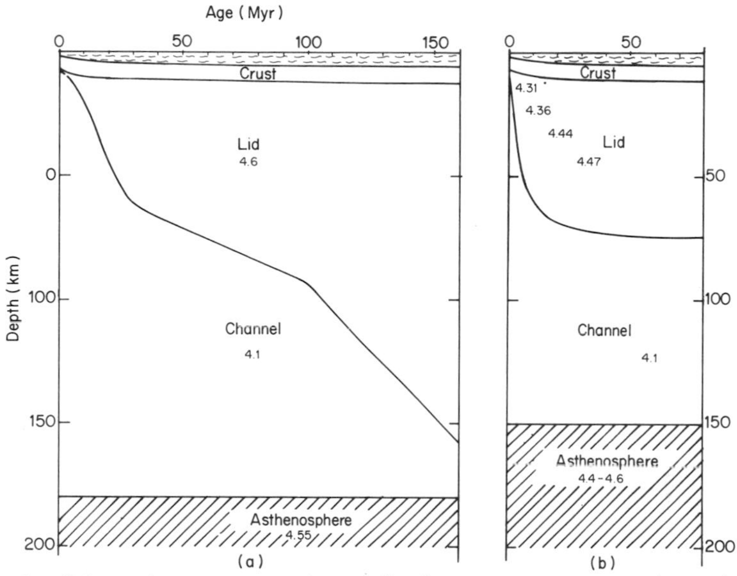

Figure 1. Schematic representation of the average properties of the lithosphere–asthenosphere system in the Pacific Ocean, as function of age. Model (a) after Leeds *et al.* (1974); model (b) after Forsyth (1975).

the low velocity layer is almost absent in the (southern) African side while it is rather well developed in the (northern) European side. In the oldest part (African side) it is possible to detect a layering in the lithosphere, indicating a decrease of *S*-wave velocity with increasing depth; in the fracture zone the *S*-wave velocity distribution versus depth is anomalous for an oceanic area. The fact that the oceanic lithosphere gets thicker and more rigid with age is in rather good agreement with the observed characteristics of deep focus earthquakes. In fact Vlaar and Wortel (1976) have shown that the penetration depth of subducted lithosphere increases with increasing lithospheric age. In this context the relevant age of a subducted part of the lithosphere is the age it had at the beginning of its descent. Since *S*-wave velocities are rather sensitive to temperature variations, it seems reasonable to accept that the oceanic lithosphere is getting not only thicker but also cooler on moving away from the ridge axis. The lithosphere cools by heat flow through the oceanic bottom. The loss of heat is reflected by the ocean bottom topography, which has been shown to be controlled by thermal contraction (Sclater and Francheteau, 1970; Sclater, Anderson and Bell, 1971). The lithosphere near the ridges is in equilibrium with the upper mantle, whereas by subsequent cooling it must become increasingly unstable, and the increasing instability with time may result in subduction. Although it is more usual that subduction takes place at the contact between an oceanic and a continental plate, spontaneous subduction is possible in oceanic regions which have reached a sufficient age. An example of this spontaneous process can be found in the Tonga–Kermadec region. On the other hand if oceanic lithosphere, which has been created recently, collides with continental lithosphere the subduction results in apparent subhorizontal underthrusting and is accompanied by predominantly shallow seismicity. This appears to be the case in the western part of North America (Vlaar and Wortel, 1976). Thus we may conclude that after its formation the oceanic lithosphere gets thicker, unless it undergoes tectonic processes, as in the case of the zone of plate contact between the Azores and Gibralter.

In this way the general characteristics of the subduction process of oceanic lithosphere and its relation to its cooling history appear to be well established as being a part of a convective cycle. Now the question arises as to why and where rifting of the lithosphere, which must occur at the onset of the convective process, will take place. The presently available data allow the formulation of a possible explanation for the transformation of continental to oceanic lithosphere.

Continental lithosphere

On land our knowledge of the elastic properties of the lithosphere in most of the shields, stable platforms, rift zones, mountains, is based mainly on surface wave dispersion studies (e.g. see Knopoff, 1972; Panza, Müller and Calcagnile, 1978). In the shields the lithosphere is rather thick, exceeding 100 km, and its thickness increases with age as indicated both by travel time residuals (Noponen, 1977; Poupinet, 1977) and by Rayleigh wave dispersion data (Brune and Dorman, 1963; Bloch, Hales and Landisman, 1969; Biswas and Knopoff, 1974; Calcagnile and Panza, 1978). The transition lithosphere–asthenosphere is not marked by a low velocity layer, with partial melting, but by a slight low velocity channel. This velocity inversion can be easily accomplished by the competitive effects of pressure and temperature without invoking melting. In the oldest shields the velocity inversion may even be absent.

The crustal part of the lithosphere can be reasonably schematized by two layer models. In the youngest shields the velocity inversion at the lithosphere–asthenosphere transition is more pronounced and in the crust an intermediate layer, with velocity lower than in the surrounding layers, starts to be present. In stable platforms the lithosphere does not exceed 100 km in thickness and the lithosphere–asthenosphere transition is marked by a well-developed low velocity layer, where partial melting is very likely to be present. The elastic properties of the lid in stable platforms and shields are very similar.

In the crust the intermediate layer reaches a thickness of more than 10 km. Thus the lithosphere of stable continents, in normal conditions, seems to cool off with age, as indicated also by the relation existing between reduced heat flow and age (Pollak and Chapman, 1977), and to thicken at the expense of the asthenospheric low velocity layer. This indicates that presently stable continents may eventually evolve into structures now observed in Precambrian shields. This hypothesis is strongly substantiated by the existence of a simple relationship between travel time residuals for *P*-waves and age of continental lithosphere, over the time span 0·2–2·0 billion years (Poupinet, 1977, 1979). Using the average upper mantle structures for shields and stable continents, determined from DSS and surface wave dispersion data, relative travel time residuals can be estimated, which are in very good agreement with the findings from travel times (Poupinet, 1977, 1979) (Fig. 2).

This evolution scheme represents a valuable foundation for a better understanding of the tectonic processes in the early history of the Earth. In fact in Precambrian times the continental blocks were very probably lying over an asthenospheric low velocity layer. This supports the hypothesis, up to now based mainly on a few palaeomagnetic observations (McElhinny and Briden, 1971), that there were few continental units moving over the surface of the Earth in the Precambrian age. The main drag source in the earlier stages of the tectonic development of our planet was probably supplied by the viscosity in the upper mantle, and not by the friction of the edges of the plates, as it is now (Knopoff, 1972). This is in agreement with the hypothesis that earlier orogenic belts did not arise from the formation and destruction of large oceans and resulting continent–continent collision, but arose from internal deformations, fairly reasonable for a primaeval thin lithosphere lying over a fairly warm asthenosphere.

If the continental lithosphere is not free to cool off, but lies under a tectonically active area, as for instance a sedimentary basin, its evolution pattern is totally different and may end up with the formation of a new oceanic crust. A possible model for the cyclic evolution scheme of the Earth's lithosphere–asthenosphere system is shown in Fig. 3. According to the comprehensive model of crustal evolution proposed by Müller (1978) three factors seem to be essential for the break-up of a continental plate: (1) the presence of an intermediate crustal layer with a *P* velocity of 6·3 to 6·4 km/s in the middle part of the crust; (2) the existence of a gravitational instability in the upper crust, i.e. the presence of the sialic low velocity zone; (3) the large-scale thermal activation and updoming of the crust–mantle boundary, with the associated formation of an "anomalous" uppermost mantle, characterized by a lid thinner than normal.

The first phase of the transformation of continental to oceanic lithosphere is represented by rifting. Among the existing rift areas we have detailed information about the properties of the crust lid and asthenosphere in the Central European rift system, and about the crust in the East African rift

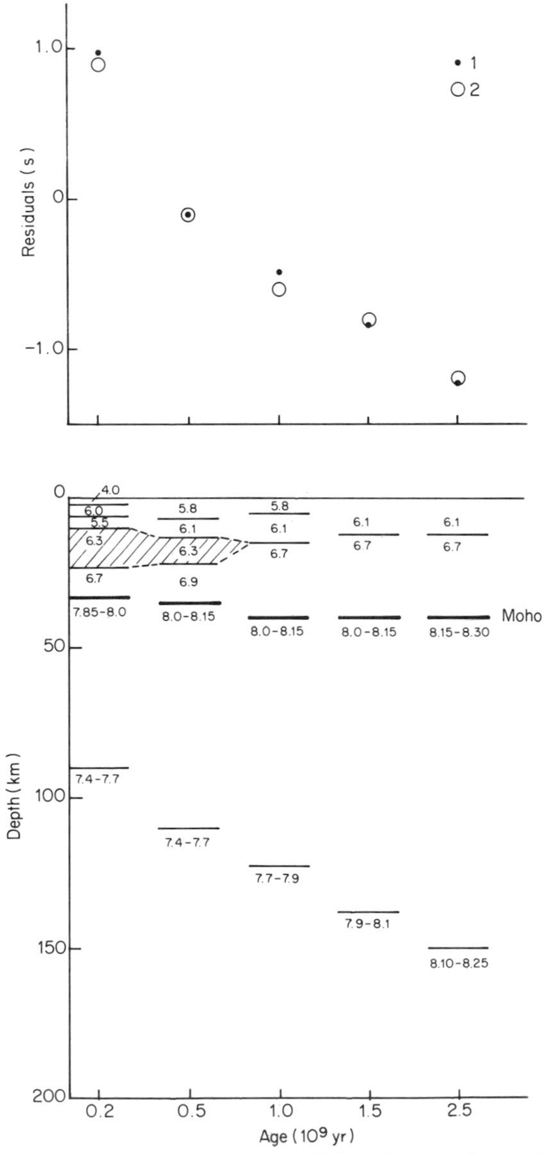

Figure 2. Schematic cross-sections of the lithosphere–asthenosphere system in Precambrian shields of different ages and stable continental platforms, as deduced from DSS (Müller, 1977) and surface wave dispersion. Numbers indicate *P*-wave velocity. The corresponding relative travel time residuals are shown in the upper part of the figure as open circles. The dots represent the average travel time residuals for vertically travelling *P*-waves, determined by Poupinet (1977, 1979) in continental areas of different ages.

system (Knopoff and Schlue, 1972; Lepine, Ruegg and Steinmetz, 1972; Müller, *et al.*, 1973; Reichenbach and Müller, 1974; Berckhemer *et al.*, 1975). From these data it can be seen that the lower and intermediate crustal layers become thinner (Hirn and Perrier, 1974) while the sialic low velocity zone increases in thickness, but the total effect is a global crustal thinning of about 20% (Müller, 1978). The sub-Moho material seems to have extremely high velocities, probably due to anisotropy (Fuchs, 1974). This high velocity sub-Moho layer must be very thin to satisfy the observed phase velocities of Rayleigh waves, and the underlying mantle is characterized by a rather anomalous *S*-wave velocity, probably reflecting the thermal alteration associated with the rift formation (Panza, Müller and Calcagnile, 1978). As a consequence of the thermal perturbation the lid thickness decreases as if the lid to the platform structure had been thinned down to nothing. The total lithospheric thickness does not exceed 60 km. As the intermediate crustal layer starts to decrease in thickness the lower crustal layer will grow proportionally. A new type of crust is formed, which is "pseudo-oceanic" in nature (Müller, 1978). An example of such a situation can be found in the Afar depression, whose most "mature" portion approaches more and more the structure of an active spreading ridge even if significant amounts of material of intermediate chemical composition are present. The transition crust–mantle is very smooth and the differentiation between lid and asthenosphere is practically lost. New oceanic crust is produced in this "ridge" phase and will move away from the ridge axis (Müller, 1978). The Balearic and Tyrrhenian basins probably represent examples of the early stage of formation of oceanic basins from the rifting of continental lithosphere (Panza and Calcagnile, 1978).

Another way in which the continental lithosphere can be consumed and recycled in the asthenosphere is through a continent–continent collision. This process has been proposed to explain the observed dispersion relations for Rayleigh waves in the Alpine area (Panza and Müller, 1978) and the travel time residuals for teleseismic events recorded on the Swiss seismic network (Baer, 1979). The model on the continental crust proposed by Müller (1977), based primarily on data from central Europe, gives the possibility of explaining the subduction of continental lithosphere. The continental crust of the stable parts is the result of the evolution of orogens into peneplains and cratons. This evolution is accomplished through a complex series of processes, such as isostatic rebound of the roots, to which is associated the peneplation, the readjustments of the isotherms, the granulitization of the lower crust with subsequent expulsion of volatiles, which migrates upward, and the migmatization and anatexis in the intermediate crust. The concentration of volatiles and partial melting are probably responsible for the reduction in rigidity of the intermediate crust, indicated by the crustal low velocity layer at a depth of about 10 km (Müller, 1977). In this "soft" layer the upper crust can be detached from the lower crust. In fact, according to the crustal model proposed by Müller (1977) the sialic low velocity zone does not represent the

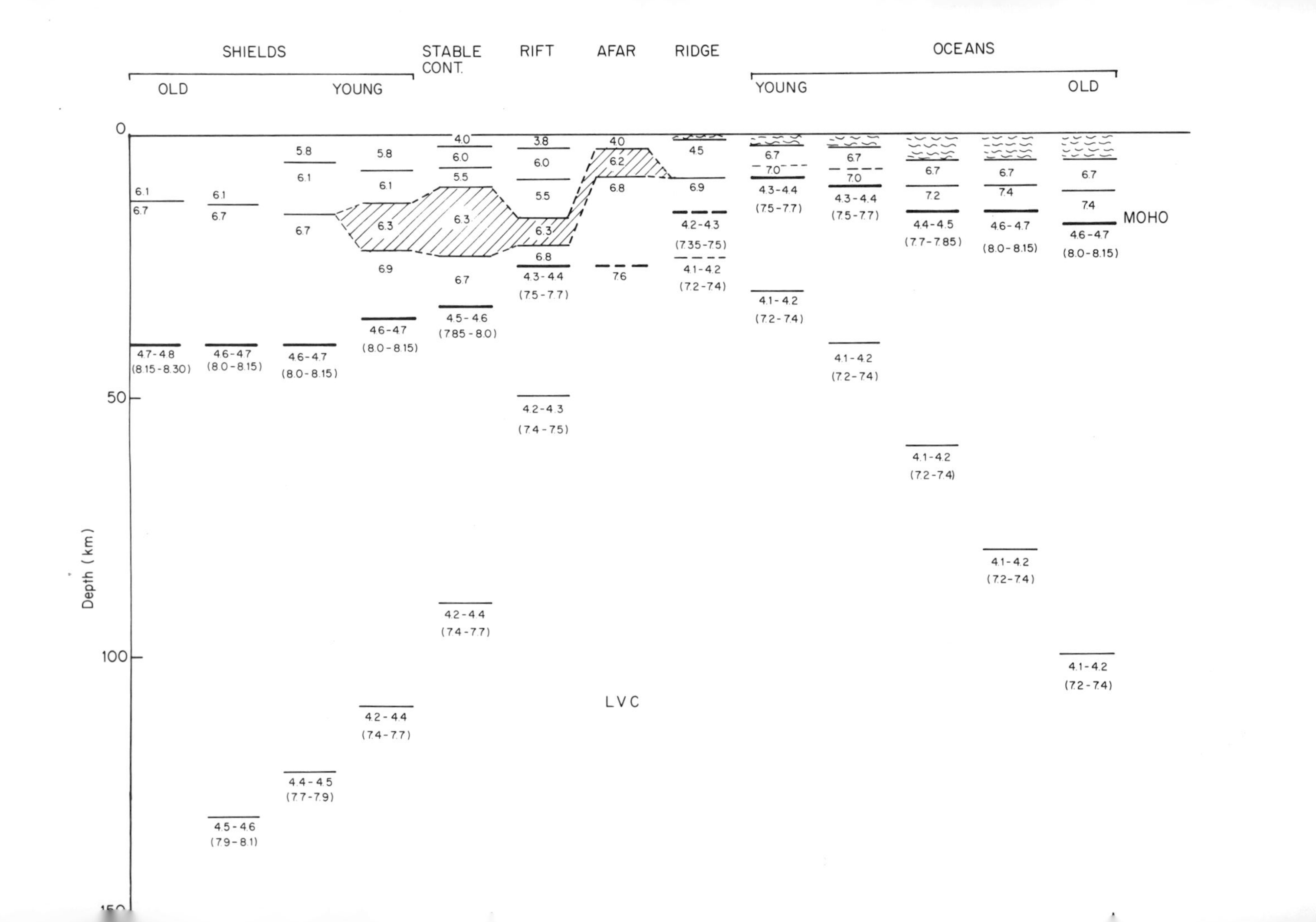
SHIELDS
OLD
YOUNG
STABLE CONT.
RIFT
AFAR
RIDGE
OCEANS
YOUNG
OLD
MOHO
LVC
Depth (km)
0
50
100
6.1
6.7
47-48
(8.15-8.30)
46-47
(8.0-8.15)
5.8
4.5-4.6
(7.9-8.1)
4.4-4.5
(7.7-7.9)
6.3
6.9
4.2-4.4
(7.4-7.7)
4.0
6.0
5.5
4.5-4.6
(7.85-8.0)
3.8
4.3-4.4
(7.5-7.7)
4.2-4.3
(7.4-7.5)
6.2
6.8
7.6
4.5
4.2-4.3
(7.35-7.5)
4.1-4.2
(7.2-7.4)
7.0
7.2
4.4-4.5
(7.7-7.85)
7.4
4.6-4.7
(8.0-8.15)

only depth level at which crustal material can be sheared off, but this can also occur at the top of the crystalline basement beneath the sedimentary cover and immediately above the Moho, as indicated by the tectonics observed in the Franco–Swiss Jura Mountains and Ivrea zone, respectively. Thus for an orogeny, which is the result of a continent–continent collision, it is possible to accept a model where the light upper crust partly forms the nappes and partly contributes to the formation of the roots, while almost all the heavy lower crust is subducted together with the lid.

Dynamical implications

Recent improvements in the solutions for the gravity fields show that ocean rises, trenches and island arcs are mass excesses while ocean basins, areas of recent glaciation and the Asian portion of the Alpine belt are mass deficiencies (Gaposhkin, 1974). The moderate magnitude of negative anomalies corresponding to areas of recent glaciation and the rapid rate of their uplift indicate the presence of a rather plastic asthenosphere.

All other features must be maintained by asthenospheric flow which, from both theory (Turcotte and Oxburgh, 1969) and observation (Heirtzler *et al.*, 1968), are essentially steady state. In a steady state flow system of incompressible material, the maintenance of a mass excess in a particular region can be accomplished by: (1) the piling up of material at the surface; (2) the replacement of less dense by more dense material at an interior interface; (3) thermal contraction; (4) transition to a denser phase; (5) petrological fractionation in which a less dense component is separated from the material before it enters the region.

The reverse of one or more of these processes is required to maintain a mass deficiency. It should be emphasized that mass transfers are necessary to create mass anomalies; densification or rarefaction alone are insufficient (Kaula, 1970).

The relationship of gravity anomalies to the flow system is not unique but depends drastically on the boundary conditions. If the upper boundary is

---

Figure 3. Summary of average characteristic crust and lid structure columns to illustrate the various phases of the evolutionary process of the Earth's lithosphere. The crustal models are essentially taken from Müller (1977). On the left is represented the possible evolution of stable continents into shields, in the centre is represented the evolutionary process from the break-up of a continent to the generation of new oceanic lithosphere, whose thickening with increasing age is represented on the right of the figure. Numbers are *P*-wave velocities in km/s as determined from DSS in the crust and *S*-wave velocities as determined from surface wave dispersion in the lid and in the asthenosphere. Numbers in parentheses are *P*-wave velocities deduced from *S*-wave velocities using the standard $\sqrt{3}$ factor.

fixed, a convective upcurrent is associated with a negative anomaly, because of its lower density (Runcorn, 1965), while if the upper boundary is free (while the lower boundary is fixed), an upcurrent is associated with a positive anomaly because the effect of the mass pushed up at the surface outweighs the density effect (Pekeris, 1935; McKenzie, 1968). In the Earth we should expect that the thicker, colder, less hydrous the lithosphere is in a particular region, the more it will behave like a fixed boundary for asthenospheric flow. On the basis of the upper mantle properties described above, to a first order approximation, we can assume that in stable continents, old oceans and subduction areas the flow system can be described well using fixed boundary conditions, while in rifts, ridges and possibly young ocean basins the flow system can be described using free boundary conditions. This schematization is not sufficient to describe realistically the flow processes in the upper mantle, which very likely has an intermediate behaviour, but solves rather clearly the apparent paradox of both tension and compression features being areas of mass excess near the surface. Thus it seems reasonable to locate convective upcurrents in tension areas, while downcurrents are very likely located in the proximity of the ocean–continent transition even if there is no subduction of oceanic lithosphere.

The structural models shown for shields indicate that upper mantle convection cannot be invoked to explain their recent movements since the mass which translates coherently seems to be at least 400 km thick. This fact, together with the observation of strong lateral velocity gradients at depths greater than 800 km beneath many subduction zones, seems to indicate that more than just the Earth's crust and upper mantle participate in whatever form of convection is responsible for plate motion (Jordan, 1975). Thus we may conclude that the dynamical evolution of the outer parts of our planet is a rather complex process in which gravitational instability and deep and shallow convection are playing interrelated roles, whose way of interaction still needs considerable investigation. The evolution scheme presented for continents and oceans does not contradict the idea that these geodynamical processes were also active in the early stages of the life of the Earth. This can be considered as support to the idea, based mainly on geological and palaeomagnetic data, that the continents did not split apart in any random manner but broke up in Phanerozoic times mainly along linear mobile belts that had undergone successive stages of tectono-thermal reactivation, in some cases traceable back in time into early Protozoic or Archaean times (Anhaeusser, 1973).

Acknowledgements

The author is greatly indebted to St. Müller for stimulating this research and for the very useful critical review of the manuscript.

References

Ampferer, O., 1906. Uber das Bewegungsbild von Faltengebirgen. *Jb. Geol. Reichsanst.* **56**, 539–622.

Anhaeusser, G. R., 1973. The evolution of the early Precambrian crust of southern Africa. *Phil. Trans. R. Soc., Ser. A* **273**, 359–388.

Baer, M., 1979. Kalibrierung des neuen stationsnetzes des schweizerischen erdberbendienstes im hinblick auf die verbesserung der lokalisierung seismicher ereignisse mit epizentralent fernunger bis 100 grad. Ph.D. Thesis-ETH Zurich, p. 46.

Berckhemer, H., Baier, B., Bartelsen, H., Behle, H., Burkhardt, H., Gebrande, H., Makris, J., Menzel, H., Miller, H. and Vees, R., 1975. Deep seismic soundings in the Afar region and on the Highland of Ethiopia. In *Afar Depression of Ethiopia* (Eds Pilger, A. and Rösler, A., 89–107, Schweizerbart-Verlag, Stuttgart.

Bird, J. M. and Dewey, J., 1970. Lithosphere plate–continental margin tectonics and the evolution of the Appalachian orogen. *Bull. Geol. Soc. Am.* **81**, 1031–1060.

Biswas, N. N. and Knopoff, L., 1974. Structure of the upper mantle under the United States from the dispersion of Rayleigh waves. *Geophys. J. R. Astr. Soc.* **36**, 515–539.

Bloch, S., Hales, A. L. and Landisman, H., 1969. Velocities in the crust and upper mantle of Southern Africa from multimode surface wave dispersion. *Bull. Seism. Soc. Am.* **59**, 1599–1629.

Brune, J. and Dorman, J., 1963. Seismic waves and Earth structure in the Canadian shield. *Bull. Seism. Soc. Am.* **53**, 167–210.

Calcagnile, G. and Panza, G. F., 1978. Crust and upper mantle structure under the Baltic shield and Barents Sea from the dispersion of Rayleigh waves. *Tectonophysics* **47**, 59–71.

Condie, K. C., 1972. A plate tectonics evolutionary model of the South Pass Archaean greenstone belt, southwestern Wyoming. 24th Int. Geol. Conf., Montreal, Canada **1**, 104–112.

Forsyth, D. W., 1975. The early structural evolution and anisotropy of the oceanic upper mantle. *Geophys. J. R. Astr. Soc.* **43**, 103–162.

Fuchs, K., 1974. Geophysical contributions to Taphrogenesis. In *Approaches to Taphrogenesis* (Eds Illies J. H. and Fuchs, K.) Schweizerbart-Verlag, Stuttgart.

Gaposhkin, E. M., 1974. Earth's gravity field to the eighteenth degree and geocentric coordinates for 104 stations from satellite and terrestrial data. *J. Geophys. Res.* **79**, 5377–5411.

Gibt, R. A., 1971. Origin of the great arc of eastern Hudson Bay: a Precambrian continental drift reconstruction. *Earth Planet Sci. Lett.* **10**, 365–371.

Gibt, R. A. and Walcott, R. I., 1971. A Precambrian suture in the Canadian shield. *Earth Planet Sci. Lett.* **10**, 417–422.

Hamilton, W., 1970. The Uralides and the motion of the Russian and Siberian platforms. *Bull. Geol. Soc. Am.* **81**, 2553–2576.

Hart, R. S. and Press, F., 1973. S_n velocities and the compostion of the lithosphere in the regionalized Atlantic. *J. Geophys. Res.* **78**, 407–414.

Heirtzler, J. R., Dickinson, G. O., Herron, E. M., Pitman, W. C. and LePichon, X., 1968. Marine magnetic anomalies, geomagnetic field reversals, and motions of the ocean floor and continents. *J. Geophys. Res.* **73**, 2119–2136.

Hirn, A. and Perrier, G., 1974. Deep seismic sounding in the Limagne graben. In *Approaches to Taphrogenesis*, (Eds Illies, J. H. and Fuchs, K.), Schweizerhart-Verlag, Stuttgart.

Jordan, T. H., 1975. Lateral heterogeneity and mantle dynamics. *Nature* **257**, 745–750.

Kaula, W. M., 1970. Earth's gravity field: relation to global tectonics. *Science* **169**, 982–948.

Knopoff, L., 1972. Observation and inversion of surface wave dispersion. *Tectonophysics* **13**, 497–519.

Knopoff, L. and Schlue, J., 1972. Rayleigh wave phase velocities for the path Addis Ababa Nairobi. *Tectonophysics* **15**, 157–163.

Laubscher, H. P., 1971. The large scale kinematics of the Western Alps and the Northern Apennines and its palinspastic implications. *Am. J. Sci.* **271**, 193–226.

Leeds, A. R., Knopoff, L. and Kausel, E. G., 1974. Variations of upper mantle structure under the Pacific Ocean. *Science* **186**, 141–143.

Lepine, J. C., Ruegg, J. C. and Steinmetz, L., 1972. Seismic profiles in the Djibouti Area. *Tectonophysics* **15**, 59–64.

McElhinny, M. W. and Briden, J. C., 1971. Continental drift during the Palaeozoic. *Earth Planet Sci. Lett.* **10**, 407–416.

McKenzie, D. P., 1968. The influence of the boundary conditions and rotation on convection in the Earth mantle. *Geophys. J. R. Astr. Soc.* **15**, 457–500.

Müller, St., 1977. A new model of the continental crust. In *The Earth's Crust, Am. Geophys. Union Geophys. Monogr.* **20**, 289–317.

Müller, St., 1978. The evolution of the Earth crust. In *Tectonics and Geophysics of Continental Rifts* (Eds Ramberg, I. B. and Neuman, E.-R.), 11–28, Reidel, Dordrecht.

Müller, St., Marillier, F. and Panza, G. F., 1979. Anomale strukturen im oberen Erdmantel Europas. 39th Annual Meeting of the Deutsch. Geophys. Gesellschaft, Kiel, April 1979.

Müller, St., Peterschmitt, E., Fuchs, K., Emter, D. and Ansorge, J., 1973, Crustal structure of the Rhinegraben area. *Tectonophysics* **20**, 381–390.

Noponen, I., 1977. The Blue road Geotraverse: the relative residual of the teleseismic *P*-wave: application to the study of the deep structure in Fennoscandia. *Geol. För. Stockh. Förh.* **99**, 32–36.

Panza, G. F. and Calcagnile, G., 1979. The upper mantle structure in Balearic and Tyrrhenian Bathyal plains and the Messinian salinity crisis. *Palaeogeography, Palaeoecology, Palaeoclimatology, Sp. Issue* **29**, 3–14.

Panza, G. F. and Müller, St., 1978. The plate boundary between Eurasia and Africa in the Alpine area. *Mem. Sci. Geol.* **33**, 43–50.

Panza, G. F., Müller, St. and Calcagnile, G., 1978. The gross features of the Lithosphere–Asthenosphere system in the European–Mediterranean area. Paper presented at the EGS–ESC Syposium: Joint interpretation of Earthquake and deep seismic sounding data, Strasbourg, 1978.

Pekeris, C. L., 1935. Thermal convection in the interior of the Earth. *Mon. Not. R. Astr. Soc. Geophys. Suppl.* **3**, 343–367.

Pollak, H. N. and Chapman, D. S., 1977. On the regional variations of heat flow, geotherms and lithospheric thickness. *Tectonophysics* **38**, 279–296.

Poupinet, G., 1977. Relation entre le temps de parcours vertical des ondes siesmiques et l'age de la lithosphere continentale. *C. R. somm. Soc. géol. Fr.* **5**, 257–259.

Poupinet, G., 1979. On the relation between *P*-wave travel time residuals and the age of continental plates. *Earth Planet. Sci. Lett.* **43**, 149–161.

Reichenbach, H. and Müller, St, 1974. Ein Krusten-Mantel-Modell für das Riftsystem um den Rheingraben, abgeleitet aus der Dispersion von Rayleigh-wellen. In *Approaches to Taphrogenesis* (Eds Illies, J. H. and Fuchs, K.), 348–354, Scweizerhart-Verlag, Stuttgart.

Runcorn, S. K., 1965. Changes in the convective pattern in the Earth's mantle and continental drift: evidence for a cold origin of the Earth. *Phil. Trans. R. Soc. Ser. A* **258**, 228–251.

Sclater, J. G., Anderson, R. N. and Bell, M. L., 1971. Elevation of ridges and evolution of the central eastern Pacific. *J. Geophys. Res.* **76**, 7888–7915.

Sclater, J. G. and Francheteau, J., 1970. The implications of terrestrial heat flow observations on current tectonic and geochemical models of the crust and upper mantle of the Earth. *Geophys. J. R. Astr. Soc.* **20**, 509–537.

Talbot, C. J., 1973. A plate tectonic model for the Archaean crust. *Phil. Trans. R. Soc. Ser. A* **273**, 413–427.

Turcotte, D. L. and Oxburgh, E. R., 1969. Convection in a mantle with variable physical properties. *J. Geophys. Res.* **74**, 1458–1474.

White, A. J. R., Jakes, P. and Christie, D. M., 1971. Composition of greenstones and the hypothesis of sea-floor spreading in the Archaean. *Sp. Pub. Geol. Soc. Aust.* **3**, 47–56.

Vlaar, N. J. and Wortel, M. J. R., 1976. Lithospheric ageing instability and subduction. *Tectonophysics* **32**, 331–351.

Yoshida, M., 1978. Group velocity distributions of Rayleigh waves and two upper mantle models in the Pacific Ocean. *Bull. Earthq. Res. Inst. Tokyo Univ.* **53**, 319–338.

An Analysis of Past Plate Motions: The South Atlantic and Indian Oceans

E. J. BARRON*, C. G. A. HARRISON

Rosenstiel School of Marine and Atmospheric Science, University of Miami, Florida, USA

Introduction

An analysis of plate tectonics as a function of time offers an additional degree of freedom to study the mechanism of plate tectonics and continental drift. In order to realize the full potential of this approach, a global tectonic model, which includes the age and geometry of past plate configurations and plate motions with respect to an absolute reference frame, is required. In addition to providing direct observational tests of various theories, this analysis can be used to investigate two other aspects of the history of the process which bear directly on the problem. (1) How variable is the intensity of the process? Is the area of sea-floor created per unit time highly variable, either due to episodes of fast spreading (Pitman, 1978) or to large variations in the length of ridge system? (2) How complicated is the geometry of the process? Are plate reorganizations, changes in the spreading rate of individual ridge segments, ridge jumps and asymmetries in crustal accretion indicative of the mechanism? In addition, a global tectonic model coupling all the lithospheric plates (at least for the whole Cenozoic) will fully test the concept of fixed hot spots, the concept of global expansion and may provide considerable insight into the various tectonic manifestations of the process (e.g. back-arc spreading).

A global tectonic model which couples the Arctic, Pacific, Indian and Atlantic Oceans has not been completed. Comprehensive single ocean analyses have been completed only for the Atlantic and Indian Oceans (Sclater, Hellinger and Tapscott, 1977; Norton and Sclater, 1979; Rabinowitz and LaBrecque, 1979). Even in the case of the Atlantic Ocean, the analyses are not straightforward (compare the reconstructions of Bullard, Everett and Smith, 1965; Pitman and Talwani, 1972; LePichon, Sibuet and Francheteau, 1977; Sclater, *et al.*, 1977). In particular, the Indian Ocean illustrates the

* Present address: National Center for Atmosphere Research, P.O. Box 3000, Boulder, Colorado 80307, USA.

variety of reconstructions which result from interpretations of available data. The Indian Ocean is an excellent illustration of the complexities of past plate motions and a comprehensive analysis is a major step toward the important goal of a global tectonic model.

A sea-floor spreading model of the Indian and South Atlantic Ocean is generated, forward in time, from 140 million years to the present, based on an initial reconstruction of the southern continents, and all available well-delineated Mesozoic and Cenozoic magnetic anomalies and DSDP site data. Continental positions are based on the timing of the initial separation of continental elements and smooth rotation vectors between magnetic anomalies. The positions of the ridges are generated based on the assumption of symmetric accretion, but are modified in cases of ridge reorganization which is required by magnetic anomaly data. The ridge shape is assumed to be the initial outline of the separating continental elements and fracture zones are the product of marginal offsets. In this manner, we attempt to model the sea-floor spreading system within the framework of the simplest possible assumptions and modify this simple system only when compelled by data.

Bathymetric maps, at 20 million year increments, are predicted using the Sclater–Francheteau age–depth relationship.

Sources of data

The initial reconstruction of the southern continents, the timing of continental break-up and the relative motions of the continents must be determined to generate a sea-floor spreading model. The sources of data are oceanic magnetic lineations, palaeomagnetic pole determinations, geologic correlations between continents, continental margin history, fracture zone lineations and geometrical fits of continental outlines.

Since Wegener (1924) and duToit (1937) presented reconstructions of Gondwanaland, the southern continental fit has been an enduring controversy. We adopt the revised reconstruction (Fig. 1) and the timing of continental break-up discussed by Barron, Harrison and Hay (1978). Previously reviewed data will not be repeated. Briefly, the controversy centres around two components: (1) the palaeoposition of Madagascar (a northerly position adjacent to Kenya, a southerly position adjacent to Mozambique, or fixed in its present position with respect to Africa) and (2) whether deformation has occurred between east and west Antarctica.

In addition to the data reviewed by Barron *et al.* (1978), Mesozoic magnetic lineations in the Mozambique Basin (Simpson *et al.*, 1979) and possible fracture zones between Madagascar and Africa (Bunce and Molnar, 1977) appear to present a coherent argument for a northerly derivation for Madagascar. However, the magnetic anomalies are so poorly delineated as to be much less than convincing. We cannot eliminate any of the three possibilities. In the reconstruction we propose that either the southerly

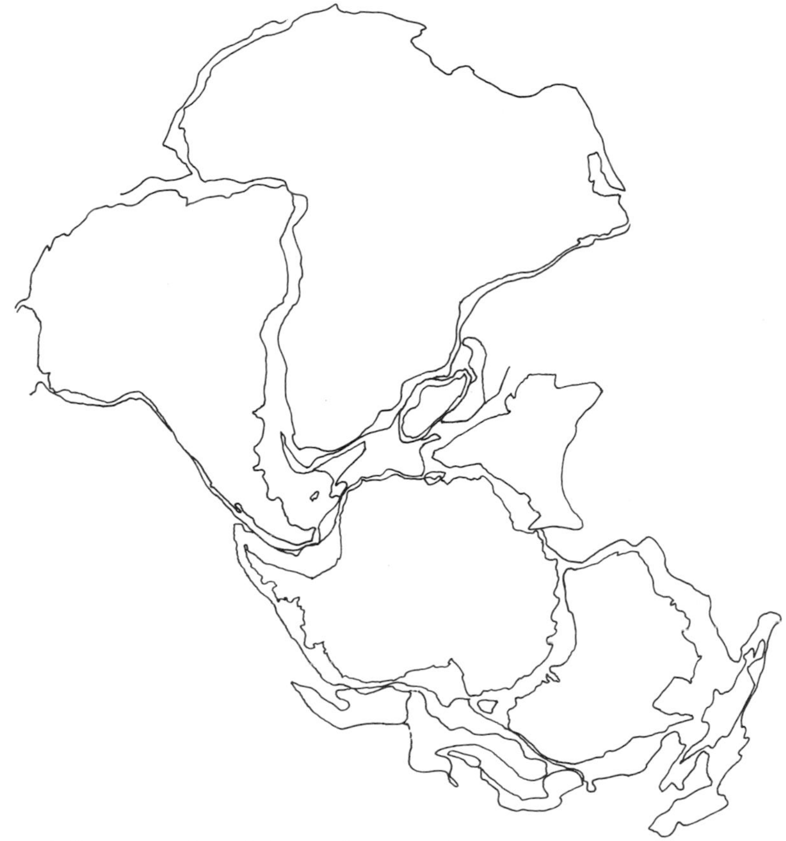

Figure 1. Reconstruction of the southern continents.

position, or a position fixed with respect to Africa, are equally acceptable geometrically (depending only on the nature of the crust of the southern Seychelles platform).

The second controversial aspect is the reconstruction of Antarctica as a single unit with Africa and South America (Fig. 1). As additional support for this hypothesis, Barron and Harrison (1979) and Harrison, Barron and Hay (1979) have demonstrated that the sea-floor spreading pattern and geologic history are consistent with this reconstruction.

The timing of continental separation is determined in one of three manners: (1) correlation of uplift, intrusion and marine sedimentation during initial rifting (e.g. Scrutton, 1973), (2) the divergence of palaeomagnetic pole positions (e.g. McElhinny, 1973) or (3) an extrapolation based on the spreading rates of the oldest available magnetic anomaly sequence (e.g. Sclater *et al.*, 1977). Based largely on data from the first category, the initial break-up into three plates, Africa–South America, India–Seychelles–Madagascar and Antarctica–Australia occurred 140 million years ago.

Magnetic lineations coupled with the strike of fracture zones are the best method of determining subsequent plate motions. The following sources of magnetic anomaly data were used to calculate the rotation vectors (Table I) for

TABLE I *Rotation vectors for the reconstruction of the Indian Ocean. A positive angle is counter clockwise*

Time (*Myr b.p.*)	*Anomaly no.*	*Latitude* (°N)	*Longitude* (°E)	*Angle* (°)
Rotations for India to Seychelles				
0– 40·0	–17	16·0	48·3	–22·4
40·0– 54·3	17–23	13·1	12·0	– 7·6
54·3– 63·7	23–28	13·1	12·0	–18·3
63·7–140·0	28–			0·0
Rotations for Seychelles to Madagascar				
0– 63·7	–28			0·0
63·7–100·0	28–	15·3	42·7	–17·9
100·0–140·0				0·0
Rotations for Madagascar to Africa				
0–100·0				0·0
100·0–140·0		44·0	3·5	– 6·0
Rotations for South America to Africa				
0– 35·6	–13	57·4	–37·5	13·4
35·6– 80·0	13–34	66·6	–37·4	20·5
80·0–108·5	34–M0	24·1	–15·7	22·1
108·5–125·5	M0–	17·9	– 9·6	4·4
125·5–140·0				0·0
Rotations for Australia to Antarctica				
0– 9·0	–5	9·7	36·5	– 6·8
9·0– 19·5	5–6	19·2	32·7	– 5·4
19·5– 27·1	6–8	16·1	29·4	– 4·0
27·1– 32·6	8–12	2·1	38·8	– 3·0
32·6– 35·6	12–13	5·0	33·2	– 1·6
35·6– 42·1	13–18	– 0·3	34·8	– 3·2
42·1– 55·0	18–	14·1	25·4	– 8·1
55·0–140·0				0·0
Rotations for Antarctica to Africa				
0– 33·6	–12	10·0	–41·0	5·0
33·6–140·0	12–	10·0	–41·0	32·7
Rotations for Arabia to Africa (Nubia)				
0– 16·0		36·5	18·0	– 6·1
16·0–140·0				0·0
Rotations for Somalia to Arabia				
0– 21·0		26·5	21·5	7·6
21·0–140·0				0·0

the southern continents. All magnetic anomaly identifications are based on the time scale of LaBrecque, Kent and Cande (1977).

India to Seychelles. The rotation of India to the Seychelles is derived from McKenzie and Sclater (1971), Whitmarsh (1972) and Schlich (1974). This closes India to the Seychelles at anomaly 28 (63·7 Myr). The present phase of spreading initiated at anomaly 17 (40·0 Myr).

Seychelles to Madagascar. The rotation of Seychelles to Madagascar is a smooth rotation from 63·7 to 100 Myr. The rotation is defined by the reconstruction of Seychelles–India to Antarctica–Africa in the initial reconstruction. The oldest anomalies available to constrain this rotation are the record of anomaly 34–79·9 Myr (Schlich, 1974). The position of the ridge crest between Madagascar and the Seychelles is not constrained by magnetic anomaly data.

Madagascar to Africa. This is a smooth rotation to bring Madagascar adjacent to Mozambique at 140 Myr. In this manner the rotations of India, Seychelles and Madagascar have been determined with respect to Africa.

South America to Africa. The rotation of South America to Africa is that of Sibuet and Mascle (1978) and we have assumed that the initial separation occurred at 125·5 Myr.

Australia to Antarctica. The rotation of Australia to Antarctica is based on the synthesis of Weissel, Hayes and Herron (1977) for the time period back to anomaly 18 (42·1 Myr). The rotation from anomaly 18 to the initial reconstruction at 55 Myr is based on the initial reconstruction of Griffiths (1974) and Laird, Cooper and Jago (1977).

Antarctica to Africa. The present day instantaneous pole (Minster *et al.*, 1974) between Antarctica and Africa, the Mesozoic pole described by Bergh (1977) based on a magnetic anomaly suite from 113 to 121 Myr and the pole of rotation derived for the reconstruction of Barron *et al.* (1978) are almost coincident. This suggests that the rotation between Antarctica and Africa has remained about a fixed pole since separation occurred (at 140 Myr). The Mesozoic pole must be modified if we attempt to fit this rotation with the poorly defined anomalies of Simpson *et al.* (1979) in the Mozambique Basin. Little is known about the rates of rotation. Norton (1976) suggested a half-spreading rate of 0·8 cm/yr for the recent past, and Bergh (1977) suggested that the Mesozoic rate was 1·75 cm/yr. We have adopted 0·8 and 1·8 cm/yr for the half-spreading rates, with the change in rate occurring at the time of anomaly 12/13. We have thus defined the rotations of Australia and Antarctica to Africa.

Arabia to Africa. The rotation pole is that of McKenzie, Davies and Molnar (1970). More complicated spreading patterns (e.g. Richardson and Harrison, 1976) are not considered.

Somalia to Arabia. The rotation pole and angle is that of McKenzie *et al.* (1970) from the work of Laughton (1966). The age of initial separation was calculated using the present spreading rate.

Assumptions of the sea-floor spreading model

The initial reconstruction and the rotations described in Table I define the past plate motions used to model the Indian and South Atlantic Oceans. In addition, we will assume that the initial shape of the ridge crest mirrors the two separating continental elements and that the generation of crust is symmetric between the two elements. The position of the ridge and the age of the sea-floor will be calculated as the continents diverge. Initially, all offsets (fracture zones) of the ridge crest will result solely from marginal offsets (from the assumption that the ridge shape mirrors the continental outline). The correspondence of fracture zones and marginal offsets and their predictive potential was noted by Francheteau and LePichon (1972). In this manner we have assumed the very simplest and most straightforward process of sea-floor generation.

Next, all available Mesozoic and Cenozoic, well-delineated, magnetic anomalies and DSDP drilling sites are plotted on respective plates. Any deviation between the initial form of the ridge crest and the observed position, or any offset between magnetic anomaly lineations is corrected prior to advancing the next step forward in time. In this manner the simple sea-floor spreading scenario is corrected to take into account any complications (ridge jumps, asymmetric spreading, etc.) which are required by the data.

Bathymetric maps are generated using the age–depth relationship of Sclater and Francheteau by calculating the positions of the 2, 20 and 50 million year isochrons. These isochrons represent the 3000, 4000 and 5000 m contours respectively.

Based on the data and assumptions described, we present one model of the sea-floor spreading pattern of the Indian Ocean which can be contrasted with another plausible reconstruction of Norton and Sclater (1979). Firstly, the models are different because of the initial configuration of the southern continents. Interestingly, an additional reconstruction has been proposed (Powell, Johnson and Veevers, 1980) which has a reconstruction of Antarctica similar to Barron *et al.* (1978), but Madagascar is placed in a northerly position.

Secondly, the relative motions are reconstructed in a particular sequence (India to Seychelles to Madagascar to Africa; Australia to Antarctica to Africa; and South America to Africa). Note that the anomalies adjacent to the ninety-east ridge and the Mesozoic anomalies adjacent to the western margin of Australia (Markl, 1974; Larson, 1977) have not been used to define the relative motion of India with respect to Australia. To reconstruct India with respect to Australia requires an accurate interpretation of the origin of the ninety-east ridge. If our model successfully reproduces the past motions and geometric configuration of the Indian Ocean, these anomalies should be predicted as the ocean basins are generated forward in time.

Thirdly, because of the assumptions used to generate this model forward in time, the result is not just a simple correlation of anomaly lineations. Instead we are constrained by geometry and tectonics on a sphere. In particular we are forced to examine the complexities of the plate motions by comparison with the most straightforward process of sea-floor generation.

Reconstructions of the Indian and South Atlantic Oceans

The bathymetric maps are plotted on a Lambert Equal Area Projection centred at 40°E and 30°S with respect to present day Africa. Using compilations of palaeomagnetic pole positions from all continents (e.g. McElhinny, 1973) and the rotations in Table I, a mean pole position for each

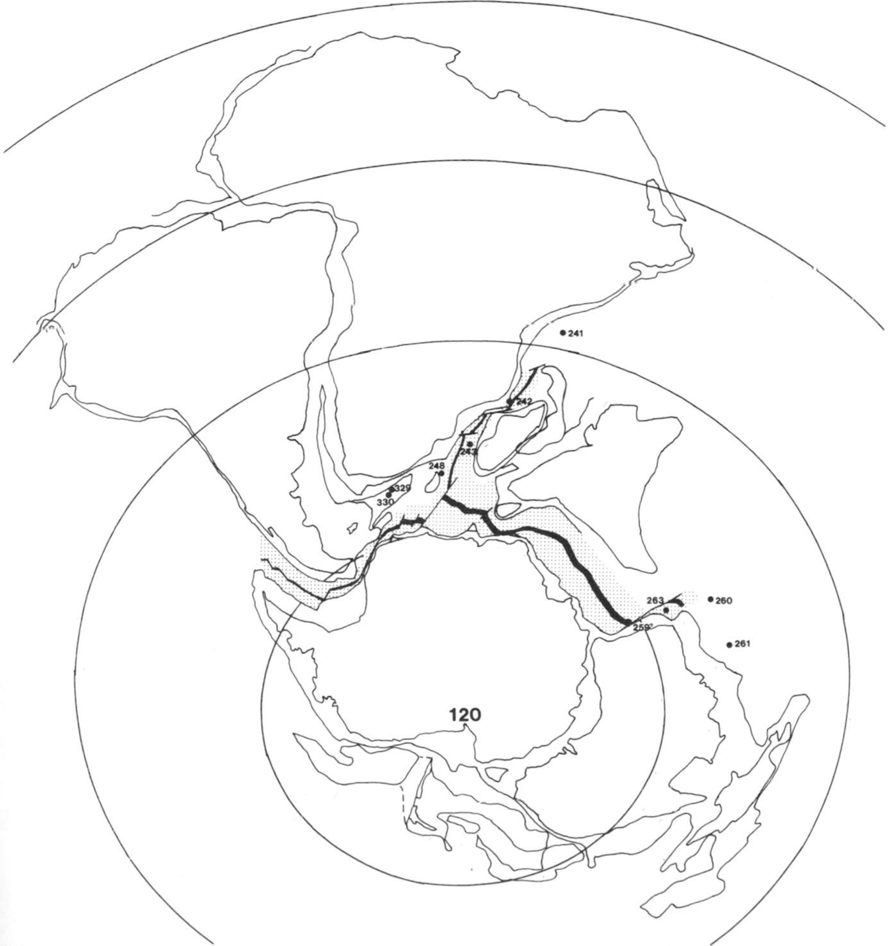

Figure 2. Reconstruction of the Indian Ocean at 120 Myr. Lambert Equal Area Projection centred at 40°E and 30°S. The 2 Myr (3000 m isobath) contours are outlined in black. Crust between 2 and 20 Myr (4000 m isobath) is denoted by a stippled pattern. DSDP sites and numbers are indicated.

time increment was calculated. Thirty degree palaeolatitude lines and the pole position (plotted as the age of the map) are given for reference. With the exception of the Falkland Plateau (3000 m isobath) the continental margin is the 2000 m isobath.

120 million years. The reconstruction at 120 million years (Fig. 2) illustrates the initial three plate configuration of the break-up of the southern continents. The 2 Myr contours (3000 m isobath) which are shaded in black, are indicative of the spreading rates on each ridge axis. The shapes of the ridge axes are the shapes of the separating continental elements and have not been modified from initial assumptions. Two sets of Mesozoic magnetic anomaly data are pertinent; suites adjacent to western Australia (Markl, 1974; Larsen, 1977) and one suite north of Dronning Maud Land (Bergh, 1977). In neither case is an adjustment of the predicted pattern required. This is especially encouraging in the case of the anomaly pattern adjacent to Australia which was not used to describe the relative motion of India with respect to Australia.

The stippled pattern delineates crust greater than 2 Myr old but less than 20 Myr old. DSDP sites which formed on crust older than 120 Myr are plotted but only rarely is a basement age available for verification. Table II lists the sites at which drilling reached basement and for which a basement age was described in the Initial Reports of the Deep Sea Drilling Project.

Site 241 is considerably older than 120 Myr (e.g. Kent, 1973). A thick wedge of sediment, approximately 4 km in thickness extends 300–400 km offshore.

TABLE II *DSDP sites and age of oldest sediments for the sites which reached basement*

DSDP site	*Basement age*	*DSDP site*	*Basement age*
14	U. Oligocene	245	L. Palaeocene
15	L. Miocene	248	L. Eocene–U. Cretaceous
16	U. Miocene	249	L. Cretaceous
17	U. Oligocene	250	Coniacian
18	U. Oligocene	251	L. Miocene
19	M. Eocene	253	M. Eocene
20	Maastrichtian	254	Tertiary
21	Maastrichtian	256	Albian
22	Maastrichtian	257	Cretaceous
212	U. Cretaceous	259	Aptian
213	U. Palaeocene	260	Albian
214	Palaeocene	261	U. Jurassic
215	Palaeocene	265	M. Miocene
216	Maastrichtian	266	L. Miocene
220	L. Eocene	267	U. Eocene
221	U. Eocene	274	L. Oligocene
235	U. Miocene	280	M. Eocene
236	L. Eocene	282	U. Eocene
239	L. Palaeocene–U. Cretaceous	283	Palaeocene
240	Miocene	355	Campanian

DSDP Leg 25 (Site 241) penetrated only 1174 m, yet reached sediments of Campanian age (Simpson *et al.*, 1974). This is difficult to explain if a northerly Madagascar derivation is assumed.

Between 140 and 120 million years a "back-arc" basin forms as the Antarctica Peninsula separates from the margin of South America. This development is discussed in detail by Harrison *et al.* (1979).

In two regions the continental outline (and therefore the strike of the ridge crest) are poorly determined. The nature of the crust in the Mozambique channel is uncertain and the extensive post-Jurassic sediment accumulation on the continental margins results in a somewhat arbitrary distinction of the ridge axis. The outline of western India, because of the progradation of the Ganges Fan and the post-Jurassic tectonic history, is smoothed. The shape of northern India prior to the formation of the Himalayas is also not specified.

It is likely that the Indian Ocean was a restricted basin at this time. Predominantly strike slip motion between parts of Antarctica and South America, and between Madagascar and Africa, probably means that these channels were not open for normal oceanic circulation. Even the large apparent opening between India and the west coast of Australia is probably not open for oceanic circulation because we have not included "greater India" in our reconstruction. Strike slip motion between the Himalayan position of India and Australia ensures that the opening if it exists at all, is *only* a shallow one.

100 million years. In the reconstruction at 100 Myr (Fig. 3) there is significant opening of the South Atlantic. This new ridge system is connected to the Indian Ocean by a large transform along the northern margin of the Falkland Plateau. The ridge system continues southward, separating the Falkland Plateau from the Agulhas–Mozambique platforms. The opening of the South Atlantic results in a four plate system and a large increase in the apparent length of total ridge system.

Two other significant developments occur during this time period. Firstly, the opening of the South Atlantic results in convergence between the Antarctic Peninsula and the Southern Andes. Subduction complexes and collision of an arc are recorded in the rock sequences of the Southern Andes (Katz, 1964, 1973; Bruhn and Dalziel, 1977). An important aspect of this reconstruction is that the separation of an arc from the southern Andes and then subsequent convergence and collision of the arc is explained simply as the result of the separation of South America from Africa. Secondly, at 100 Myr, sea-floor spreading between Madagascar and Africa ceases and Madagascar becomes part of the African plate.

In Fig. 3, crust older than 20 Myr (4000 m isobath) and younger than 50 Myr (5000 m isobath) is denoted by a pattern of parallel lines. The oldest sea-floor at this period is 40 Myr.

It is clear from this map that openings at least 3 km deep occur to the SW, N and E of the Indian Ocean.

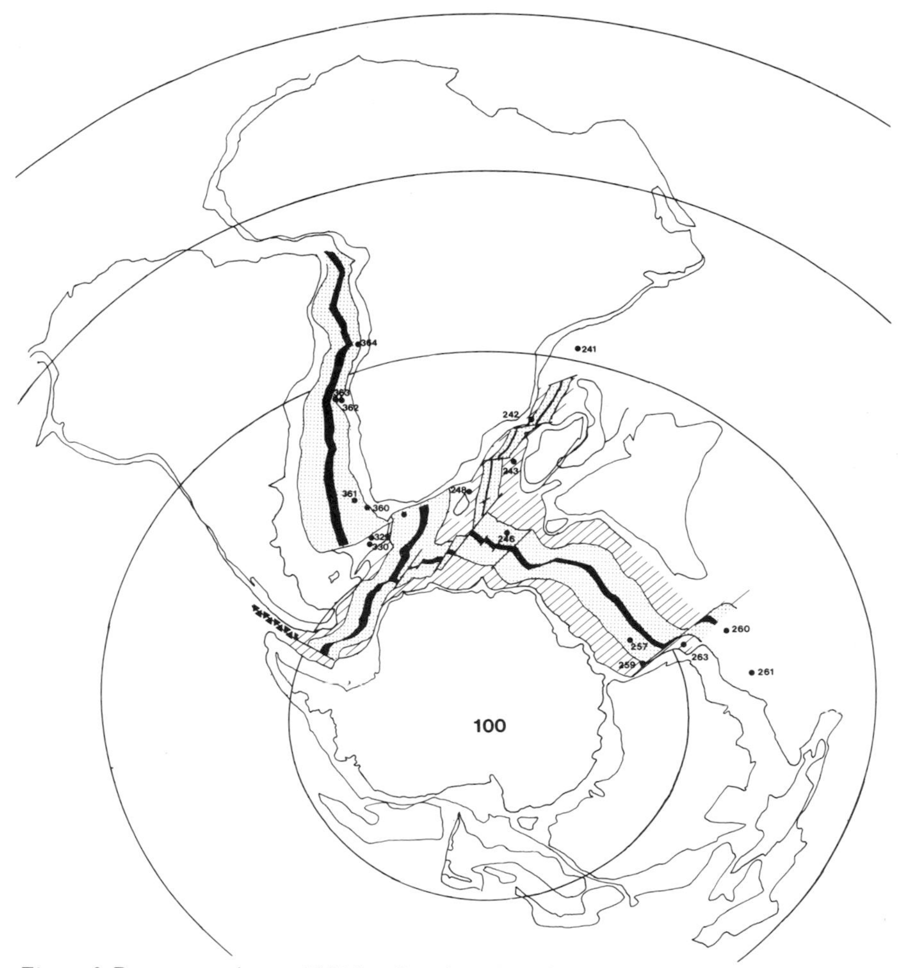

Figure 3. Reconstruction at 100 Myr. Lambert Equal Area Projection centred at 40°E and 30°S. The 2 Myr (3000 m isobath) contours are outlined in black. Crust between 2 and 20 Myr (4000 m isobath) is denoted by a stippled pattern. Crust between 20 and 50 Myr (5000 m isobath) is denoted by a lined pattern. DSDP sites and numbers are indicated.

80 million years. In the reconstruction at 80 Myr (Fig. 4) the ridge system shifts from a position between Madagascar and Africa to a position separating the Madagascar–Africa plate from the India–Seychelles plate. Anomaly 34 (79·7 Myr) is observed by Bergh and Norton (1976) along the Prince Edward Fracture Zone (south of Madagascar) and by Schlich (1974) in the Madagascar Basin. The Schlich anomalies define the southern segment of the newly formed ridge crest (between the southern tip of India and Antarctica). Anomaly 34, identified by Bergh and Norton (1976), has the same strike as the ridge axis generated at 80 Myr but is displaced northward. This adjustment is

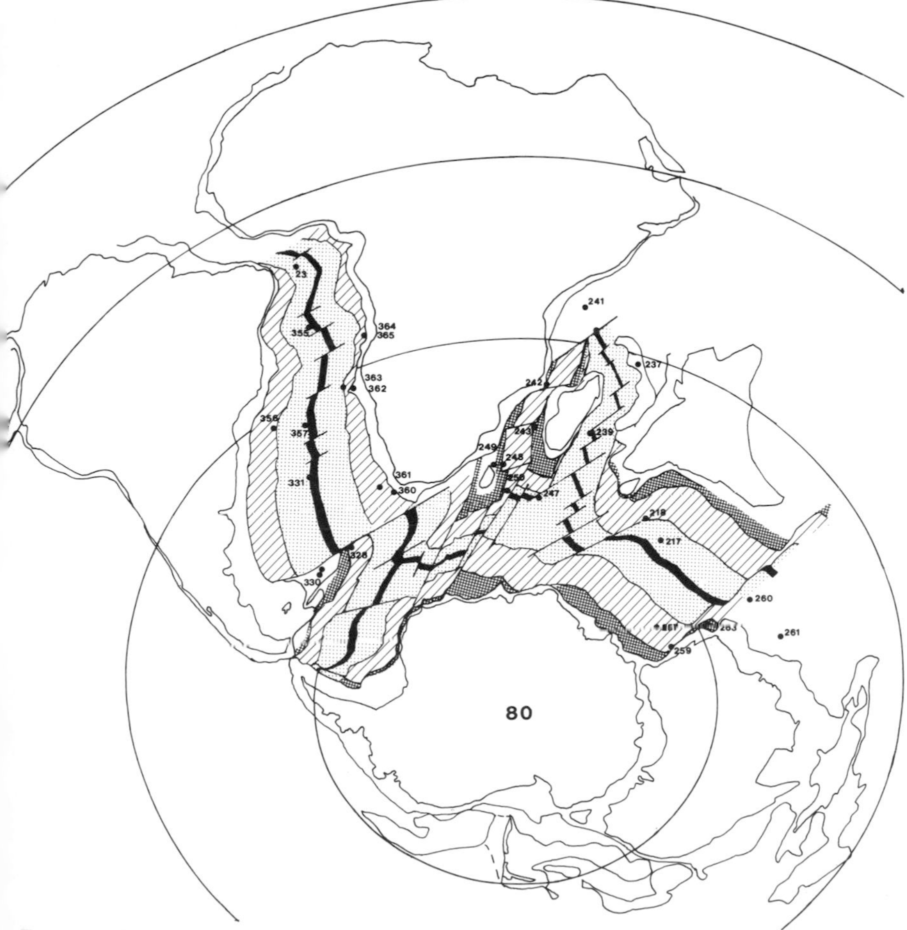

Figure 4. Reconstruction at 80 Myr. Lambert Equal Area Projection centred at 40°E and 30°S. The 2 Myr (3000 m isobath) contours are outlined in black. Crust between 2 and 20 Myr (4000 m isobath) is denoted by a stippled pattern. Crust between 20 and 50 Myr (5000 m isobath) is denoted by a lined pattern. Crust older than 50 Myr is denoted by a cross-hatched pattern. DSDP sites and numbers are indicated.

associated with the reorganization of the ridge system as the Seychelles separates from Madagascar.

Note that the relative motions of Antarctica and South America have resulted in a continuous end-to-end link of the Southern Andes and the Antarctica Peninsula. This probably has caused a break in the deep water circulation between the Indian Ocean and the eastern Pacific.

The age of the sea-floor formed during the initial continental break-up exceeds 50 Myr (5000 m isobath) and is denoted by a cross-hatched pattern.

60 million years. At 63·7 Myr, Seychelles becomes fixed with respect to Africa, resulting in a shift of the spreading axis as India separates from the

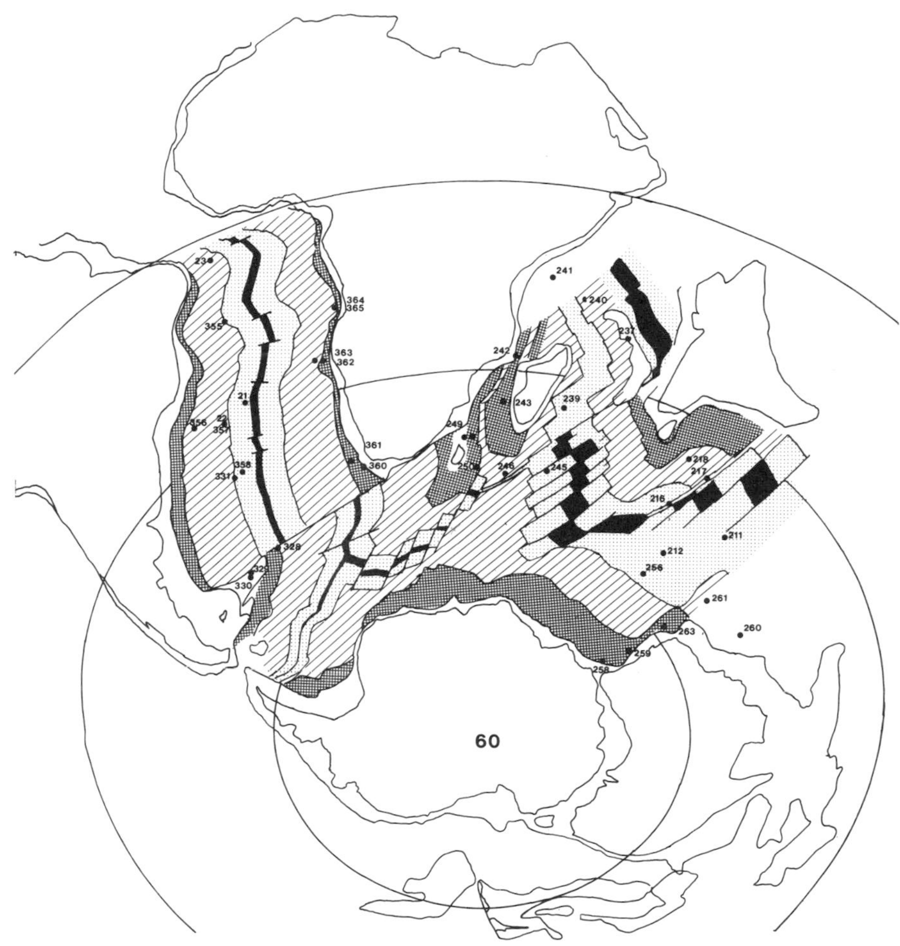

Figure 5. Reconstruction at 60 Myr. Lambert Equal Area Projection centred at 40°E and 30°S. The 2 Myr (3000 m isobath) contours are outlined in black. Crust between 2 and 20 Myr (4000 m isobath) is denoted by a stippled pattern. Crust between 20 and 50 Myr (5000 m isobath) is denoted by a lined pattern. Crust older than 50 Myr is denoted by a cross-hatched pattern. DSDP sites and numbers are indicated.

Seychelles (Fig. 5). At this time, the ridge system separating Africa and Antarctica is coupled to the India–Antarctica spreading system largely by a transform fault.

At anomaly 33 time (71·6 Myr) the magnetic lineation data become much more extensive. Most of the ridge axis surrounding India is now matched with observations. As illustrated by the width of two million year old crust, this time period represents the fastest rate of crustal generation.

The first evidence is available of a significant shift in the ridge position separating India and Australia. This reorganization could have occurred at any time between the Mesozoic anomaly sequence of Markl (1974) and anomaly

33 time. If the Albian age of site 256 is correct this reorganization may have occurred near the time of the separation of the Seychelles from Madagascar. There is no directly apparent reason for the spreading axis shift.

One interesting aspect of the time period between 80 and 60 Myr is that a small amount of compression occurs along a fracture zone. DSDP sites 216 and 217 plot on the zone of compression. These two sites were drilled on the present day ninety-east ridge. The compression along this fracture zone is a simple result of the change in the direction of sea-floor spreading (as Seychelles–India separated from Madagascar and then as India separated from the Seychelles) as might be predicted from a straightforward analysis of tectonics on a sphere (e.g. Menard and Atwater, 1968). The question is only whether the ridge system, in response to the compression, will readjust significantly enough to prevent the formation of a clearly compressional feature. The opposite readjustment, a "leaky" transform, is an increasingly common observation.

At 60 Myr, the relative motion of South America and Antarctica is such that we preserve the continuity of the Southern Andes and the Antarctic Peninsula.

The South Atlantic ridge system continues to be offset by a major transform north of the Falkland Plateau (Agulhas Fracture Zone). At the present, this ridge system is characterized by multiple offsets, but there are no data which suggest that the predicted pattern should be modified at this time.

40 million years. The separation of Australia from Antarctica 55 Myr ago initiates a five plate sea-floor spreading pattern (Fig. 6). This system continues until approximately 39 million years at which time India and Australia form a single plate. The youngest magnetic anomaly that records sea-floor spreading between India and Australia is anomaly 17 at age 39 Myr (Sclater and Fisher, 1974).

Until 40 Myr ago, compression continues along the fracture zone which is now marked by the ninety-east ridge. In addition to DSDP sites 216 and 217, sites 214 and 253 plot on the zone of predicted compression.

If the outline of the Kerguelen plateau is fixed with respect to Antarctica and Broken Ridge is fixed with respect to Australia, the two features are immediately adjacent across the Australia–Antarctica Ridge. Both features are on crust for which a range of 60–40 Myr is predicted. The oldest anomaly north of Kerguelen is anomaly 18 (41 Myr) (Schlich and Patriat, 1971). There exists considerable speculation as to the age and origin of the feature. It is unlikely that Kerguelen is a continental fragment (Watkins *et al.*, 1974; Houtz, Hayes and Markl, 1977). If seismic interpretations are correct, the age of the sediment on the plateau may be as old as Cenomanian (Houtz *et al.*, 1977).

The sea-floor spreading rate on the India–Antarctica and India–Seychelles Ridges decreases by half at 40 Myr. The change in rate from approximately 100–180 mm/yr to only about 50 mm/yr is typically considered to indicate the initial collision between Asia and India (Molnar and Tapponnier, 1975). The change in rate by anomaly 17 (39 Myr) and the cessation of spreading between

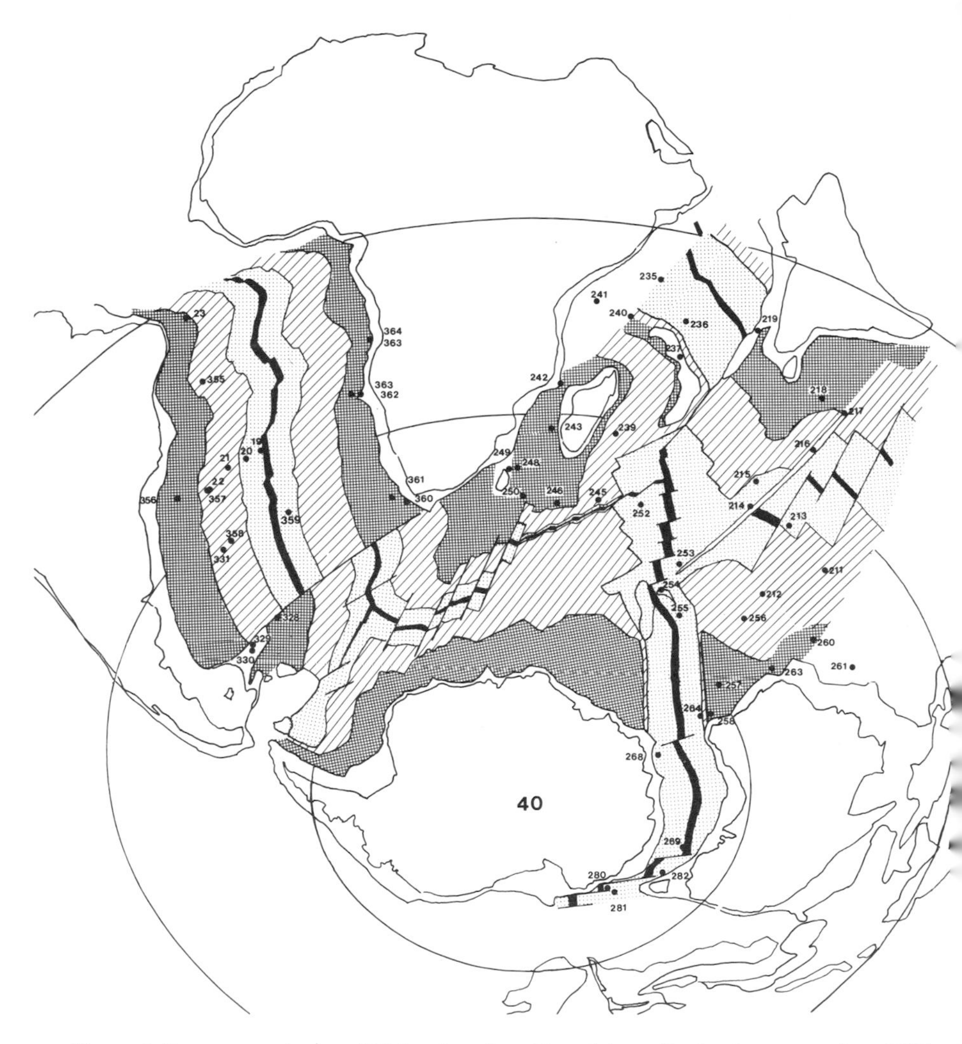

Figure 6. Reconstruction at 40 Myr. Lambert Equal Area Projection centred at 40°E and 30°S. The 2 Myr (3000 m isobath) contours are outlined in black. Crust between 2 and 20 Myr (4000 m isobath) is denoted by a stippled pattern. Crust between 20 and 50 Myr (5000 m isobath) is denoted by a lined pattern. Crust older than 50 Myr is denoted by a cross-hatched pattern. DSDP sites and numbers are indicated.

Australia and India at approximately the same time initiate the modern phase of sea-floor spreading in the Indian Ocean.

20 million years. The phase of sea-floor spreading initiated at anomaly 17 time continues with relatively little modification (Fig. 7). The initial opening of the Gulf of Aden, calculated using the present spreading rate, occurred at 21 Myr.

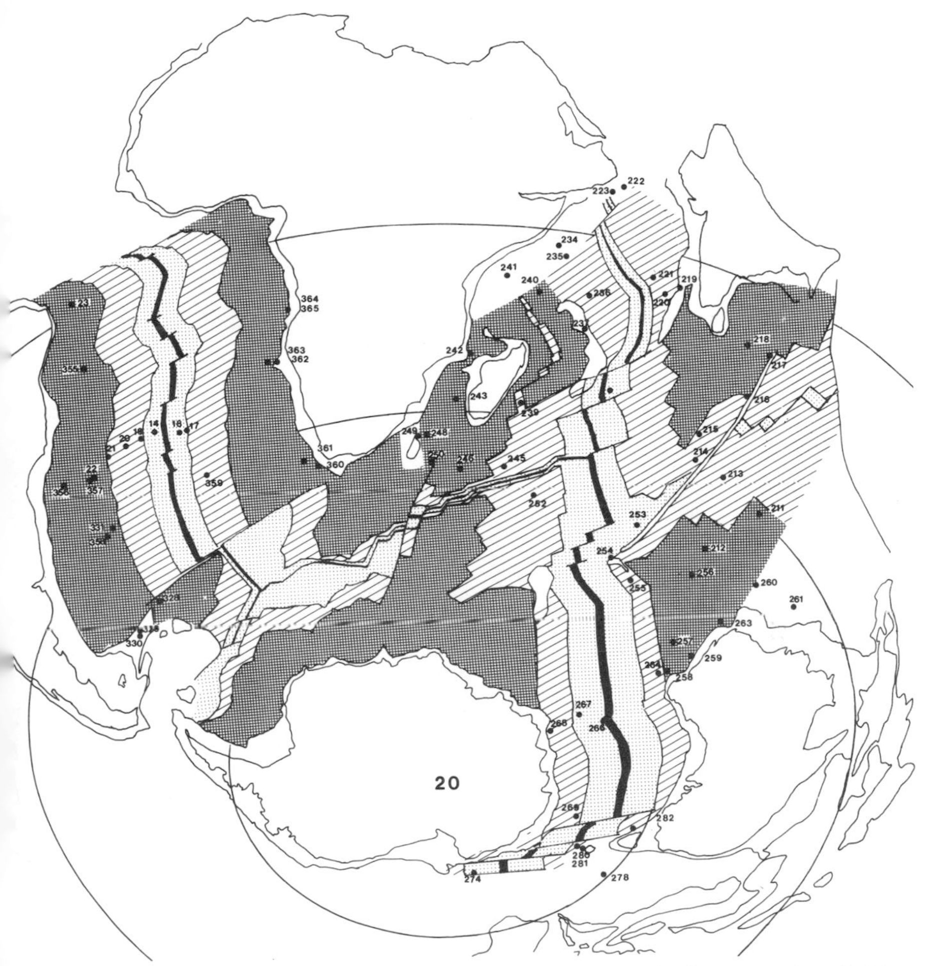

Figure 7. Reconstruction at 20 Myr. Lambert Equal Area Projection centred at 40°E and 30°S. The 2 Myr (3000 m isobath) contours are outlined in black. Crust between 2 and 20 Myr (4000 m isobath) is denoted by a stippled pattern. Crust between 20 and 50 Myr (5000 m isobath) is denoted by a lined pattern. Crust older than 50 Myr is denoted by a cross-hatched pattern. DSDP sites and numbers are indicated.

The separation of the Southern Andes and the Antarctic Peninsula is a post-Cretaceous process (Barker and Griffiths, 1972) for which there exist clearly defined anomalies at least to number 20 (44·8 Myr) (Barker, 1970). No attempt was made specifically to model the evolution of the Scotia Sea.

The present day. Figure 8 is a bathymetric map generated for the present day excluding sediment cover. Plate motions from anomaly 34 time (79·6 Myr) are well defined. Since the oldest contour is a 5000 m isobath (50 Myr) we would, therefore, expect this map to be very similar to a modern bathymetric map.

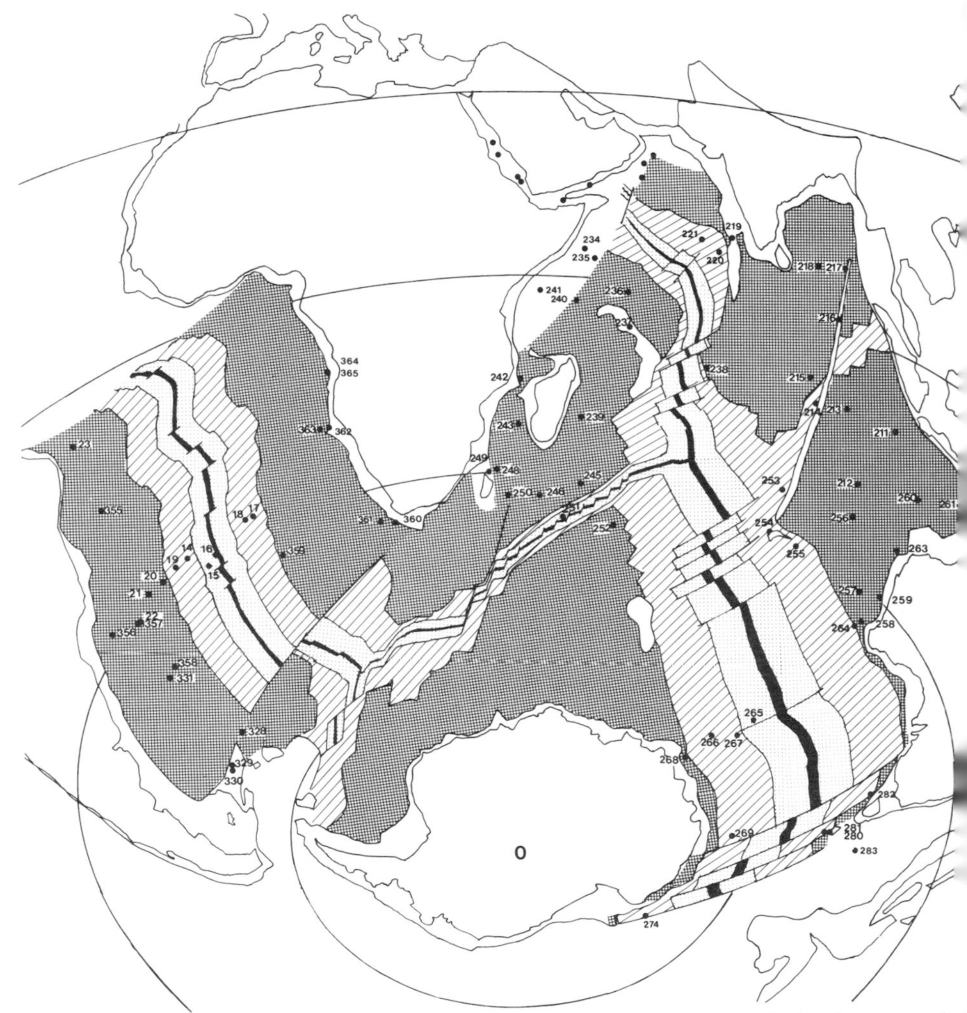

Figure 8. Reconstruction for the present day. Lambert Equal Area Projection centred at 40°E and 30°S. The 2 Myr (3000 m isobath) contours are outlined in black. Crust between 2 and 20 Myr (4000 m isobath) is denoted by a stippled pattern. Crust between 20 and 50 Myr (5000 m isobath) is denoted by a lined pattern. Crust older than 50 Myr is denoted by a cross-hatched pattern. DSDP sites and numbers are indicated.

We have thus defined the evolution of the Indian Ocean and South Atlantic Ocean from 140 Myr to the present.

The origin of the ninety-east ridge

The ninety-east ridge has been interpreted as a horst (Francis and Raitt, 1967), as a result of convergence (LePichon and Heirtzler, 1968), as a "leaky" fracture zone (Sclater and Fisher, 1974) and as the trace of one or more hot spots

(Luyendyk and Rennick, 1977; Pierce, 1978). The ninety-east ridge is a fracture zone, with the crust to the east becoming younger to the south, and the crust to the west becoming younger to the north (Sclater and Fisher, 1974). The ridge, which becomes older to the north, fits the Sclater–Francheteau age–depth curve. The five DSDP sites (Davies *et al.*, 1974; von derBorch *et al.*, 1974) reached shallow littoral and pyroclastic basal sediments. The age of the ninety-east ridge is mid-Cretaceous to early Oligocene. Palaeolatitude studies (Pierce, 1978) indicate that the ridge formed at a point relatively fixed in latitude (48–53°S). This is considered the most important constraint for a hot spot origin. The ridge has an asymmetric cross-section with a steep east facing scarp; an observation which gave rise to the hypothesis of convergence (Le Pichon and Heirtzler, 1968). Based on crustal models, constrained by limited seismic and gravity data, Bowin (1973) concluded that the small free air gravity anomaly was inconsistent with either a horst or a compressive feature.

The model we have presented of the Indian Ocean correctly predicts the ninety-east ridge in both time and space, as a compressive feature. This requires a reanalysis of the seismic and gravity data. This feature would not be apparent in a reconstruction of India and Australia based on the magnetic anomalies in the basins adjacent to the ridge. Sclater and Fisher (1974) interpret the anomaly pattern as parallel lineations which are offset by the ninety-east ridge. The compression is apparent in Figs 5 and 6 because the ridge crest is advanced forward in time using the ridge crest of the previous interval. This results in compression and a predicted position of the spreading ridge (to the east of the ninety-east ridge) which is very slightly subparallel to the anomaly pattern of Sclater and Fisher (1974). A geometric feature of tectonics on a sphere is that the appropriate change in spreading direction will yield compression or a "leaky" ridge axis or fracture zone. Despite this observation, compressive features have not been noted as a product of sea-floor spreading. This may only mean that the ridge axes readjust sufficiently to prevent formation of a distinct feature (in cases where the result is not subduction of one oceanic slab beneath another). Alternatively the ninety-east ridge may be the "type" example of the processes which occur from slight compression along a fracture zone. The "fixed" palaeolatitude of the source does not seem to be a constraint, as surprisingly, the spreading ridge axis in our model maintains a position close to that of the palaeomagnetic studies (Pierce, 1978). We question the hypothesis that the ninety-east ridge is an absolute frame of reference for the reconstruction of the Indian Ocean, which would be especially coincidental as it is also a fracture zone representative of India's relative motion.

Discussion

This analysis of the past plate motions of the Indian and South Atlantic Ocean was generated, forward in time, based on an initial reconstruction of the southern continents and the available magnetic anomaly data. We have assumed that the ridge system corresponds to the shape of separating continental margins and that there is a correspondence between marginal offsets and fracture zones. Changes in this configuration result from the reorganization of plate motions (i.e. changes in the relative rotations of the continental elements). The analysis of the sea-floor spreading system is governed by the geometric constraints of tectonics on a sphere. This is the simplest sea-floor spreading scenario, which is then only modified if required by data.

In contrast to the history of the Atlantic Ocean, the evolution of the Indian Ocean is highly complex. Continental break-up occurs throughout the history of the Indian Ocean, at widely different intervals from the initial break-up of the southern continents to the recent opening of the Red Sea and the divergence of Somalia and Africa. The Indian Ocean is characterized by numerous crustal fragments of diverse origins (e.g. Kerguelen Plateau, Broken Ridge, the Seychelles and the ninety-east ridge). The pattern of sea-floor evolution is also modified by a major continental collision, between India and Asia. Many aspects of this analysis are of particular significance.

To a large degree the position of ridge crests and fracture zones can be determined based on simple assumptions, such as symmetric accretion and the idea that the shapes of ridge systems mirror corresponding continental outlines. The deviations from predicted offsets and positions of ridge crests are associated with changes in the direction of plate motion from the initial pattern.

The separation of India from Africa is particularly interesting. The relative motion is quite simple but consists of a series of continental separations, first from Africa, then Madagascar and finally from the Seychelles. This series of continental separations is quite distinct from the Atlantic history. Although the general pattern of separation between India and Africa appears quite uniform in our reconstruction, each continental separation along India's trailing margin is associated with a small change in the pole of rotation of India. In each case this results in a modification of the sea-floor spreading pattern and also is responsible for the formation of other complex features such as the ninety-east ridge. The most recent change in the spreading pattern between India and Africa is associated with another factor, the convergence of Asia and India.

The history of the ridge axis between Africa and Antarctica is distinctly different. Apparently the relative motion can be described by a single pole of rotation over a period of 140 Myr with the only variation being a change in the rate of crustal accretion.

The general pattern of plate motion appears to be fairly simple, but the sequence of continental separations and small modifications in the direction of sea-floor spreading result in a highly complex evolution. The evolution of the Southern Andes (see Harrison *et al.*, 1979) is an excellent illustration of how relatively simple plate motions result in a complex geologic history. This is again illustrated by the formation of the ninety-east ridge as a product of compression which is predicted from small changes in the relative motion of India with respect to Australia.

Acknowledgements

This research was sponsored by NSF Grant OCE78–26679 and by Phillips Petroleum. James Sloan and Jill Whitman compiled and digitized the DSDP data.

References

Barker, P. F., 1970. Plate tectonics of the Scotia Sea region. *Nature* **228**, 1293–1296.

Barker, P. F. and Griffiths, D. H., 1972. The evolution of the Scotia Ridge and Scotia Sea. *Phil. Trans. R. Soc.* **A271**, 151–183.

Barron, E. J. and Harrison, C. G. A., 1979. Reconstructions of the Campbell Plateau and the Lord Howe Rise. *Earth Planet. Sci. Lett.* **45**, 87–92.

Barron, E. J., Harrison, C. G. A. and Hay, W. W., 1978. A revised reconstruction of the southern continents. *Trans. Am. Geophys. Union* **59**, 436–449.

Bergh, H. W., 1977. Mesozoic sea floor off Dronning Maud Land, Antarctica. *Nature* **269**, 686–687.

Bergh, H. W. and Norton, I. O., 1976. Prince Edward Fracture Zone and the evolution of the Mozambique Basin. *J. Geophys. Res.* **81**, 5221–5239.

Bowin, C., 1973. Origin of the Ninety-East Ridge from studies near the Equator. *J. Geophys. Res.* **78**, 6029–6043.

Bruhn, R. L. and Dalziel, I. W. D., 1977. Destruction of the Early Cretaceous marginal basin in the Andes of Tierra del Fuego. In *Island Arcs, Deep Sea Trenches and Back-arc Basins* (Eds Talwani, M. and Pitman, W. C., III), 395–405, Amer. Geophys. Union, Washington, D.C.

Bullard, E. C., Everett, J. E. and Smith, A. G., 1965. The fit of the continents around the Atlantic. *Phil. Trans. R. Soc.* **A258**, 41–51.

Bunce, E. T. and Molnar, P., 1977. Seismic reflection profiling and basement topography in the Somali Basin: possible fracture zones between Madagascar and Africa. *J. Geophys. Res.* **82**, 5305–5312.

Davies, T. A., Luyendyk, B. P., *et al.*, 1974. *Initial Reports of the Deep Sea Drilling Project, V. 26, Washington*, US Government Printing Office.

duToit, A. L., 1937. *Our Wandering Continents*, Oliver and Boyd, Edinburgh.

Francheteau, J. and LePichon, X., 1972. Marginal fracture zones as structural framework of continental margins in South Atlantic Ocean. *Bull. Am. Ass. Petrol. Geol.* **56**, 991–1007.

Francis, T. J. G. and Raitt, R. W., 1967. Seismic refraction measurements in the southern Indian Ocean. *J. Geophys. Res.* **72**, 3015–3041.

Griffiths, J. R., 1974. Revised continental fit of Australia and Antarctica. *Nature* **249**, 336–337.

Harrison, C. G. A., Barron, E. J. and Hay, W. W., 1979. Mesozoic evolution of the Antarctic Peninsula and the southern Andes. *Geology* **7**, 374–378.

Houtz, R. E., Hayes, D. E. and Markl, R. G., 1977. Kerguelen plateau bathymetry, sediment distribution and crustal structure. *Marine Geol.* **25**, 95–130.

Katz, H. R., 1964. Strukturelle verhaltnisse in den sudlichen Patagonischen Anden und deren beziehung zue Antarktis: eine diskussion. *Geologische Runds.* **54**, 1195–1213.

Katz, H. R., 1973. Contrasts in tectonic evolution of orogenic belts in the south-east Pacific. *R. Soc. N.Z. J.* **3**, 333–362.

Kent, P. E., 1973. Mesozoic history of the east coast of Africa. *Nature* **238**, 147–148.

LaBrecque, J. L., Kent, D. V. and Cande, S. C., 1977. Revised magnetic polarity time scale for Late Cretaceous and Cenozoic time. *Geology* **5**, 330–335.

Laird, M. G., Cooper, R. A. and Jago, J. B., 1977. New data on the lower Palaeozoic sequence of nothern Victoria Land, Antarctica, and its significance for Australian–Antarctic relations in the Palaeozoic. *Nature* **265**, 107–110.

Larson, R. L., 1977. Early Cretaceous break-up of Gondwanaland off western Australia. *Geology* **5**, 57–60.

Laughton, A., 1966. The birth of an ocean. *New Scientist* **27**, 218–220.

LePichon, X. and Heirtzler, J. R., 1968. Magnetic anomalies in the Indian Ocean and sea-floor spreading. *J. Geophys. Res.* **73**, 2101–2117.

LePichon, X., Sibuet, J.-C. and Francheteau, J., 1977. The fit of the continents around the north Atlantic. *Tectonophysics* **38**, 169–209.

Luyendyk, B. P. and Rennick, W., 1977. Tectonic history of aseismic ridges in the eastern Indian Ocean. *Bull. Geol. Soc. Am.* **88**, 1347–1356.

McElhinny, M. W., 1973. *Palaeomagnetism and Plate Tectonics*, Cambridge University Press, New York.

McKenzie, D., Davies, D. and Molnar, P., 1970. Plate tectonics of the Red Sea and East Africa. *Nature* **226**, 1–6.

McKenzie, D. and Sclater, J. G., 1971. The evolution of the Indian Ocean since the Late Cretaceous. *Geophys. J. R. Astr. Soc.* **24**, 437–528.

Markl, R. G., 1974. Evidence for the break-up of eastern Gondwanaland by the early Cretaceous. *Nature* **251**, 196–200.

Menard, H. W. and Atwater, T. M., 1968. Changes in direction of sea-floor spreading. *Nature* **219**, 463–467.

Minster, J. B., Jordan, T. H., Molnar, P. and Haines, E., 1974. Numerical modelling of instantaneous plate tectonics. *Geophys. J. R. Astr. Soc.* **36**, 541–576.

Molnar, P. and Tapponnier, P., 1975. Cenozoic tectonics of Asia: effects of a continental collision. *Science* **189**, 419–426.

Norton, I. O., 1976. The present motion between Africa and Antarctica. *Earth Planet. Sci. Lett.* **33**, 219–230.

Norton, I. O. and Sclater, J. G., 1979. A model for the evolution of the Indian Ocean and the break-up of Gondwanaland. *J. Geophys. Res.* **84**, 6803–6830.

Pierce, J. W., 1978. The northward motion of India since the Late Cretaceous. *Geophys. J. R. Astr. Soc.* **52**, 277–311.

Pitman, W. C., III, 1978. Relationship between eustasy and stratigraphic sequences of passive margins. *Bull. Geol. Soc. Am.* **89**, 1389–1403.

Pitman, W. C., III and Talwani, M., 1972. Sea-floor spreading in the north Atlantic. *Bull. Geol. Soc. Am.* **83**, 619–646.

Powell, C. McA., Johnson, B. D. and Veevers, J. J., 1980. A revised fit of east and west Gondwanaland. *Tectonophysics* **63**, 13–30.

Rabinowitz, P. D. and LaBrecque, J., 1979. The Mesozoic south Atlantic Ocean and evolution of its continental margins. *J. Geophys. Res.* **84**, 5973–6002.

Richardson, E. S. and Harrison, C. G. A., 1976. Opening of the Red Sea with two poles of rotation. *Earth planet. Sci. Lett.* **30**, 135–142.

Schlich, R., 1974. Sea-floor spreading history and deep sea drilling results in the Madagascar and Mascarene basins, Western Indian Ocean. In *Initial Reports of the Deep Sea Drilling Project, V. 25, Washington* (Eds Simpson, E. S. W. *et al.*), 663–678. U.S. Government Printing Office.

Schlich, R. and Patriat, P., 1971. Anomalies magnetiques de la branche est de la dorsale medio-indienne entre les isles Amsterdam et Kerguelen. *C. r. Acad. Sci. Fr.*, **272**, 773–776.

Sclater, J. G. and Fisher, R. L., 1974. Evolution of the East Central Indian Ocean, with emphasis on the tectonic setting of the ninety-east ridge. *Bull. Geol. Soc. Am.* **85**, 683–702.

Sclater, J. G., Hellinger, S. and Tapscott, C., 1977. The palaeobathymetry of the Atlantic Ocean from the Jurassic to the present. *J. Geol.* **85**, 509–552.

Scrutton, R. A., 1973. The age relationship of igneous activity and continental break-up. *Geol. Mag.* **110**, 227–234.

Sibuet, J.-C. and Mascle, J., 1978. Plate kinematic implications of Atlantic equatorial fracture zone trends. *J. Geophys. Res.* **83**, 3401–3421.

Simpson, E. S. W., Schlich, R., *et al.*, 1974. *Initial Reports of the Deep Sea Drilling Project, V. 25 Washington*, 87–138, U.S. Government Printing Office.

Simpson, E. S. W., Sclater, J. G., Parsons, B., Norton, I. and Meinke, L., 1979. Mesozoic magnetic lineations in the Mozambique Basin. *Earth Planet. Sci. Lett.* **43**, 260–264.

von derBorch, C. C., Sclater, J. G., *et al.*, 1974. *Initial Reports of the Deep Sea Drilling Project, V. 22, Washington*, U.S. Government Printing Office.

Watkins, N., Gunn, B., Nougier, J. and Baksi, A., 1974. Kerguelen: continental fragment or oceanic island? *Bull. Geol. Soc. Am.* **85**, 201–212.

Wegener, A., 1924. *The Origins of Continents and Oceans*, Dover, London.

Weissel, J. K., Hayes, D. E. and Herron, E. M., 1977. Plate tectonic synthesis; the displacements between Australia, New Zealand and Antarctica since the Late Cretaceous. *Marine Geol.* **25**, 231–277.

Whitmarsh, R. B., 1972. Some aspects of plate tectonics in the Arabian Sea. In *Initial Reports of the Deep Sea Drilling Project, V. 23, Washington*, (Eds Whitmarsh, R. B. *et al.*), 527–536, U.S. Government Printing Office.

Small-Scale Convection in the Upper Mantle and the Isostatic Response of the Canadian Shield

R. STEPHENSON, C. BEAUMONT

Departments of Geology and Oceanography, Dalhousie University, Halifax, Nova Scotia, Canada

Introduction

The consensus of opinion of most Earth scientists is that the motion of lithospheric plates is driven by some form of thermal convection which derives its energy from primordial heat and/or heat produced by radioactive elements distributed in the mantle. There are, however, few observations to provide independent evidence of the existence and nature of the hypothesized mantle circulation. In this respect some attention has been given to long wavelength gravity anomalies and their relationship to topography on the Earth's surface of similar dimension. It has been assumed that the topography may be a result of plate flexure or uplift dynamically supported by forces associated with the mantle convection (e.g. McKenzie, 1967). Anderson, McKenzie and Sclater (1973), for example, showed that differences in bathymetric depth and gravity anomalies at active mid-ocean ridges are correlated in a way similar to that predicted by numerical models of convection in a Newtonian fluid (McKenzie, Roberts and Weiss, 1974). More recent studies in the North Atlantic (Sclater, Lawyer and Parsons, 1975) and in the Central Pacific (Watts, 1976) have tended to support their conclusions. In contrast, Cochran and Talwani (1977) disputed the existence of a consistent direct correlation between long wavelength gravity anomalies and bathymetry throughout the world's oceans. They argued that the lithosphere must be strongly decoupled from the main body of the asthenosphere since the gravity anomalies do presumably have their source beneath the lithosphere. An alternative approach to the problem of the relationship between gravity anomalies and topography which may clarify the uncertainties in its interpretation is one in which the relationship is determined systematically in the wavenumber (**k**) domain in terms of a linear transfer function (or admittance), $Q(\mathbf{k})$. McKenzie (1977) calculated the behaviour of $Q(\mathbf{k})$ as it depends on the Rayleigh number, the degree of internal heating, viscosity variations, and the depth of the convecting layer. He did not take into account the effect on Q of the defection of an abrupt density change interface within the lithosphere, such as the Mohorovičič discontinuity, during its flexure by the forces derived from convection. Furthermore, the comparison of observed to McKenzie's theoretical values of Q is obviously complicated by the possible presence of topography on the surface of the lithosphere not associated with deformation caused by underlying convection. McKenzie and Bowin (1976), for example, attempted to detect the effects of convection in observations of gravity and bathymetry made along two profiles in the Atlantic Ocean but found that the observed $Q(\mathbf{k})$ could best be explained by isostatic compensation within the lithosphere to simple surface loading by topography. No deformation of the plate by convection was detected.

Convective flow in the Earth may occur with two distinct horizontal length scales: (1) large-scale mass circulation of the lithosphere plates themselves in combination with some form of return flow at an as yet undetermined depth in the mantle and (2) small-scale Rayleigh–Bénard convection in the upper mantle which provides the mechanism of heat transport to the base of the lithosphere evident under the older parts of oceans and under continents. It was the effect of the smaller scale of convection which McKenzie and Bowin (1976) attempted to detect; its hypothetical existence is based on theoretical analyses of the efficiency of heat transport in convective systems (Richter, 1973; McKenzie and Weiss, 1975) and on experimental results (Richter and Parsons, 1975). For regions younger than approximately 70 Myr mean depth varies as the square root of age ($t^{1/2}$), an observation which can be satisfactorily explained in terms of the oceanic lithosphere behaving as a simple cooling boundary layer in the large-scale convective flow regime. In older regions the mean oceanic depth is less than predicted by $t^{1/2}$ suggesting the presence of an efficient heat transfer to the base of the lithosphere by small-scale convection. Parsons and McKenzie (1978) have modelled the onset of small-scale convection in terms of the development of a thermal instability beneath the oceanic lithosphere, assumed to be a mechanically rigid boundary layer, as it thickens with age.

Depending on the Rayleigh number (Ra) of the convecting layer and the velocity of the upper boundary (the lithospheric plate), theory (Richter, 1973) and experiment (Richter and Parsons, 1975) demonstrate that the small-scale circulation may eventually take the form of longitudinal rolls aligned with axes parallel to the direction of shear between the plate and the underlying layer, shown schematically in Fig. 1(a). For $Ra = 10^6$, an estimate based on the results of numerical models compared to observations of heat flux (McKenzie *et al.*, 1974), Richter and Parsons (1975) suggest, on the basis of scaled experimental results, that the formation of longitudinal rolls would take from 20 to 50 Myr to several hundred million years for absolute plate velocities in the range 10–2 cm/yr. More complex patterns of small-scale convection cells,

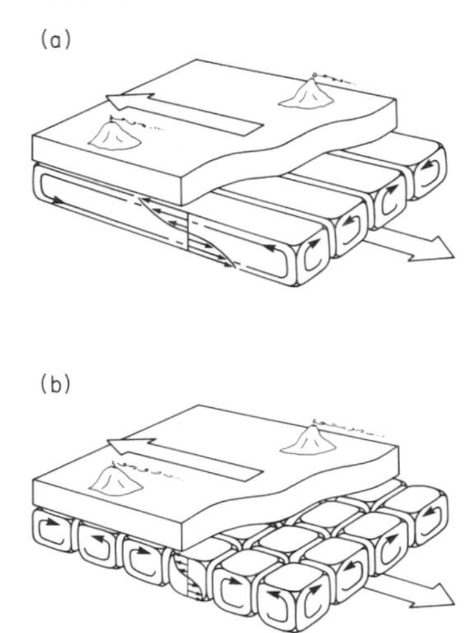

Figure 1. (a) Schematic drawing of sub-lithospheric convection cells in the form of longitudinal rolls aligned with axes parallel to the shear between the lithospheric plate and the underlying asthenosphere; (b) schematic drawing of sub-lithospheric convection cells having a bimodal configuration. The directions of absolute plate motion and mantle return flow are indicated by the large open arrows but note that these directions are not necessarily antiparallel; flow in a vertical section of one cell is shown by the smallest arrows; the volcanoes serve to illustrate the direction of absolute plate motion and that the lithosphere is loaded at its surface as well as its base. Note that these schematic drawings are not drawn to any realistic scale.

such as a bimodal configuration (Fig. 1(b)), may be possible if the age versus plate velocity constraint is not met or if Ra is larger.

A test for the presence of small-scale convection in the upper mantle

In a plate model of the lithosphere, isostatic compensation of surface loads is effected by flexure of the plate. The form of the compensation is revealed by the wavenumber domain relationship between the load and deflection, measurable as the topography ($H_T(\mathbf{k})$), and its associated gravity signature ($G_T(\mathbf{k})$), and can be approximated by the linear transfer function ($Q_T(\mathbf{k})$) (Fig. 2(a)) such that

$$Q_T(\mathbf{k}) = \frac{G_T(\mathbf{k}) + N(\mathbf{k})}{H_T(\mathbf{k})} \tag{1}$$

$Q_T(\mathbf{k})$, generally referred to as the isostatic response function, is a special case of the admittance, mentioned previously, between topography and gravity, in which it is assumed the only load causing flexure of the lithosphere is that of tectonically uplifted surface topography.

In order to estimate $Q_T(\mathbf{k})$ it is assumed that the topography can be measured perfectly, whereas $G_T(\mathbf{k})$ is subject to geological noise ($N(\mathbf{k})$) that is

(a)

$H_0(\mathbf{k}) \rightarrow H_T(\mathbf{k}) \rightarrow \boxed{Q_T(\mathbf{k})} \rightarrow G_T(\mathbf{k}) \rightarrow G_0(\mathbf{k})$

(b)

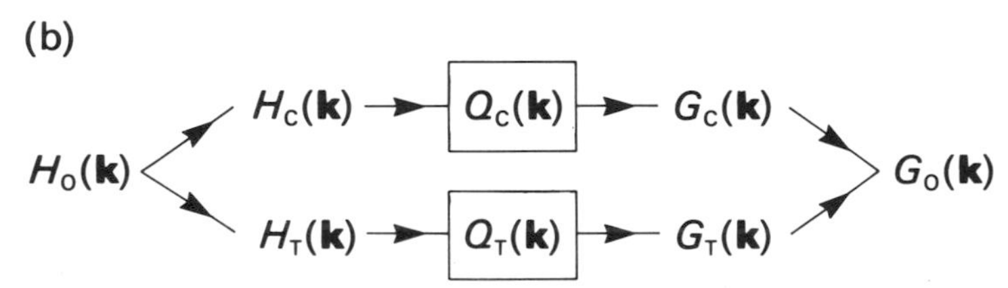

Figure 2. Linear models employed in the interpretation of observed topographic and gravity data; symbols are explained in the text. (a) A model in which the lithosphere is loaded only by tectonically induced topography on its surface; (b) a model in which the lithosphere is also loaded at its base by forces associated with convection.

mainly caused by lateral density variations in the upper crust. If the noise is uncorrelated with topography, $Q_T(\mathbf{k})$, the isostatic response function, is best estimated by

$$Q(\mathbf{k}) = \hat{S}_{GH}(\mathbf{k})/\hat{S}_H(\mathbf{k}), \tag{2}$$

Where $\hat{S}_{GH}(\mathbf{k})$ is a smoothed estimate of the cross spectrum of the gravity and topography and $\hat{S}_H(\mathbf{k})$ is a smoothed estimate of the power spectrum of the topography. It is the smoothing, accomplished by ensemble averaging, which minimizes the random noise (Munk and Cartwright, 1966). It is also useful to determine the coherence squared ($\gamma^2(\mathbf{k})$) of the system, a measure of the portion of the gravity field which is caused by the topography. In the presence of random noise

$$\gamma^2(\mathbf{k}) = \frac{n(\hat{S}_{GH}(\mathbf{k})\hat{S}^*{}_{GH}(\mathbf{k})/\hat{S}_G(\mathbf{k})\hat{S}_H(\mathbf{k})) - 1}{\mathrm{n}-1}, \tag{3}$$

where * indicates complex conjugate, $\hat{S}_G(\mathbf{k})$ is the smoothed estimate of the power in the gravity signal, and n is the number of data in the ensemble. The ensemble of data which is averaged (Eqn (2)) comprises all of those which fall within a given wavenumber band or annulus, symmetric about the origin, on the assumption that the lithosphere is directionally isotropic in its response to a point load and that Q and γ^2 can therefore be estimated as functions of wavenumber modulus $|\mathbf{k}|$. For the same reason, $Q(|\mathbf{k}|)$ is expected to be a real function, though no *a priori* assumption to this effect is required if it is estimated using Eqn (2).

Estimates of Real $[Q(|\mathbf{k}|)]$ in continental regions have been found for the United States (Lewis and Dorman, 1970), Australia (McNutt and Parker, 1978), and East Africa (Banks and Swain, 1978) and have been inverted using local (Dorman and Lewis, 1972) and regional (Banks, Parker and Huestis,

1977) models of isostatic compensation to reveal possible density structures and rheological properties of the continental lithosphere.

However, consider a model in which the lithosphere is loaded at its base by forces associated with small-scale upper mantle convection as discussed previously in addition to surficial topography. This class of model has two independent transfer functions (Fig. 2(b)), which are assumed to be linear, such that the observed admittance is

$$Q(\mathbf{k}) = \frac{G_T(\mathbf{k}) + G_C(\mathbf{k}) + N(\mathbf{k})}{H_T(\mathbf{k}) + H_C(\mathbf{k})} = \frac{Q_T(\mathbf{k})H_T(\mathbf{k}) + Q_C(\mathbf{k})H_C(\mathbf{k}) + N(\mathbf{k})}{H_O(\mathbf{k})}. \quad (4)$$

$G_C(\mathbf{k})$ and $H_C(\mathbf{k})$ are the components of the observed gravity and topography signals caused by convective forces and $Q_C(\mathbf{k})$ is their transfer function. Similarly, $Q_T(\mathbf{k})$ relates tectonic topography $H_T(\mathbf{k})$ to its induced gravity signal $G_T(\mathbf{k})$. $H_O(\mathbf{k})$, the observed topography, is obviously the sum of $H_C(\mathbf{k})$ and $H_T(\mathbf{k})$. Note that $H_T(\mathbf{k})$, as in Eqn (1), comprises the topographic load itself as well as a deflection of the lithosphere in response to that load. The estimate of the admittance, $Q(\mathbf{k})$, is derived as before by Eqn (2) and shall continue to be referred to as the isostatic response function even though there may be dynamic forces supporting the lithosphere. If the lithosphere responds isotropically to point loads applied both from above and below both $Q_T(\mathbf{k})$ and $Q_C(\mathbf{k})$ are real functions but $Q(\mathbf{k})$ is complex since $H_C(\mathbf{k})$ and $H_T(\mathbf{k})$ are unlikely to have the same spatial phase. In general, $H_C(\mathbf{k})$ will only be non-zero for those wavenumbers, $\mathbf{k}$, for which convection cells exist. Elsewhere, $H_C(\mathbf{k}) = 0$ and $Q(\mathbf{k})$ will be a true estimate of $Q_T(\mathbf{k})$ (Eqn (4)). Furthermore, if convection is in the form of rolls aligned with axes parallel to the shear between the lithospheric plate and the underlying asthenosphere (Fig. 1(a)) or has a bimodal configuration with similar orientation (Fig. 1(b)), apparently anomalous values of $Q_T(\mathbf{k})$ will be observed only in those directions which are normal to the modal directions of convective flow.

We therefore test the presence of small-scale convection in the upper mantle by examining whether the isostatic response function of the Canadian Shield is real and isotropic. We choose the Canadian Shield because it is tectonically very old and therefore has a small $H_T(\mathbf{k})$ and because the gravity data, furnished by the Earth Physics Branch, Ottawa, Canada, provide extensive coverage.

Isostatic response of the Canadian Shield

Estimates of $Q(\mathbf{k})$ and $\gamma^2(\mathbf{k})$ have been computed for two overlapping but relatively rotated portions of the Canadian Shield: one with dimensions 3200 km by 1600 km (area I, Fig. 3), and the other 3000 km by 1500 km (area II, Fig. 3), each containing approximately 62 000 evenly distributed data. These were averaged within 50 km by 50 km cells on a map having a Lambert conformal conic projection and the gridded values, having been tapered

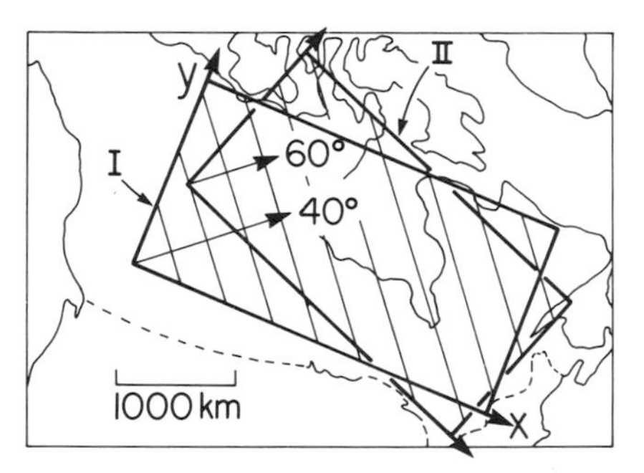

Figure 3. Location of the regions for which $Q(\mathbf{k})$ and $\gamma^2(\mathbf{k})$ were computed; the "A" direction of anomalous isostatic response (see text) is shown by the arrows labelled 40° and 60°. The parallel diagonal lines are explained in the text.

around the edges in order to suppress noise introduced by the finite data area, were transformed to the wavenumber domain by means of a two-dimensional FFT algorithm. A spherical harmonic representation of the Earth's gravity field to degree and order 22, computed from the Goddard Earth Model 10 (Lerch *et al.*, 1977), was removed prior to the analysis. Q and γ^2 as functions of $|\mathbf{k}|$ for the free air gravity field of area I of the Canadian Shield (Fig. 3) are shown in Fig. 4. We have examined the amplitude of Q because we did not want to make the *a priori* physical assumption that Q is solely a real function.

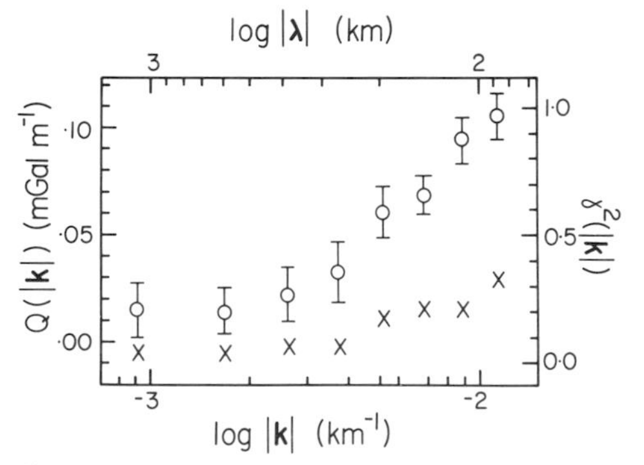

Figure 4. The amplitude of Q (open circles), which represents the free air gravity anomaly in mGal arising from a periodic load of amplitude 1 m, with error bars corresponding in length to two standard errors; and γ^2 (crosses), which measures the proportion of the gravity field caused by the topography, as functions of wavenumber modulus $|\mathbf{k}|$ for data in area I.

In fact the imaginary part of $Q(|\mathbf{k}|)$ is very small, such that it is always of the order of the standard error and usually less. The removal of the GEM10 geoid representation of the gravity field does not affect the estimates of Q at wavelengths of approximately less than 600 km. Rather, it assures that for the very long wavelengths, as $|\mathbf{k}|$ approaches zero, so also does Q: the theoretically expected result for a lithosphere plate responding isostatically to periodic topographic loading (e.g. Walcott, 1976). It is implicity assumed that these long wavelength gravity anomalies in the GEM10 representation are derived from sources beneath the lithosphere. Conversely, the computed values of the isostatic response function at short wavelengths, as $|\mathbf{k}|$ increases, approach 0·112 mGal/m, which is simply the Bouguer correction, assuming a topographic load density of 2670 kg/m^3, and is the theoretically expected result. $\gamma^2(|\mathbf{k}|)$ also behaves as expected: at long wavelengths it is essentially nil showing that topographic features are compensated; as the wavelength becomes shorter γ^2 increases somewhat, either because features are not compensated or because the compensation lies too deeply in the crust to be detected. The fact that the coherence remains as small as it does at the shortest wavelength for which there is a non-aliased estimate ($\gamma^2 = 0{\cdot}21$ at $|\mathbf{k}|^{-1} = 112$ km) is indicative of the large amount of the gravity signal caused by near surface crustal density variations (geological "noise"). Q as a function of wavenumber modulus $|\mathbf{k}|$ for area II (Fig. 3), though not illustrated here, is similar to that for area I. Q for both areas has also been computed as a function of $\mathbf{k}$ by averaging within wavenumber annuli through an azimuthal arc length of 30° (Figs 5(a), 6(a)). These results are comparable to those for $Q(|\mathbf{k}|)$, Fig. 4, insofar as the values increase from zero to more than 0·10 mGal/m from the longest to shortest wavelengths. Associated standard errors are relatively large as shown by the extensive zones (diagonally hatched) in which the values are not greater than zero by more than one standard error. However, anomalous values in the areas labelled A and B contrast substantially with those at equivalent wavelengths in Fig. 4 and with those computed for other regions (Lewis and Dorman, 1970; Banks and Swain, 1978; McNutt and Parker, 1978). Furthermore, relatively large imaginary components in the estimates of the isostatic response function appear to contribute to the anomalously high amplitudes in areas A and B. To judge better the significance of these anomalies, which suggest a directionally anisotropic $Q(\mathbf{k})$, the hypothesis $\mathscr{H}_0$: $Q'_n = Q_{mn}$ was tested using a F-distributed statistic formulated through the principle of extra sums of squares (Draper and Smith, 1967). The Q_{mn}, for annulus n, were computed from data in the 30° arc centred on azimuth m; Q'_n were computed from the ensemble of remaining data in the annulus. For areas A and B of data set I and area B of data set II (Figs 5(b), 6(b)) the probability that the process which induces the isostatic response is the same as in other directions at the appropriate wavelength is less than 5%. The effects of geological noise of course cannot totally be discounted. Futhermore, although there is no significant (arbitrarily defined as greater than 0·5) anomaly in γ^2 at B in either of the data sets (Figs 5(c), 6(c)), the free air gravity and topography signals at A are exceptionally coherent, particularly in data set I. At similar wavelengths, 200–600 km, coherence predicted by reasonable rheological models of the lithosphere in response to surface loading as well as shown by results displayed in Fig. 4 and observations made by others (Lewis and Dorman, 1970; Banks and Swain, 1978; McNutt, 1978) is very much smaller or is non-existent.

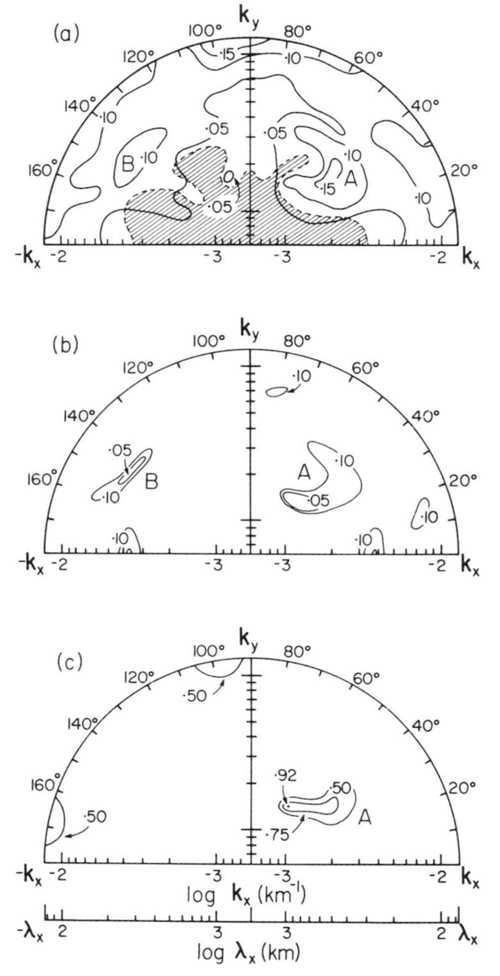

Figure 5. (a) The amplitude of Q in mGal/m; (b) the probability that the hypothesis $\mathscr{H}_0$ (see text) is true, equivalent to the level of significance at which $\mathscr{H}_0$ can be rejected; and (c) γ^2 as functions of $\mathbf{k}$ for data in area I. Note that data along the $-\mathbf{k}_x$ and $\mathbf{k}_x$ axes are symmetric about the origin; the $\mathbf{k}_x$ and $\mathbf{k}_y$ directions are analogous to the spatial x and y coordinates (Fig. 3); the amplitude of Q minus its associated standard error is zero or less in the diagonally hatched region; features labelled A and B are discussed in the text.

Quantitative interpretation of the individual gravity and topography power and cross spectra is difficult because they are not expected to be directionally isotropic even in the absence of convection and because there is a strong trend to increasing power at long wavelengths. The anomalous $Q(\mathbf{k})$ in areas A and B appear, however, to be more the result of an increase in the correlation between $H_0(\mathbf{k})$ and $G_0(\mathbf{k})$ than of changes in the individual power spectra. The probability that this increased correlation is caused by highly correlated

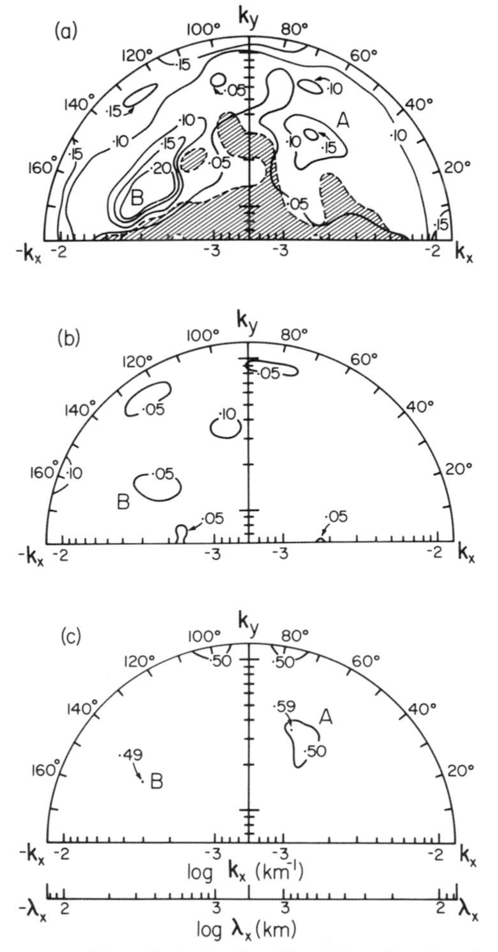

Figure 6. (a), (b) and (c). As for Fig. 5 but for data in area II.

random geological noise is $<5\%$, as demonstrated by the F-test, if the noise is directionally isotropic.

Discussion

The reason for analysing two overlapping data sets was simply to check for internal consistency of the results and to avoid accepting those which were artifacts of the method of analysis. Because each data set contains a large subset of the other the results for each were expected to be similar; merely rotated with respect to the coordinate axes. This is partly the case: anomaly A in $Q(\mathbf{k})$ for set I, for example, is centred approximately on the 40° azimuth whereas for set II (Fig. 6(a)) it centres on 60°. The 20° rotation is appropriate to the orientation of the two data sets since the anomaly reflects a spatial perturbation normal to its wavenumber domain azimuth (cf. the parallel diagonal lines in Fig. 3). Anomalies A and B, elsewhere in the results of $Q(\mathbf{k})$ and $\gamma^2(\mathbf{k})$ in the two data sets, are also located accordingly. The F-test, as it has been formulated, however, appears not to have been a totally appropriate statistical judge of the significance of features A and B in data set II since feature B is so pervasive at the anomalous wavelengths. In this respect it should be noted that the estimates of the amplitude of $Q(\mathbf{k})$ in anomaly B for both data sets are, in fact, equivalent within bounds established by their respective standard errors. More importantly, the wavelength of anomaly B in set I is centred at approximately 200 km whereas in set II it ranges from 200 to nearly 400 km. This frequency shift may be partly because the raw spectra are digitized on a Cartesian grid; since the spatial dimensions of the two data sets are not the same neither are the frequencies for which spectral estimates exist. Furthermore, the Cartesian gridded spectral estimates for one data set compared to the other may be thought of as having been rotated in some cases through the polar annular boundaries which during a rotation remain stationary. The wavelengths of the annuli were chosen to be the same for each data set. Mostly, however, the differences in the results for the two data sets are attributable to having too few data in each spectral estimate to sufficiently reduce the high level of noise which is derived both from the finite nature of the data sets and from variations in the geology of continental crust.

On the basis of the seemingly correct rotations of the salient features of $Q(\mathbf{k})$ and $\gamma^2(\mathbf{k})$ between the two data sets, we tentatively accept the validity of our results which show that the isostatic response of the Canadian Shield is significantly anisotropic. This could be the result of a grossly anisotropic lithosphere, an interpretation for which we know no supporting evidence, either observational or theoretical; alternatively we suggest it may be indicative of small-scale sub-lithospheric convective flow as explained earlier. If so, the presence of two anomalous directions (features A and B in the estimates of $Q(\mathbf{k})$) which are approximately perpendicular to one another suggests that the form of the convection is bimodal (Fig. 1(b)). The approximate orientation of one of the characteristic directions (corresponding to anomaly A in the data) is shown by the parallel diagonal lines in Fig. 3; the wavelength of both modes of convection cell is in the range of 600–200 km suggesting that penetration to the 650 km mantle phase transition does not occur even if the cells have a unitary aspect ratio.

Further interpretation of the results in terms of the properties of the convecting layer is not possible because $Q_C(\mathbf{k})$ cannot be accurately estimated unless $H_T(\mathbf{k}) << H_C(\mathbf{k})$ (Eqn (4)), a condition that requires a billiard ball Earth in the absence of convection. Even a knowledge of $Q_T(\mathbf{k})$ is not useful because the observed topography cannot be partitioned into its convective and tectonic components. Furthermore, convectively induced anomalies in $Q(\mathbf{k})$ will almost always be obscured in data from areas of significant tectonic topography (Eqn (4)). This may explain why the anomaly was less distinct in the third data set, which encompassed a region twice as large as the Canadian Shield set I and included a considerable portion of the United States; this data set had a relatively larger $H_T(\mathbf{k})$.

Alternatively, the shear flow between lithosphere and asthenosphere is not necessarily uniformly parallel under any single plate. Preliminary numerical models of net flow based on the relative motions and geometry of plates (Chase, 1979) suggest that this may be the case for the North American plate. Under the Canadian Shield, however, the mean net flow vectors calculated by Chase are aligned in a direction approximately parallel to that indicated by the isostatic response data (Fig. 3). Shear stress vectors at the base of the lithosphere in the region of the Canadian Shield computed from kinematic models of large-scale mantle flow by Hager and O'Connell (1979) are similarly oriented. It should be noted, however, that in both these models the directions of the assumed plate motion and the computed return asthenospheric flow thereof are not antiparallel beneath the Canadian Shield and that the consequences of this type of flow regime on small-scale convection are unknown. The correspondence of one of the characteristic directions in a bimodal configuration of convection with the direction of shear between the lithosphere and asthenosphere is, of course, consistent with the theory of small-scale convection discussed previously. The plan form of convection will depend on the properties of the convecting layer as well as the velocity of the overlying plate and the duration of the applied shear (Richter and Parsons, 1975). The age of sea-floor spreading in the North Atlantic does not necessarily provide any insight regarding the persistence of the direction of the shear flow beneath the Canadian Shield; therefore the likelihood of a bimodal configuration of convection under the Canadian Shield cannot with certainty be tested against the theoretical predictions. However, a bimodal configuration cannot be rejected on the basis of our knowledge or the origin of spreading and the velocity of the North American plate.

Acknowledgement

Data and financial assistance were provided by the Earth Physics Branch, Department of Energy, Mines and Resources, Ottawa.

References

Anderson, R. N., McKenzie, D. P. and Sclater, J. G., 1973. Gravity, bathymetry and convection in the Earth. *Earth Planet. Sci. Lett.* **18**, 391–407.

Banks, R. J., Parker, R. L. and Huestis, S. P., 1977. Isostatic compensation on a continental scale: local versus regional mechanisms. *Geophys. J. R. Astr. Soc.* **51**, 431–452.

Banks, R. J. and Swain, C. J., 1978. The isostatic compensation of East Africa. *Proc. R. Soc.* **A364**, 331–352.

Chase, C. G., 1979. Asthenospheric counterflow: a kinematic model. *Geophys. J. R. Astr. Soc.* **56**, 1–18.

Cochran, J. R. and Talwani, M., 1977. Free-air gravity anomalies in the world's oceans and their relationship to residual elevation. *Geophys. J. R. Astr. Soc.* **50**, 495–552.

Dorman, L. M. and Lewis, B. T. R., 1972. Experimental isostasy, 3. Inversion of the isostatic Green function and lateral density changes. *J. Geophys. Res.* **77**, 3068–3077.

Draper, N. R. and Smith, H., 1967. *Applied Regression Analysis*, Wiley, New York.

Hager, B. H. and O'Connell, R. J., 1979. Kinematic models of large-scale flow in the Earth's mantle. *J. Geophys. Res.* **84**, 1031–1048.

Lerch, F. J., Klosko, S. M., Laubscher, R. E. and Wagner, C. A., 1977. Gravity model improvement using GEOS-3 (GEM 9 and 10). Goddard Space Flight Center Document X-921-77-246.

Lewis, B. T. R. and Dorman, L. M., 1970. Experimental isostasy, 2. An isostatic model for the U.S.A. derived from gravity and topographic data. *J. Geophys. Res.* **75**, 3367–3386.

McKenzie, D. P., 1967. Some remarks on heat flow and gravity anomalies. *J. Geophys. Res.* **72**, 6261–6273.

McKenzie, D. P., 1977. Surface deformation, gravity anomalies and convection. *Geophys. J. R. Astr. Soc.* **48**, 211–238.

McKenzie, D. P. and Bowin, C., 1976. The relationship between bathymetry and gravity in the Atlantic Ocean. *J. Geophys. Res.* **81**, 1903–1915.

McKenzie, D. P. and Weiss, N., 1975. Speculations on the thermal and tectonic history of the Earth. *Geophys. J. R. Astr. Soc.* **42**, 131–174.

McKenzie, D. P., Roberts, J. M. and Weiss, N. O., 1974. Convection in the Earth's mantle: towards a numerical simulation. *J. Fluid Mech.* **62**, 465–538.

McNutt, M. K., 1978. Continental and oceanic isostasy. Ph.D. thesis, University of California, San Diego.

McNutt, M. K. and Parker, R. L., 1978. Isostasy in Australia and the evolution of the compensation mechanism. *Science* **199**, 773–775.

Munk, W. H. and Cartwright, D. E., 1966. Tidal spectroscopy and prediction. *Phil. Trans. R. Soc.* **259**, 533–581.

Parsons, B. and McKenzie, D., 1978. Mantle convection and the thermal structure of the plates. *J. Geophys. Res.* **83**, 4485–4496.

Parsons, B. and Sclater, J. G., 1977. An analysis of the variation of ocean floor bathymetry and heat flow with age. *J. Geophys. Res.* **82**, 803–827.

Richter, F. M., 1973. Convection and the large-scale circulation of the mantle. *J. Geophys. Res.* **78**, 8735–8745.

Richter, F. M. and Parsons, B., 1975. On the interaction of two scales of convection in the mantle. *J. Geophys. Res.* **80**, 2529–2541.

Sclater, J. G., Lawyer, L. A. and Parsons, B., 1975. Comparison of long wavelength residual elevation and free air gravity anomalies in the North Atlantic and possible implications for the thickness of the lithospheric plate. *J. Geophys. Res.* **80**, 1031–1052.

Walcott, R. I., 1976. Lithosphere flexure, analysis of gravity anomalies, and the propagation of seamount chains. In *The Geophysics of the Pacific Ocean Basin and its Margin, Am. Geophys. Union Monograph Ser.*, Vol. 19 (Eds Sutton, G. H., Manghnani, M. H. and Moberly, R.), 431–438, Am. Geophys. Union, Washington DC.

Watts, A. B., 1976. Gravity and bathymetry in the Central Pacific Ocean. *J. Geophys. Res.* **81**, 1533–1553.

Watts, A. B., 1978. An analysis of isostasy in the world's oceans, 1. Hawaiian–Emperor Seamount Chain. *J. Geophys. Res.* **83**, 5989–6004.

Q_β^{-1} Models for the East Pacific Rise and the Nazca Plate

J. A. CANAS*, B. J. MITCHELL

Department of Earth and Atmospheric Sciences, Saint Louis University, Missouri, USA

A. M. CORREIG

Departmento de Fisica de la Tierra y del Cosmos, Facultad de Fisica, Universidad de Barcelona, Spain

Introduction

Seismic surface wave velocities have been used broadly to obtain information about the elastic properties of the Pacific Ocean floor. Kausel, Leeds and Knopoff (1974), Forsyth (1975), Yoshii (1975), Schlue and Knopoff (1977) and Yu and Mitchell (1979) obtained surface wave velocity models that in general increase with increasing age of the lithosphere. These studies also show a systematic increase in thickness of the lithosphere and a decrease in the thickness of the asthenosphere with age.

Although amplitude attenuation data of surface waves provide important information about the anelasticity of the Earth, only a small number of studies have been carried out in the Pacific Ocean, mainly because it is more difficult to obtain good amplitude data than it is to obtain information about velocities. Several adverse factors may influence amplitude data. These include lateral refraction, multipath propagation, scattering, and mode conversion. Earlier work on attenuation across the Pacific Ocean floor has been done by Tasi and Aki (1969) who used a single path and Mitchell *et al.* (1976) who studied several paths over a broad region of the Pacific using both Rayleigh and Love waves. Canas and Mitchell (1978) obtained attenuation coefficients as a function of the age for three regions (0–50 Myr, 50–100 Myr, and >100 Myr) of the Pacific. Correig and Mitchell (1980) obtained attenuation coefficients for the north-eastern Pacific, Nazca plate and the East Pacific

* Present address: Universidad C. de Madrid, Facultad de C. Fisicas, Cátedra de Geofisica, Madrid, Spain.

Rise showing that attenuation coefficients decrease rapidly with increasing age in the eastern Pacific.

Using inversion theory (Backus and Gilbert, 1967, 1968, 1970), Mitchell (1976) obtained models of Q_β^{-1} using Rayleigh and Love waves. The favoured model was that obtained from the inversion of the Rayleigh attenuation data alone. His average model shows that the low-Q zone is situated approximately at depths between 60 and 200 km. Canas and Mitchell (1978) also applied inversion theory to obtain regionalized Q_β^{-1} models of the Pacific plate for the regions 0–50 Myr, 50–100 Myr and > 100 Myr. They showed that Q_β in the low-Q zone increases with increasing age of the sea-floor. The data also suggest that a systematic increase in thickness of the lithosphere occurs with increasing age of the sea-floor.

In the present study we apply inversion theory to the attenuation data obtained by Correig and Mitchell (1980) for the Nazca Plate and the East Pacific Rise to obtain average Q_β^{-1} models for both regions. These models are then compared with the region 0–50 Myr (Canas and Mitchell, 1978) to obtain differences in the anelastic properties of the Earth under these regions.

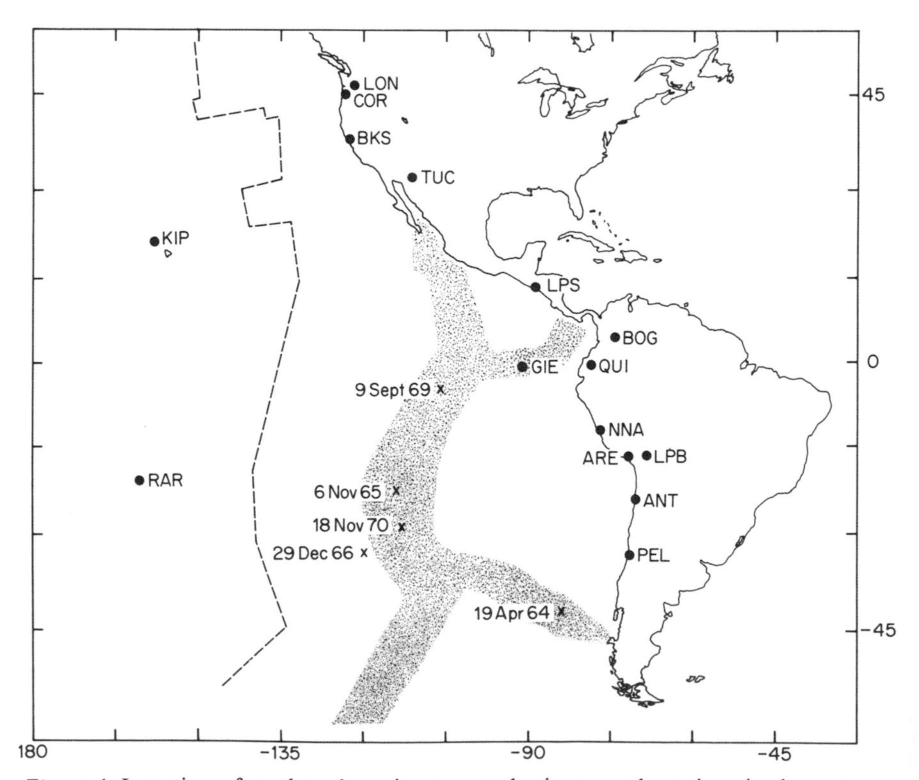

Figure 1. Location of earthquake epicentres and seismograph stations in the eastern Pacific. The dashed line indicates the 50 Myr isochron and the shading denotes a region which is less than about 10 Myr.

Attenuation data

Figure 1 shows the location of the WWSSN stations and epicentres used by Correig and Mitchell (1980) to obtain attenuation data for the Nazca Plate and the East Pacific Rise. They used two new methods: (1) the reference station method in which amplitudes are compared at several stations with the amplitude at one selected reference station; and (2) an iterative method that compares observed and theoretical spectral amplitudes for a given region. The main advantages of these two methods are that they can be applied to small regions in which the methods applied by Tsai and Aki (1969), Mitchell (1976) and Canas and Mitchell (1978) could not be used.

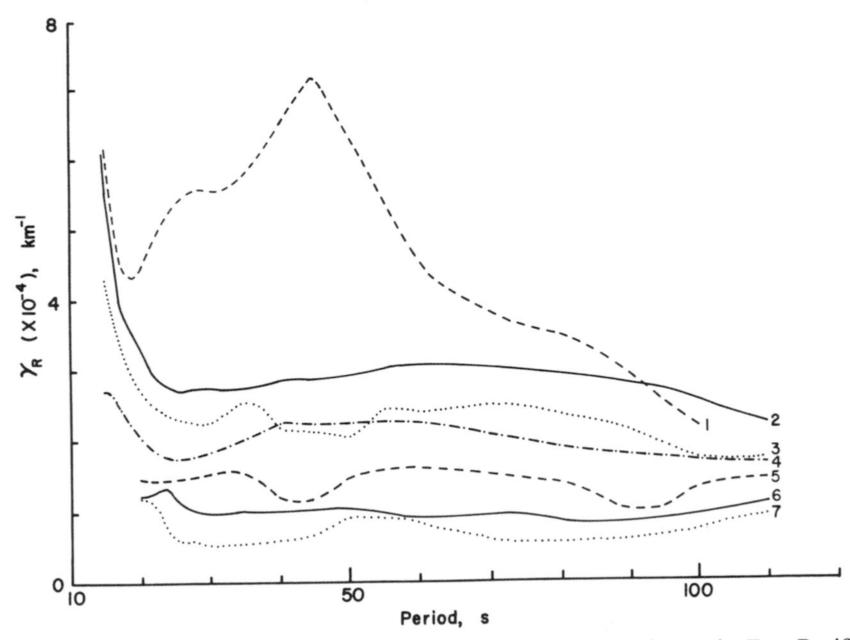

Figure 2. Comparison of Rayleigh wave attenuation coefficients for (1) the East Pacific Rise, (2) the Nazca Plate, (3) the combined Cocos and Nazca Plates, (4) the north-eastern Pacific (Correig and Mitchell, 1980), and values for regions of the Pacific plate which are (5) 0–50 Myr, (6) 50–100 Myr and (7) > 100 Myr in age (Canas and Mitchell, 1978.

Figure 2 shows the attenuation coefficient values obtained by Correig and Mitchell (1980) together with those values obtained by Canas and Mitchell (1978) for the Pacific plate. It is apparent from the data that the values of the attenuation coefficients for the Eastern Pacific are higher than the values obtained by Canas and Mitchell (1978) for the Pacific plate.

Data inversion

Inversion theory (Backus and Gilbert, 1967, 1968, 1970) in its stochastic form (Wiggins, 1972) as applied by Mitchell (1976) and Canas and Mitchell (1978) to the Pacific plate is now applied to the attenuation data of the Nazca Plate and the East Pacific Rise (Fig. 2) to obtain Q_{β}^{-1} models for both regions. We use the formulation of Anderson, Ben-Menahem and Archambeau (1965) assuming that the internal friction of the shear and compressional waves ($Q_{\beta}^{-1}, Q_{\alpha}^{-1}$) is independent of frequency and that shear wave internal friction is twice as large as compressional wave internal friction.

The layered earth models used to obtain the necessary partial derivatives

$$\frac{\partial C_{\mathrm{R}}}{\partial \alpha_i} \quad \text{and} \quad \frac{\partial C_{\mathrm{R}}}{\partial \beta_i}$$

to perform the inversion are the 0–20 Myr model of Yu and Mitchell (1979) for the East Pacific Rise and the average model of the eastern Pacific of Correig (1977) for the Nazca Plate. C_{R}, α_i and β_i are the Rayleigh phase velocity, compressional and shear wave velocities, respectively, in layer i.

Modern inversion theory allows us to determine resolving kernels, the widths of which give us an idea of the layer thicknesses which can be resolved, and standard deviations, which give an indication of the reliability of the model.

Attenuation data for the East Pacific Rise obtained by Correig and Mitchell (1980) consist of only a single observation and therefore no standard deviations are given by those authors. It is also apparent from Fig. 2 that these

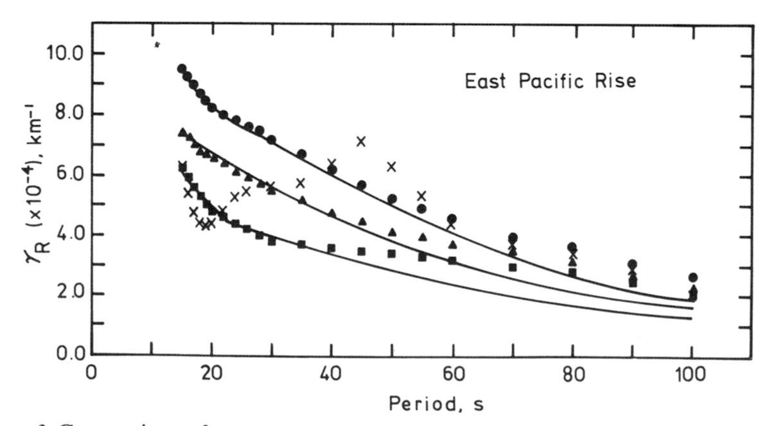

Figure 3. Comparison of assumed attenuation coefficients with theoretical values (solid lines) corresponding to the inversions for the East Pacific Rise. The crosses indicate the observational attenuation coefficients. Circles, triangles and squares indicate theoretical values for three models which approximate different portions of the complex data set for this region.

data could be affected by several of the difficulties described in the introduction, especially between periods of 20 and 60 s. To obtain the range over which values of Q_β^{-1} can vary along the East Pacific Rise, we have considered three sets of attenuation coefficients (Fig. 3) that span the range of the observational ones. We have also assumed that the standard deviations of each set are the same as those of the Nazca Plate (Fig. 4).

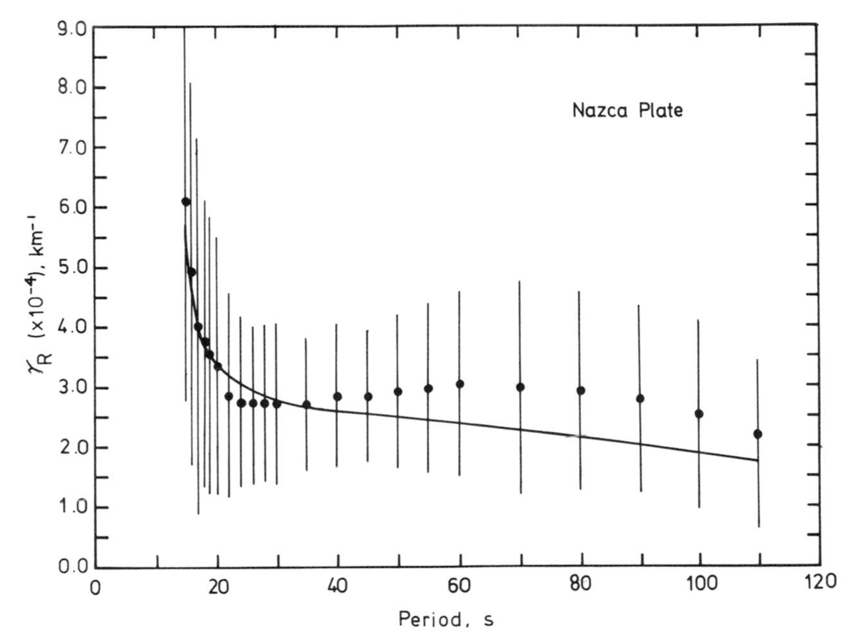

Figure 4. Comparison of attenuation coefficients with theoretical values (solid line) corresponding to the inversion for the Nazca Plate.

Until more attenuation data for the East Pacific Rise become available, we think that our approach is a reasonable one since we are trying to obtain a range of possible Q_β^{-1} values for the East Pacific Rise.

In Figs 3 and 4, we have also plotted the theoretical attenuation coefficients corresponding to the final Q_β^{-1} models that are described in the following section.

Results

An application of the inversion procedure discussed in the previous section yields the Q_β^{-1} model for the East Pacific Rise which is shown in Fig. 5. The corresponding resolving kernels are shown in Fig. 6. We can see that the resolving kernels become broader (and poorer) as the depth increases; the best resolution is therefore at shallower depths.

In Fig. 5 we have plotted the three different inversions carried out for the East Pacific Rise. The range of acceptable Q_β^{-1} models is given by the

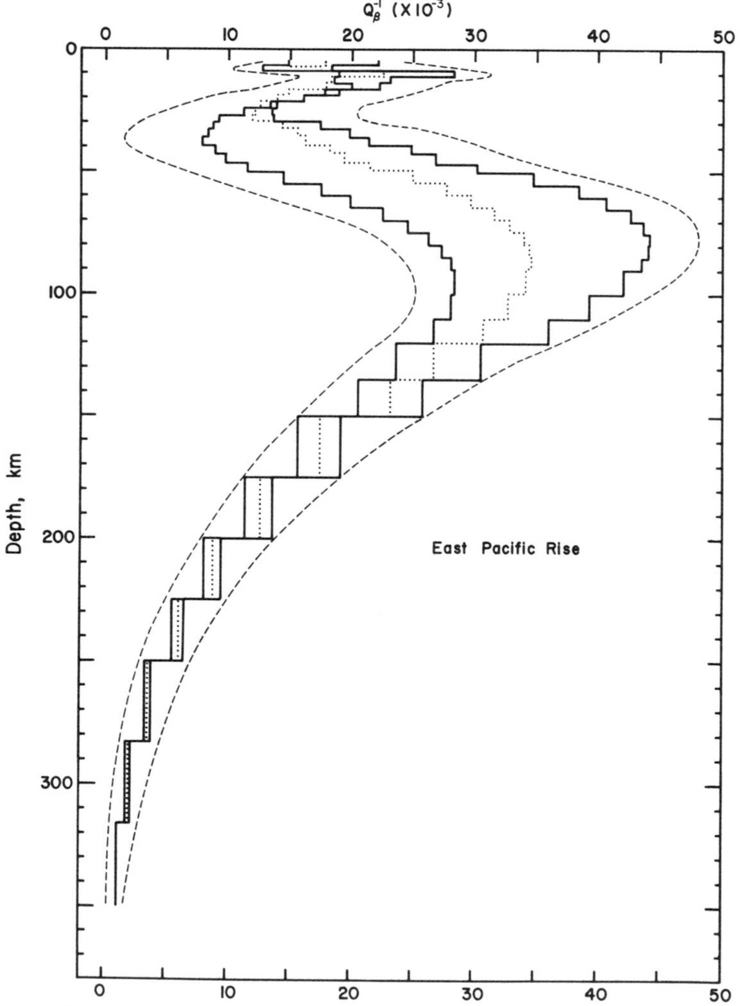

Figure 5. Q_β^{-1} models resulting from the three inversions for the East Pacific Rise. The dashed lines indicate the standard deviations.

intersection of the results of the three different inversions. Since the long-period data are likely to be more reliable than the short-period data, we believe that the model obtained from those data (circles in Fig. 3) is likely to be the most nearly correct one. Therefore the correct model is likely to be the low-Q

(or high Q_β^{-1}) portion of the range of models in Fig. 5. The following conclusions are suggested by that figure:

(1) A low-Q zone appears to occur at about 10 km depth in the lithosphere. This result is not definitive, however, because of the broad resolving kernels for this model.

(2) Although Q_β^{-1} values have large standard deviations, it is possible to delineate a strong low-Q zone beginning at around 25–35 km depth.

(3) Q_β^{-1} has its maximum value (or Q_β its minimum) at about 75 km depth.

(4) Q_β^{-1} at a depth of 200 km has approximately the same value as that which occurs in the lithosphere. This result, although tentative because of the broad resolving kernels at 200 km, provides an estimate for the depth at which the bottom of the asthenosphere occurs.

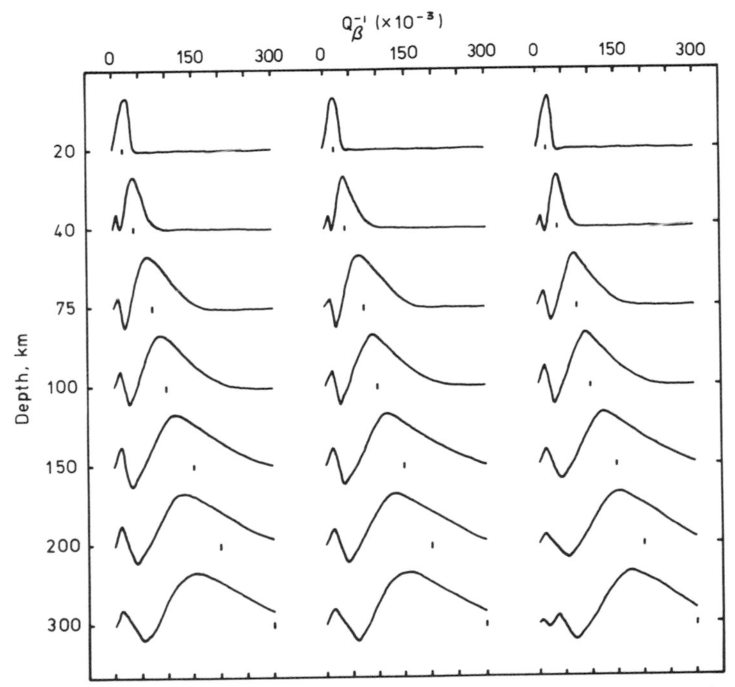

Figure 6. Resolving kernels for the models of Fig. 5. The depth to which each of them refers is indicated by the number beside, and by the vertical dash below, each kernel.

In Fig. 7 we have plotted a Q_β^{-1} model for the Nazca Plate and in Fig. 8 we have plotted the corresponding resolving kernels. From Fig. 7 we can observe the following:

(1) The low-Q zone in the mantle seems to begin at about 35 km depth.

(2) Q_β^{-1} has its maximum value (or Q_β its minimum) at about 95 km depth.

(3) As in the East Pacific Rise, there is indication that the bottom of this low-Q zone can be situated at around 200 km depth.

We have compared the models for the East Pacific Rise region (< 10 Myr) and the Nazca Plate (~40 Myr) with that obtained by Canas and Mitchell

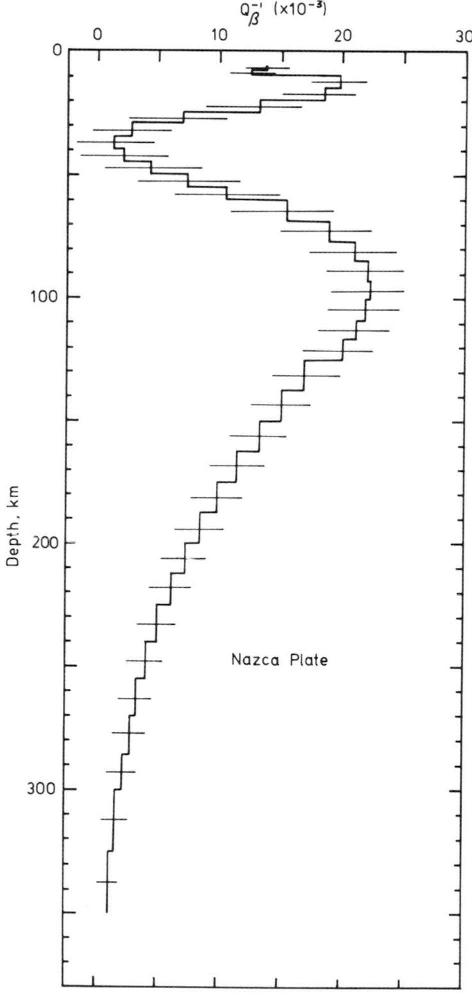

Figure 7. Q_β^{-1} models and standard deviations (horizontal bars) for the Nazca Plate.

(1978) for the region 0–50 Myr of the Pacific plate (Fig. 9). A comparison of these models leads to the following results:

(1) All three models contain a low-Q zone. The highest values of Q_β^{-1} occur in the youngest region (East Pacific Rise, < 10 Myr in age), intermediate values occur in the Nazca Plate (0–40 Myr) and the lowest values occur in the 0–50 Myr region of the Pacific plate.

(2) Although there is overlap of standard deviations among the Q_β^{-1} models, it appears that the lithosphere becomes thicker as the age of the Pacific sea-floor increases.

(3) The bottom of the low-Q zone occurs at a depth of about 200 km in all models. Because the resolving kernels are quite broad at that depth, however, it cannot be stated unequivocably that the asthenosphere bottoms uniformly at that depth.

(4) All three models suggest a low-Q zone in the lithosphere. Because this feature is narrower than the resolving kernels, however, its existence must be considered questionable.

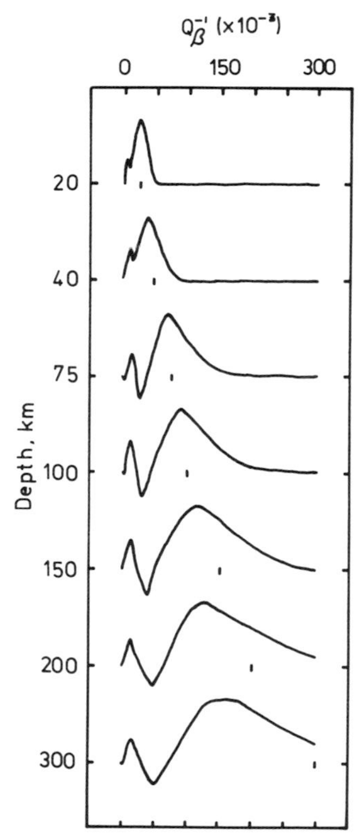

Figure 8. Resolving kernels for the model of Fig. 7.

This new study supports the results of Canas and Mitchell (1978) that as the age of the ocean floor increases the lithosphere increases in thickness and Q_β^{-1} values are smaller. Our derived models suggest that the lower limit of the low-Q zone is situated at about 200 km depth in all regions; if this result is correct, the low-Q zone becomes smaller and less pronounced as the age of the sea-floor increases.

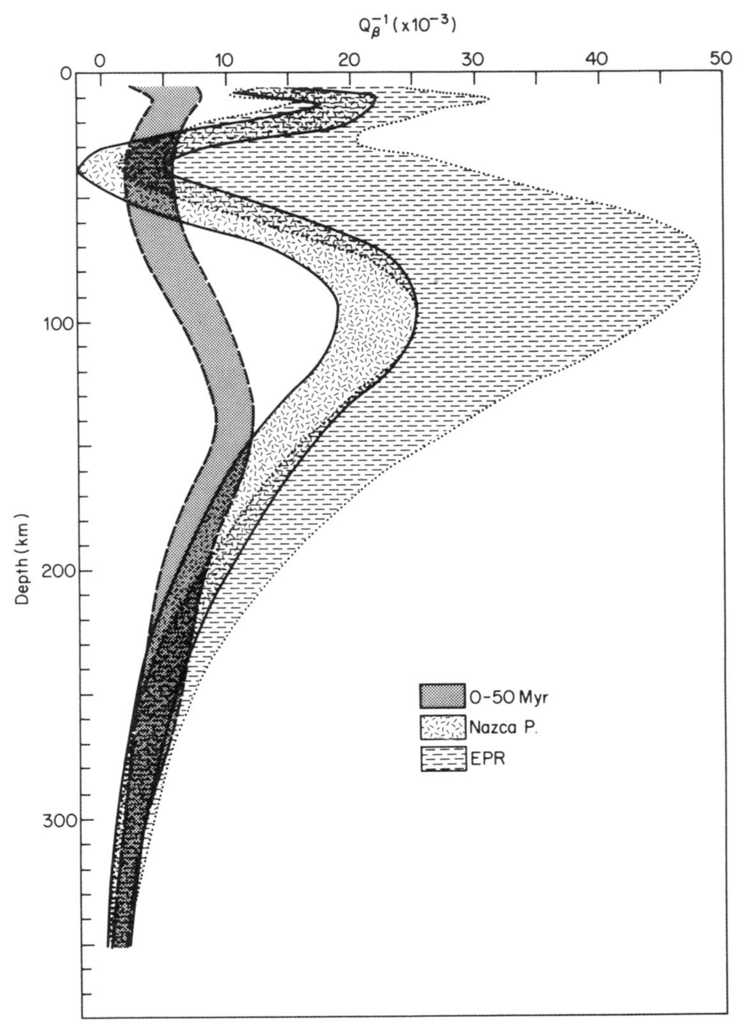

Figure 9. Comparison of the Q_β^{-1} models for the East Pacific Rise and the Nazca Plate with a model for the region 0–50 Myr of the Pacific plate (Canas and Mitchell, 1978). The shaded area indicate estimates of the standard deviation distribution with depth.

Conclusions

Inversion of the attenuation data for the East Pacific Rise and Nazca Plate shows a clear low-Q zone. In the East Pacific Rise (<10 Myr) it is thicker and has considerably higher Q_β^{-1} values than it has in the Nazca Plate (~ 40 Myr). The low-Q region extends to a depth of about 200 km and the top of the low-Q zone is deeper in the older region (35 km) than in the younger one (25–35 km).

Comparison of these results with the region 0–50 Myr of the Pacific plate (Canas and Mitchell, 1978) shows that as the age of the sea floor increases the low-Q zone is less pronounced and becomes thinner with a lower limit situated at about 200 km depth.

Acknowledgement

This research was supported by the National Science Foundation, Division of Earth Sciences, under Grant EAR 77-14482.

References

Anderson, D. L., Ben-Menahem, A. and Archambeau, C. B., 1965. Attenuation of seismic energy in the upper mantle. *J. Geophys. Res.* **70**, 1441–1448.

Backus, G. and Gilbert, F., 1967. Numerical applications of a formalism for geophysical inverse problems. *Geophys. J. R. Astr. Soc.* **13**, 247–276.

Backus, G. and Gilbert, F., 1968. The resolving power of gross Earth data. *Geophys. J. R. Astr. Soc.* **16**, 169–205.

Backus, G. and Gilbert, F., 1970. Uniqueness in the inversion of gross earth data. *Phil. Trans. R. Soc., Ser. A*, **266**, 123–192.

Canas, J. A. and Mitchell, B. J., 1978. Lateral variation of surface wave anelastic attenuation across the Pacific. *Bull. Seism. Soc. Am.* **68**, 1637–1650.

Correig, A. M., 1977. Estudi del mecanisme focal dels terratremols a partir d'ones sismiques. Aplicacio a terratremols de les dorsals de l'Atlantic i oriental del Pacific. Ph. D. Thesis, University of Barcelona.

Correig, A. M. and Mitchell, B. J., 1980. Regional variation of Rayleigh wave attenuation coefficients in the eastern Pacific. *Pure Appl. Geophys.* **118**, 831–846.

Forsyth, D. W. 1975. A new method for the analysis of multi-mode surface-wave dispersion; application to Love-wave propagation in the east Pacific. *Bull. Seism. Soc. Am.* **65**, 323–342.

Kausel, E. G., Leeds, A. R. and Knopoff, L., 1974. Variations of Rayleigh wave phase velocities across the Pacific Ocean. *Science* **186**, 139–141.

Mitchell, B. J., 1976. Anelasticity of the crust and upper mantle beneath the Pacific Ocean from the inversion of observed surface wave attenuation. *Geophys. J.* **46**, 521–533.

Mitchell, B. J., Leite, L. W. B., Yu, Y. K. and Herrmann, R. B., 1976. Attenuation of Love and Rayleigh waves across the Pacific at periods between 15 and 110 seconds. *Bull. Seism. Soc. Am.* **66**, 1189–1201.

Schlue, J. W. and Knopoff, L.. 1977. Shear-wave polarization anisotropy in the Pacific basin. *Geophys. J.* **49**, 145–165.

Tsai, Y. B. and Aki, K., 1969. Simultaneous determination of seismic moment and attenuation of surface waves. *Bull. Seism. Soc. Am.* **59**, 275–287.

Wiggins, R. A., 1972. The general linear inverse problem: implication of surface waves and free oscillations for Earth structure. *Rev. Geophys. Space Sci.* **10**, 251–285.

Yoshii, T., 1975. Regionality of group velocities of Rayleigh waves in the Pacific and thickening of the plate. *Earth Planet. Sci. Lett.* **25**, 305–312.

Yu, G. and Mitchell, B. J., 1979. Regionalized shear wave velocities of the Pacific from the inversion of observed surface wave dispersion. *Geophys. J. R. Astr. Soc.* **57**, 311–341.

Can Membrane Tectonics be Used to Explain the Break-Up of Plates?

S. J. FREETH

Department of Geology, University College of Swansea, UK

Introduction

A major question in the field of plate tectonics concerns the primary cause of the break-up of plates. Once active ridges and subduction zones have been established, around a plate, many would agree that explaining the continued movement of that plate presents no insuperable problems. However, there is as yet no generally acceptable model to explain how intra-continental rifts are initiated and what processes operate during the initial spreading phase. I suspect that stresses generated by membrane tectonics may provide at least part of the answer to this question.

The basic concepts of membrane tectonics

The Earth is an oblate spheroid and consequently the curvature of a plate is modified as it moves over the Earth's surface. At the very simplest level a plate moving towards the equator must decrease its radius of curvature and consequently be subjected to compressional stress in the peripheral part and tension in the centre (Fig. 1). The basic concept of membrane tectonics was put forward and subsequently developed in a series of papers by Turcotte and Oxburgh. They were able to demonstrate that under certain circumstances the stresses generated could be large enough to propagate lithospheric fractures. The Hawaii island chain (Turcotte and Oxburgh, 1973; Turcotte, 1974) and East Africa (Oxburgh and Turcotte, 1974; Turcotte and Oxburgh, 1976) were proposed as examples of membrane tectonics in action. Further examples of membrane tectonics have subsequently been suggested, and I have pointed out that the change from a compressional tectonic regime in West Africa to an extensional regime coincides closely with the time at which the centre of Africa crossed the equator (Freeth, 1978a, b).

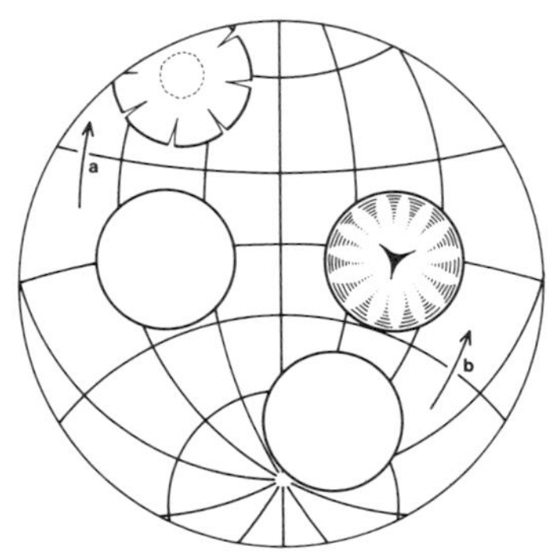

Figure 1. Sketch illustrating the basic concept of membrane tectonics, based on the original diagram of Oxburgh and Turcotte (1974). Plates moving away from the equator (a) suffer peripheral extension and central compression whilst plates moving towards the equator (b) suffer peripheral compression and central extension.

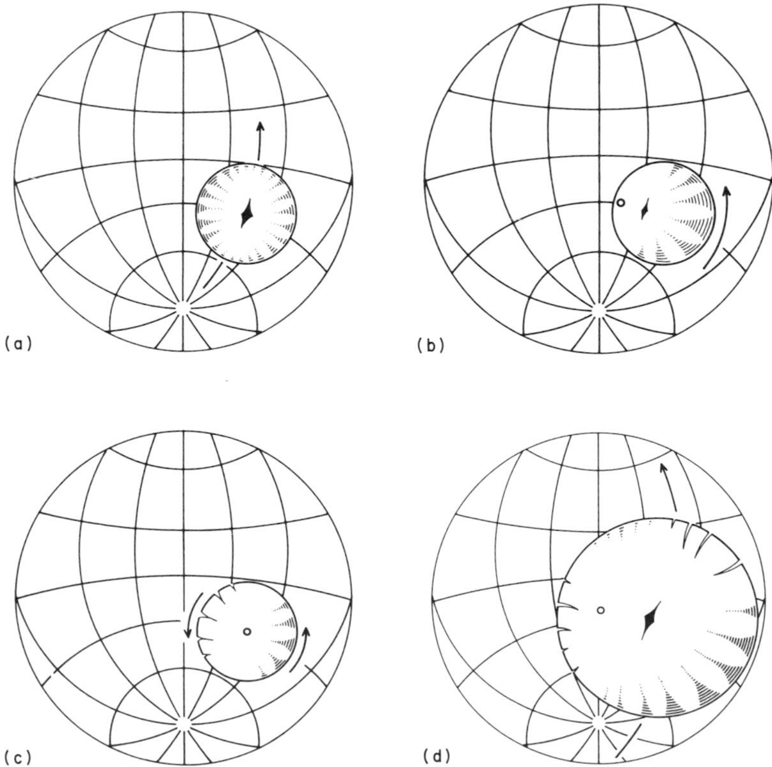

Figure 2. Diagrammatic representations of the relative stresses generated in circular plates as they move across the surface of the earth.

(a) A medium sized plate moving along a line of longitude towards the equator; (b) a medium sized plate moving towards the equator and rotating about a point near its circumference; (c) a medium sized plate rotating about its centre; (d) a large plate rotating about a point near its circumference.

In the papers cited above, quantitative estimates of the membrane stresses generated by plate movement were all based on equivalent circular plates being subjected to a uniform change of radius of curvature. It was realized, from the first attempts to analyse membrane stresses in plates, that the shape of a plate would influence the distribution of stresses within a plate (Turcotte, 1974). A particularly obvious example is the way in which the major re-entrant to the north and east of the Gulf of Guinea might be expected to modify stress trajectories in the African plate (Freeth, 1978b). The extent to which the latitude dependence of rate of change of radius of curvature would significantly modify the stress generation in different parts of large plates does not seem to have been fully appreciated initially. More seriously, it has only recently been pointed out (Freeth, 1979) that as a consequence of the fact that plates tend to spin across the surface of the Earth, rather than move along lines of longitude, the amount and even the sense of membrane stress generation may vary markedly from place to place. This effect is illustrated diagrammatically in Fig. 2.

If significant progress is to be made, then a way must be found to analyse quantitatively the generation and distribution of membrane stresses. There are two fairly obvious ways of tackling this problem. Perhaps the best solution would be to utilize the "finite element analysis" techniques to construct a quantitative model. The preparation of a geologically realistic model is complex, to say the least, and although our own preliminary results are most encouraging we feel that it will be some time before they are ready for publication. The other approach involves a *quantitative* determination of the stress generated at particular points within a plate followed by a *qualitative* assessment of the resulting membrane stress generation domains. This approach, which has already been applied with some success to the African plate (Freeth, 1979), is outlined in the following section.

The calculation of membrane stress generation domains

The membrane stresses generated at any particular point within a plate are proportional to the change radius of curvature at that point.

Using the formula for the radius of curvature (R) of the geoid as a function of latitude (γ):

$$R = a(1-\varepsilon+2\varepsilon\sin^2\gamma),$$

where a = equatorial radius of the Earth (6378 km) and ε = ellipticity of the Earth (0·00335).

It therefore follows that the change in radius of curvature (ΔR) at any particular point moving from an initial latitude of γ_0 to a latitude of γ will be:

$$\Delta R = 2a\varepsilon(\sin^2\gamma-\sin^2\gamma_0)$$

By omitting the constants ($2a\varepsilon$), reversing the sign (so that peripheral tensions follow the normal geological convention of tension being negative) and incorporating a scaling factor (10^2) a measure of stress generated by change in latitude (σ_γ) can be set up (Freeth, 1979).

$$\sigma_\gamma = (\sin^2\gamma_0 - \sin^2\gamma) \times 10^2 \ .$$

If the values of σ_γ for any particular time interval are plotted for a number of points within a plate, then domains of membrane stress generation can be outlined and used to assess the way in which a plate might be expected to deform.

West Africa – an example of membrane stresses in action

The West African rift system situated, as it is, at the apex of a major re-entrant and extending deep into the African plate, is ideally situated to preserve the exact timing and extent of phases of compression in that part of the African plate. Similarly the neighbouring Cameroun Volcanic Line preserves an unambiguous record of recent extension in the same part of the African plate. Details of the timing of these events are presented in Fig. 3.

Membrane stress generation domains can be calculated for the continental part of the African plate, for 20 Myr time intervals, using the technique described in the previous section. The results obtained have been set alongside the observed deformation in Fig. 3. Since a detailed discussion of each time interval has already been published (Freeth, 1979), I will restrict myself to two particularly obvious examples.

The values for latitude generated membrane stress (σ_γ) during the period 60 Myr to 40 Myr are presented in Fig. 4(a). During this period Africa rotated, relative to the geoid, in an anticlockwise direction about a point in West Africa. Three stress domains were generated: a major negative domain in the north-east of the Afro–Arabian plate, a major positive domain in south and east Africa and a smaller positive domain in north-west Africa. The resultant effect of the location of a domain of peripheral expansion to the north-east of the West African rift system and domains of peripheral compression to the south and to the west is relatively easy to visualize. Relative extension in the north-east would cause the other two domains to move closer together and give rise to compression across the Lower Benue Rift. The anticipated phase of compression during the period 60 Myr to 40 Myr corresponds closely with the observed phase of compression which lasted from mid-Palaeocene to mid-Eocene times (Fig. 3).

The values for latitude generated membrane stress (σ_γ) during the last 20 Myr are presented in Fig. 4(b). During this period Africa moved steadily northward by between 8° and 9°, with only very minor clockwise rotation. Two stress domains were therefore generated, a negative domain to the north of the average position of the equator and a positive domain to the south. The effect of interaction between these two domains is easy to visualize. Peripheral

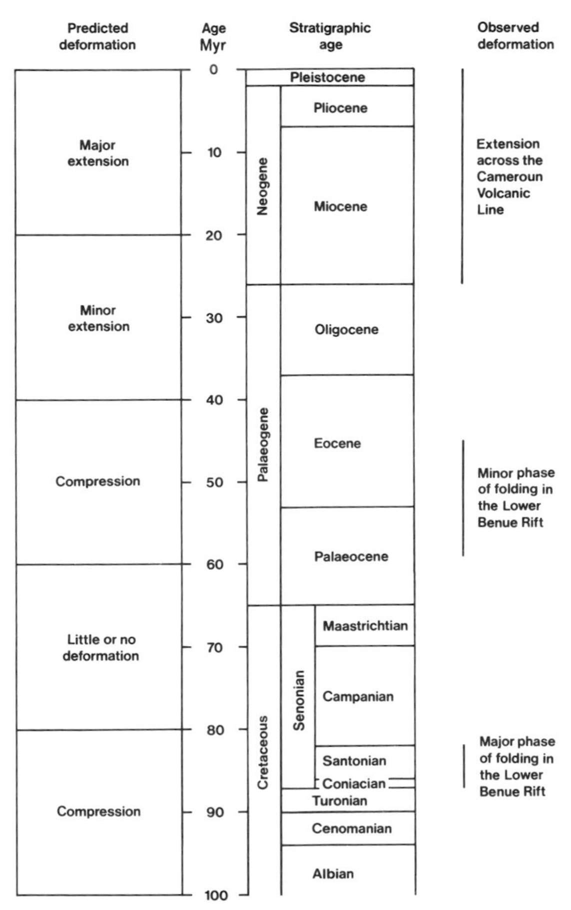

Figure 3. Comparison between the deformation observed in West Africa and the deformation predicted from an examination of the membrane stress domains generated during 20 Myr time intervals from 100 Myr to the present day. Dating of extension across the Cameroun Volcanic Line is based on radiometric evidence from the volcanics. Dating of the phases of deformation within the Lower Benue Rift is based on stratigraphic evidence. This figure has been modified from a previous version (Freeth, 1979) to take into account the revised Cretaceous time scale adopted by Rawson *et al.* (1978).

tension over the whole of the northern domain would tend to open up fractures particularly where stresses were concentrated, such as in the Gulf of Guinea re-entrant. Interaction between the two opposing domains would further enhance the tendency for a major fracture to open up in this area.

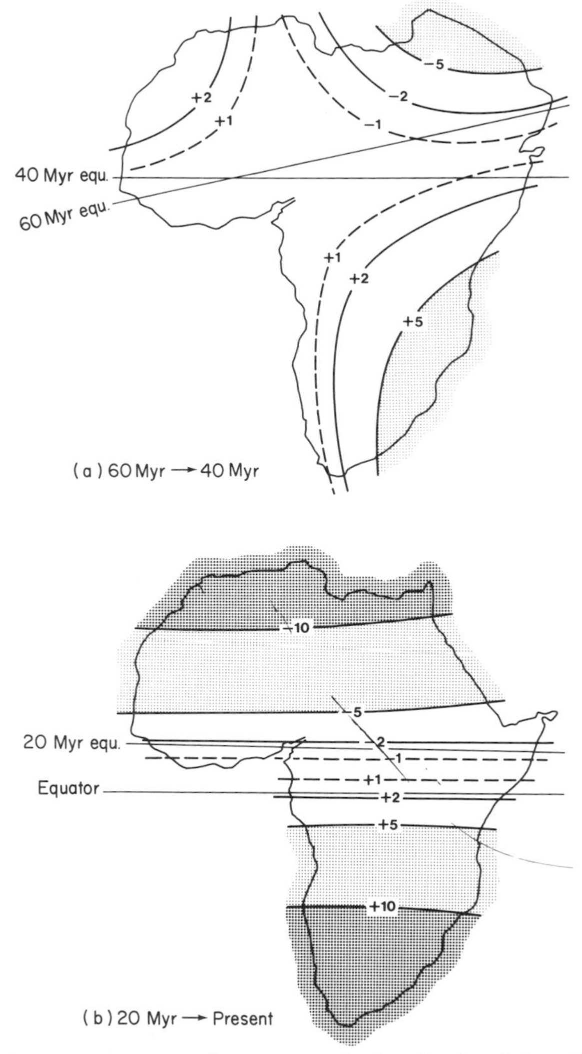

Figure 4. Membrane stress generation patterns during the time intervals (a) 60 Myr to 40 Myr and (b) 20 Myr to the present day. Values for latitude generated stress (σ_γ) were calculated at the intersection points of the present day 10° latitude–longitude grid and then contoured. The palaeocontinental positions of Africa are based on the maps of Smith and Briden (1977).

Similar, though less marked, stress domains were set up during the previous 20 Myr. The anticipated phase of tensions corresponds with the initiation of activity along the Cameroun Volcanic Line, about 25 Myr ago, and its continued extension up to the present day.

Gondwanaland – a possible example of membrane stresses causing plate break-up

It is now generally accepted that the break-up of Gondwanaland occurred about 125 Myr ago (Burke and Whiteman, 1973; Ladd, Dickson and Pitman, 1973; Smith and Briden, 1977). Certainly little room for doubt remains concerning the timing of rifting in the South Atlantic (Herz, 1977) though discussion will probably continue for some time as to the exact nature and timing of the development of rifts between Africa, Antarctica and India. Many authors would also accept that the onset of uplift and volcanic activity predates the opening of intra-continental rifts by perhaps 20 to 25 Myr (Vine and Hess, 1971; Burke and Whiteman, 1973; Scrutton, 1973).

I am not entirely happy that the development of major rifts is not a much longer term process. It is difficult to find an objective measure of tectonic activity which can be related to the initiation of rifting. Igneous activity is intimately associated with rifting and the level of igneous activity probably provides the best available record of tectonic activity during the early history of a rift. Unfortunately there are few, if any, parts of the world which have been studied in the sort of detail which would allow an accurate graph to be drawn of changes in the level of igneous activity with time. In the absence of such detailed information I believe that a simple compilation of radiometric ages is the nearest we can get to an objective measure of the timing and level of tectonic activity associated with the initiation of rifting.

Obviously such an approach has major limitations: any compilation of radiometric ages will immediately be biased by the presence of a particularly active person or group working on a stratigraphically restricted group of rocks or within a geographically restricted area. Even accepting these limitations the compilation of 429 K–Ar ages and 23 Rb–Sr isochron ages from Gondwanaland presented in Fig. 5 has a number of extremely interesting features. If this histogram is taken at face value then it would appear that there has been some igneous activity in one part or other of Gondwanaland from at least 220 Myr ago until well after the break-up of Gondwanaland. There would also appear to be peaks of activity at around 165 Myr and 125 Myr.

If the data is broken down in terms of country of origin the peak at around 165 Myr would appear to be due not to a peak of activity in any particular part of the plate, but to the fact that for almost every country for which data is available there are some dates in the range 160 to 170 Myr. On the other hand, the peak at 125 Myr is due primarily to the large number of K–Ar dates obtained from a relatively restricted group of rocks in southern Brazil by Amaral *et al.* (1966; 1967) and Melfi (1967). If we take Gondwanaland as a whole, it seems probable that tectonic activity built up to a peak at around 165 Myr and then continued at a relatively high level until after the break-up of the continent about 125 Myr ago.

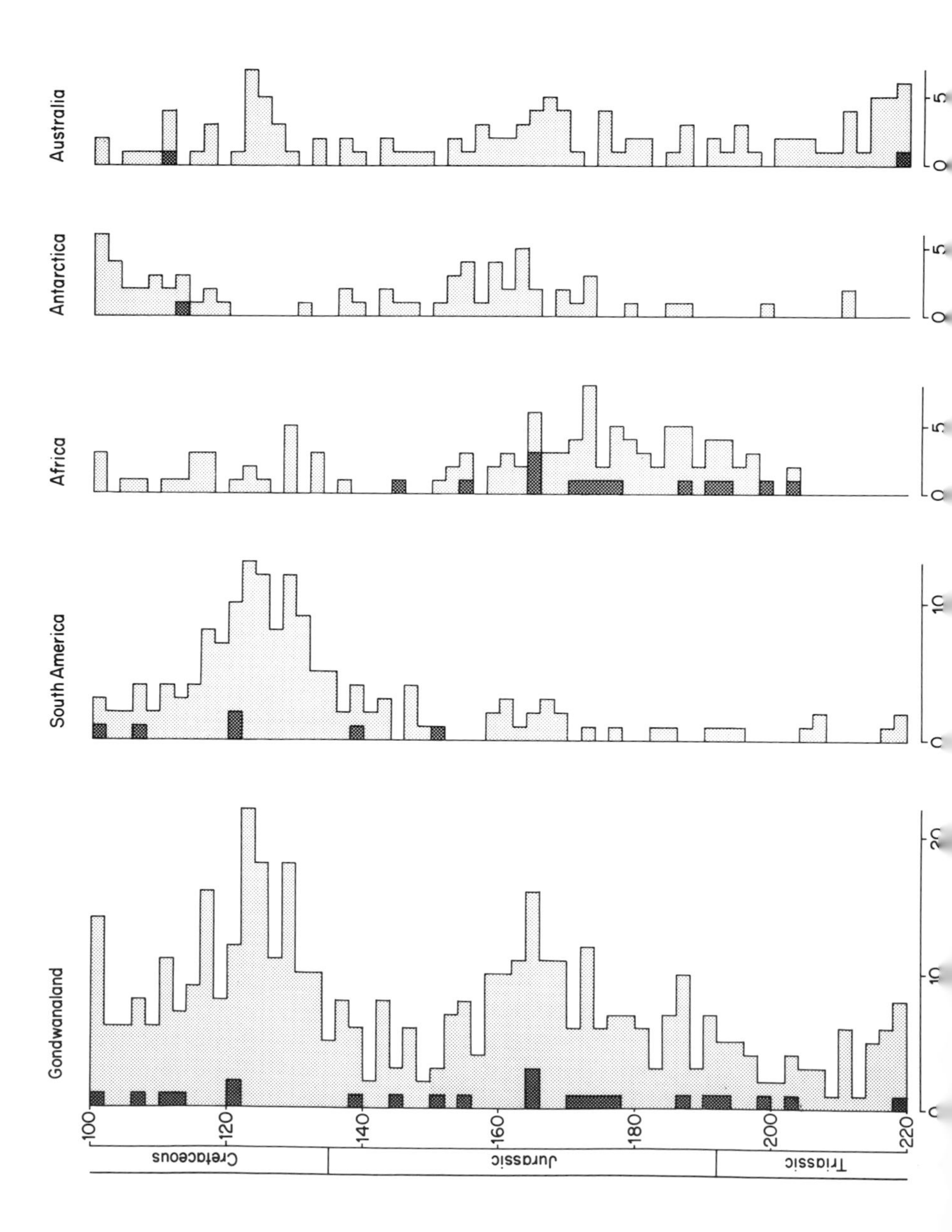
Australia
Antarctica
Africa
South America
Gondwanaland
100
120
140
160
180
200
220
Cretaceous
Jurassic
Triassic

The membrane stresses generated as a direct consequence of the movement of Gondwanaland, relative to the geoid, present a possible explanation for the long phase of igneous activity leading up to its eventual break-up.

The values for latitude-generated membrane stress (σ_γ) during the period 220 Myr to 200 Myr are presented in Fig. 6(a). During this period Gondwanaland moved towards the north with little or no rotation relative to the geoid. The amount of movement was relatively large and consequently a domain of very large positive membrane stresses was generated in the centre of the plate (Fig. 6(a)). As stated previously, a positive domain is one with peripheral compression and central tension. Assuming that the palaeomagnetic data (Smith and Briden, 1977) on which these calculations were based is reasonably accurate, it is difficult to believe that a positive membrane stress domain of such magnitude would not have resulted in the opening up of a rift system somewhere near the centre of the plate. There are relatively few radiometric ages from central Gondwanaland for the period 220 Myr to 200 Myr. Many of those which do exist come from southern Brazil, Argentina and Nigeria (i.e. near the centre of the positive stress domain) but the data, particularly in the absence of any information for northern Brazil, is too sparse to warrant any firm conclusion. During the same period a major negative stress domain, i.e. a domain of peripheral extension, extended across the north-eastern part of Australia (Fig. 6(a)). It is therefore extremely interesting to note that all the igneous activity, reflected by the radiometric dates shown for this period in Fig. 5, is from south-eastern Queensland. In other words, the most active igneous province is located on the Gondwanaland continental margin in the centre of a domain of peripheral extension.

During the period 200 Myr to 180 Myr Gondwanaland rotated in a clockwise direction about a pole located somewhere in northern India. The domain of positive membrane stresses which this movement generated extended over most of South America and over the southern part of Africa (Fig. 6(b)), and domains of negative membrane stresses extended over much of Australia and over part of north-west Africa. The consequence of these stress domains should have been the continued extension of rifts in the centre of the positive domain, continued igneous activity in north-eastern Australia and possibly the opening up of peripheral rifts near the north-west African continental margin. In practice during this period there was major igneous

---

Figure 5. Histograms of radiometric ages from Gondwanaland and its constituent parts. Only K–Ar ages (shown in light tone) and Rb–Sr isochron ages (shown in dark tone) have been used. The ages plotted were culled from 54 papers and their distribution according to country of origin is as follows: Surinam (K–Ar 3), Brazil (K–Ar 107), Chile (K–Ar 23 and Rb–Sr 5), Argentina (K–Ar 18), Uruguay (Rb–Sr 1), Morocco (K–Ar 10), Sierra Leone (K–Ar 4 and Rb–Sr 1), Liberia (K–Ar 8), Nigeria (K–Ar 12 and Rb–Sr 9), Tanzania (K–Ar 3), Namibia (K–Ar 17), Malawi (K–Ar 4), Zimbabwe (K–Ar 12 and Rb–Sr 3), Mozambique (K–Ar 4), South Africa (K–Ar 22 and Rb–Sr 1), Australia (K–Ar 114 and Rb–Sr 2), Antarctica (K–Ar 68 and Rb–Sr 1).

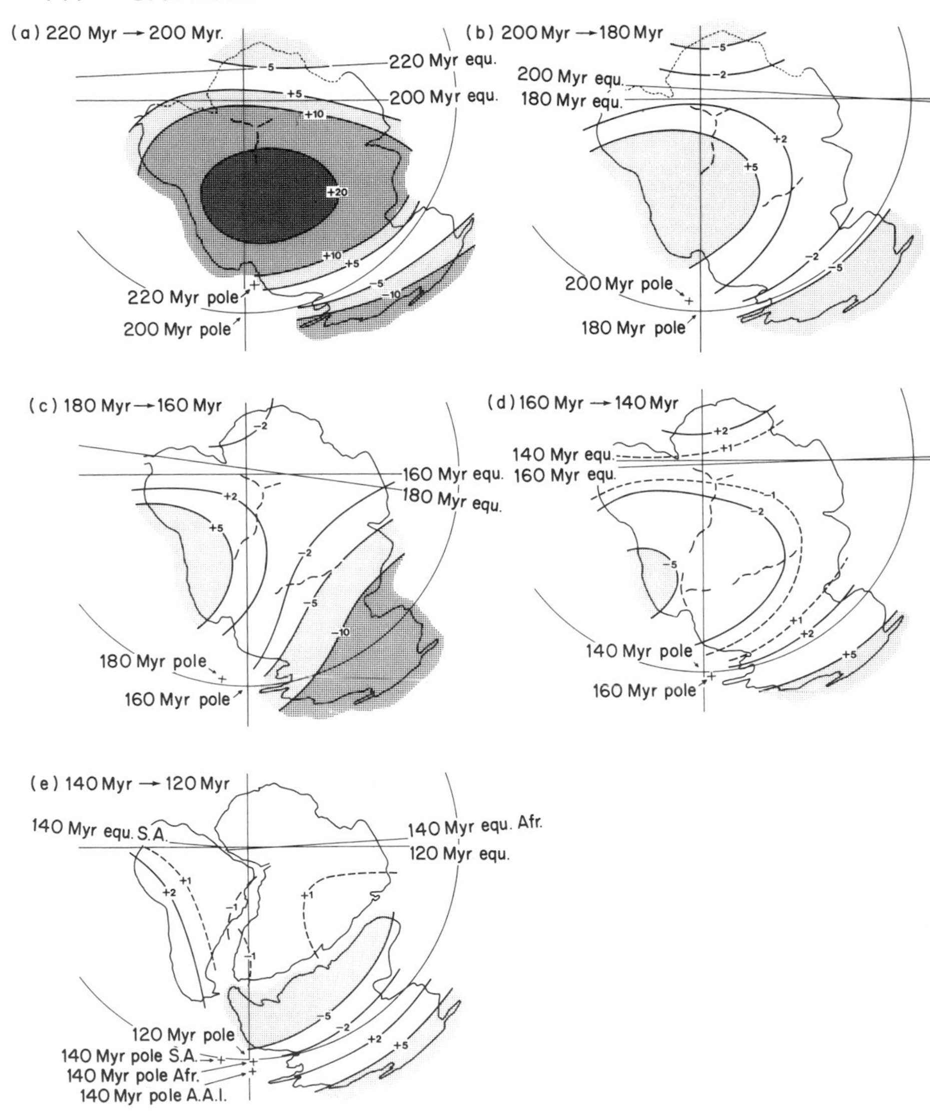

Figure 6. Membrane stress generation patterns during 20 Myr time intervals from 220 Myr to 120 Myr. Values for latitude generated stress (σ_y) were calculated at the intersection points of a present day 10° latitude–longitude grid and then contoured. The palaeocontinental positions of Gondwanaland are based on the maps of Smith and Briden (1977). The locations of the postulated rifts discussed in the text are marked on with a heavy dashed line.

activity, judging by the record of radiometric ages, in South Africa and Zimbabwe. This period also marked the decline of igneous activity in Queensland (Fig. 5) and the onset of activity in Morocco. In the latter case although late Triassic-early Jurassic volcanic activity is quite common little radiometric dating work has yet been done and although the only radiometric dates of which I am aware lie in a group between 181 Myr and 184 Myr, the actual age range was presumably considerably greater.

From 180 Myr to 160 Myr Gondwanaland continued to rotate in a clockwise direction, relative to the geoid, the pole of rotation during this period being located in central Chad. The membrane stress generation domains were similar to those in the previous period but shifted significantly towards the west (Fig. 6(c)), the positive domain being centred in the south-west of South America and the main negative domain being far more intense and extending over Australia, much of Antarctica and part of north-east Africa. During this period igneous activity continued in South Africa and Zimbabwe and a number of radiometric ages have also been obtained from Argentina. This period also marked the onset of a major phase of igneous activity in Antarctica and in the south-eastern part of Australia.

A complete change in the direction of movement occurred around 160 Myr ago, during the period 160 Myr to 140 Myr. Gondwanaland rotated in an anticlockwise direction about a pole located in the north-east of Arabia. As a consequence the positive domain which had been located over South America and southern Africa for the previous 60 Myr was replaced by a negative (i.e. peripherally extensive) domain (Fig. 6(d)). It would seem reasonable to anticipate that a negative domain centred in southern South America would result in a rift propagating into the continent from the major re-entrant in the plate margin located at the point where South America can be fitted against Antarctica. Assuming that the intra-continental rifts, postulated previously, actually existed, it is reasonable to postulate that the peripheral rift(s) would continue to propagate until it (they) joined up with the intra-continental rifts, thus separating Gondwanaland into three major fragments; South America, Africa and Australia–Antarctica–India. The switch from a peripherally compressive to a peripherally extensive regime in southern South America and southern Africa corresponds with the apparent decrease in igneous activity mentioned earlier. The effective end of activity in southern Africa occurs at about 160 Myr and the start of the major phase of igneous activity which leads up to the opening of the south Atlantic occurs at about 150 Myr in southern Brazil.

The period 140 Myr to 120 Myr marked the break-up of Gondwanaland. During this period Africa remained almost stationary, South America rotated slightly in a clockwise direction relative to the geoid and Australia–Antarctica–India moved away to the south-east. Peripherally extensive membrane stress generation domains were located in the south-west of Africa and in the east of South America (Fig. 6(e)); it is particularly interesting to note that these were the main areas of igneous activity in the respective continents. The stress domains during this period (Fig. 6(e)) and during the two successive 20 Myr periods were peripherally compressive over most of Australia and peripherally extensive over most of Antarctica. The radiometric age record shows a build-up of activity from 120 Myr onwards in Antarctica and a decrease of activity in Australia.

Crucial to the whole scenario outlined above is the early history of the South Atlantic. Marine sediments of Albian age have been identified in the Lower Benue Rift (Burke, Dessaugavie and Whiteman, 1973) and of Aptian age on the western side (Asmus and Ponte, 1973) and on the eastern side (Franks and Nairn, 1973) of the South Atlantic Rift. The earliest marine deposits on either side of the South Atlantic are underlain by a sequence of continental deposits. Recent reviews of the available information (Asmus and Ponte, 1973; Franks and Nairn, 1973) favour the view that these continental deposits accumulated in an intra-cratonic basin during Upper Jurassic and Lower Cretaceous times. However, in the absence of evidence to the contrary one could speculate that the basin might have started to develop much earlier. One minor piece of evidence which lends strong support to the suggestion that active rifting near the centre of Gondwanaland had already started by middle Jurassic times is the presence of basic volcanic rocks on the Ghana continental shelf which have been dated at 160–165 Myr (personal communication from Cudjoe and Khan quoted by Delteil *et al.*, 1974).

One of the most striking features of the pre-Albian sediments of the South Atlantic is the presence of thick sequences of evaporites. Although subsurface exploration for oil has proved the presence of salt in areas such as the Gabon Basin (Brink, 1974) the full extent of evaporite distribution can only be inferred from the distribution of structures identified from the geophysical record as salt diapirs (Burke, 1975; Leyden *et al.*, 1976). "Salt" diapirs have been mapped on either side of the South Atlantic, from the Niger Delta–Recife Plateau in the north to the Walvis Ridge–Rio Grande Rise in the south. Very little can be said with any degree of certainty concerning the age range of the evaporites. In the Gabon Basin the youngest evaporites are intercalated with carbonates of Aptian age (Brink, 1974) and have been interpreted as being lagoonal in origin. The age and origin of the oldest evaporites is at present a completely open question.

It has been pointed out by Burke (1975) that the South Atlantic salt deposits originated at latitudes at which evaporation probably exceeded rainfall. Most authors (Asmus and Ponte, 1973; Franks and Nairn, 1973; Burke, 1975; Roberts, 1975; Leyden *et al.*, 1976) favour a model in which sea water spills over the Walvis Ridge–Rio Grande Rise and evaporates in a shallow basin.

If the membrane stress patterns developed in the way suggested earlier, then the South Atlantic rifts might be expected to have developed in the following stages:

(1) About 200 Myr ago three rifts started spreading out from a triple junction located near the centre of a major positive membrane stress domain (Fig. 6(a)).

(2) From about 200 Myr to about 160 Myr the rifts propagated away from the triple junction and spread laterally. Since the centre of the stress domain

was migrating steadily southward the rate of propagation and spreading across the southern rift would have been greater than on the other two rifts. During this stage the rifts would have been entirely intra-continental.

(3) About 150 Myr ago, due to the reversal of the pattern of membrane stress domains (Fig. 6(d)), the intra-continental rifts would have ceased to spread and a new rift would have propagated in from the continental margin thus separating South America from Africa.

Nothing in the sequence of events outlined above is incompatible with the known geology of the area. During the early stages of rifting, the rift margins would be elevated, drainage would be away from the rift and consequently little input of sediments would be expected for the surrounding area. Any rift system far from the nearest ocean and at a latitude of about 10° to 30° south would probably be occupied by saline lakes. A restricted sediment supply and a high salt content over a period of up to 60 Myr would fit the scanty evidence available for the thin sequence of pre-Aptian sediments on both sides of the South Atlantic. During the later stages as the rift propagating in from the continental margin joined up with the intra-continental rift system limited amounts of water presumably spilled over the divide giving the restricted marine environment observed in the sediments of Albian age.

Conclusions

A case can be made for a correlation between the sequence of geological events leading up to the break-up of Gondwanaland and domains of membrane stress generation resulting from movement of the Gondwana plate relative to the geoid. An even more convincing case can be made out for a similar correlation between membrane stress generation domains and the deformation observed in West Africa during the last 100 Myr.

Clearly much work remains to be done before the cases which have been outlined can be proved (or disproved) to anyone's satisfaction. Very little is yet known for certain about the early history of the South Atlantic rift system, detailed mapping along the whole of the continental margin together with information from offshore oil exploration will presumably rectify this in due course. Although the geological history of the Lower Benue Rift is now reasonably well established, there are as yet no accurate strain measurements for any of the three major phases of deformation or for the subsequent extension across the Cameroun Volcanic Line.

Domains of membrane stress generation can be calculated, from palaeomagnetic data, and compared in a qualitative way with observed deformation. The next stage in the development of membrane tectonic theory will be the construction of a numerical model which can translate movement relative to the geoid into fully quantitative patterns of stress distribution in actual lithospheric plates. Work on such a model, using finite element analysis, is currently in hand and we hope to make our initial results available in the near future.

References

Amaral, G., Bushee, J., Cordani, U. G., Kawashita, K. and Reynolds, J. H., 1967. Potassium–argon ages of alkaline rocks from southern Brazil. *Geochim. Cosmochim. Acta* **31**, 117–142.

Amaral, G., Cordani, U. G., Kawashita, K. and Reynolds, J. H., 1966. Potassium–argon dates of basaltic rocks from southern Brazil. *Geochim. Cosmochim. Acta* **30**, 159–189.

Asmus, H. E. and Ponte, F. C., 1973. The Brazilian marginal basins. In *The Ocean Basins and Margins* (Eds Nairn, A. E. M. and Stehli, F. G.), Vol. 1, The South Atlantic, 87–133, Plenum Press, New York.

Brink, A. H., 1974. Petroleum geology of Gabon basin. *Bull. Am. Assoc. Petrol. Geol.* **58**, 216–235.

Burke, K., 1975. Atlantic evaporites formed by evaporation of water spilled from Pacific, Tethyan, and Southern oceans. *Geology* **3**, 613–618.

Burke, K., Dessaugavie, T. F. J. and Whiteman, A. J., 1972. Geological history of the Benue Valley and adjacent areas. In *African Geology, Ibadan 1970* (Eds Dessauvagie, T. F. J. and Whiteman, A. J.), 187–205, Univ. Ibadan.

Burke, K. and Whiteman, A. J., 1973. Uplift, rifting and the break-up of Africa. In *Implications of Continental Drift to the Earth Sciences* (Eds Tarling, D. H. and Runcorn, S. K.), 735–755, Academic Press, London.

Delteil, J.-R., Valery, P., Montadert, L., Fondeur, C., Patriat, P. and Mascle, J., 1974. Continental margins in the northern part of the Gulf of Guinea. In *The Geology of Continental Margins* (Eds Burk, C. A. and Drake, C.L.), 297–311, Springer-Verlag, New York.

Franks, S. and Nairn, A. E. M., 1973. The equatorial marginal basins of West Africa. In *The Ocean Basins and Margins* (Eds Nairn, A. E. M. and Stehli, F. G.), Vol. 1, The South Atlantic, 301–350, Plenum Press, New York.

Freeth, S. J., 1978a. Tectonic activity in West Africa and the Gulf of Guinea, since Jurassic times – an explanation based on membrane tectonics. *Earth Planet. Sci. Lett.* **38**, 298–300.

Freeth, S. J., 1978b. A model for tectonic activity in West Africa and the Gulf of Guinea during the last 90 Myr based on membrane tectonics. *Geol. Rdsch.* **67**, 675–687.

Freeth, S. J., 1979. Deformation of the African plate as a consequence of membrane stress domains generated by post-Jurassic drift. *Earth Planet. Sci. Lett.* **45**, 93–104.

Herz, N., 1977. Timing of spreading in the South Atlantic: information from Brazilian alkalic rocks. *Bull. Geol. Soc. Am.* **88**, 101–112.

Ladd, J. W., Dickson, G. O. and Pitman III, W. C., 1973. The age of the South Atlantic. In *The Ocean Basins and Margins* (Eds Nairn, A. E. M. and Stehli, F. G.), Vol. 1, The South Atlantic, 555–573, Plenum Press, New York.

Leyden, R., Asmus, H., Zembruscki, S. and Bryan, G., 1976. South Atlantic diapiric structures. *Bull. Am. Ass. Petrol. Geol.* **60**, 196–212.

Melfi, A. J., 1967. Potassium–argon ages for core samples of basaltic rocks from southern Brazil. *Geochim. Cosmochim. Acta.* **31**, 1079–1089.

Oxburgh, E. R. and Turcotte, D. L., 1974. Membrane tectonics and the East African Rift. *Earth Planet. Sci. Lett.* **22**, 133–140.

Rawson, P. F., Curry, D., Dilley, F. C., Hancock, J. M., Kennedy, W. J., Neale, J. W., Wood, C. J. and Worssam, B. C., 1978. A correlation of Cretaceous rocks in the British Isles. *Sp. Rep. Geol. Soc. Lond.* No. 9.

Roberts, D. G., 1975. Evaporite deposition in the Aptian South Atlantic Ocean. *Marine Geol.* **18**, M65–M72.

Scrutton, R. A., 1973. The age relationship of igneous activity and continental break-up. *Geol. Mag.* **110**, 227–234.

Smith, A. G. and Briden, J. C., 1977. *Mesozoic and Cenozoic Palaeocontinental Maps*, Cambridge University Press.

Turcotte, D. L., 1974. Membrane tectonics. *Geophys. J. R. Astr. Soc.* **36**, 33–42.

Turcotte, D. L. and Oxburgh, E. R., 1973. Mid-plate tectonics. *Nature* **244**, 337–339.

Turcotte, D. L. and Oxburgh, E. R., 1976. Stress accumulation in the lithosphere. *Tectonophysics* **35**, 183–199.

Vine, F. J. and Hess, H. H., 1971. Sea-floor spreading. In *The Sea* (Ed. Maxwell, A. E.), Vol. 4, New concepts of sea-floor spreading evolution, 587–622, Wiley, New York.

Self-Driven Motions of Plates and Descending Slabs

G. SCHUBERT

Department of Earth and Space Sciences, University of California, Los Angeles, California, USA

Although the tectonic plates are an integral part of a global mantle convection system, there is a widespread tendency among geophysicists to isolate the plates from the rest of the mantle circulation and then to inquire into the nature of the forces – slab pull, ridge push or gravitational sliding, basal traction, etc. – driving the plates (Forsyth and Uyeda, 1975; Harper, 1975; Chapple and Tullis, 1977; Davies, 1978). An unfortunate consequence of this approach is that it leads to seemingly endless and unproductive debates about whether the surface plates drive the motions in the rest of the mantle or vice versa. We will discuss how the plates can be regarded as driving both themselves and the motions in the rest of the mantle. However, the plates would not move, or even exist, if not for the Earth's heat which must be removed from its interior and brought to the surface. In this sense, the mantle drives the plates, for it is the past and present radiogenic heating throughout the mantle which is the ultimate energy source for all the motions.

To understand how the plates can be thought of as driving both themselves and the rest of the mantle one need only consider the boundary layer nature of high Rayleigh number or vigorous thermal convection of an ordinary Boussinesq fluid between parallel plates (Turcotte and Oxburgh, 1967; Olson and Corcos, 1980) or within a spherical shell (Schubert and Zebib, 1980; Zebib, Schubert and Straus, 1980). If heating is from below, a circulation will develop in which fluid moves along the lower boundary and acquires heat, rises, moves along the upper boundary and cools, and finally sinks. Heating and cooling in the horizontal legs of the cell occur in thin thermal boundary layers. The rising and descending limbs of the cell are correspondingly thin. Most of the fluid occupies a nearly isothermal, slowly circulating core region. Because the temperature variations are confined to the thermal boundary layers and the rising and descending plumes, so are the buoyancy forces. One could calculate the forces driving the flow by integrating the buoyancy forces per unit volume in the lower thermal boundary layer and in the rising plume, or in the upper thermal boundary layer and in the descending plume. These forces can be thought of as applying a torque to the isothermal core and driving its circulation against resistive viscous forces. One could isolate the upper thermal boundary layer and descending plume and calculate the forces driving these portions of the flow. Clearly, the driving forces will reside within these elements themselves. Thus the boundary layers and plumes drive both their own motions and those of the interior because the buoyancy forces in the convective circulation are confined to these regions. If fluid is heated wholly from within, there will be no lower thermal boundary layer or narrow rising plume, but the upper thermal boundary layer and the descending plume will exist. The forces acting on these entities and the forces they exert on the remainder of the fluid can be calculated, as before, as due to the buoyancy of these elements themselves.

The lithospheric plates are the upper thermal boundary layers of the mantle convection system and the descending slabs its sinking plumes (Turcotte and Oxburgh, 1967). As discussed just above, the plates and descending slabs drive themselves and the rest of the mantle because the buoyancy forces of the mantle convection system reside within these elements. However, one should not lose sight of the nature of the plates and descending slabs as integral parts of the entire system and the radioactives distributed throughout the mantle as the ultimate source of energy for the motions. Slab pull (Elsasser, 1969; McKenzie, 1969; Turcotte and Schubert, 1971; Griggs, 1972; Schubert, Yuen and Turcotte, 1975) and ridge push (Lliboutry, 1969; Jacoby, 1970) or gravitational sliding due to lithospheric cooling, contraction, thickening, and subsidence (Hales, 1969; Lister, 1975; Hager, 1978), etc., are interrelated aspects of the whole convecting system. To isolate these effects and attribute special significance to each as a separate entity does not properly reflect their origin as interconnected parts of a single convecting system.

The plates and slabs can be looked upon as the sources of the buoyancy forces for mantle convection whether or not there are narrow upwelling regions and lower thermal boundary layers. The mantle is undoubtedly heated largely from within, but some heat is also certainly emanating from the core. Stacey (1977a) has estimated that the heat flow across the core–mantle interface is about 10% of the surface heat loss. Thus there is likely to be a thermal boundary layer at the core–mantle interface, but its thickness, and the average temperature drop across it, compared to the corresponding quantities for the upper thermal boundary layer, are uncertain. The numerical calculations of Zebib *et al.* (1980) and Schubert and Zebib (1980) suggest that the thickness of the boundary layer at the bottom of the mantle is comparable to that of the surface boundary layer. If this is indeed the situation, then the average temperature drop across the lower boundary layer could be substantial even if relatively little heat, in terms of surface heat flow, enters the mantle from the core because of the relatively small surface area of the core as compared with that of the Earth. For example, if 10% of the surface heat loss crosses the core–mantle boundary then the average temperature drop across the lower boundary layer could be as large as 40% of the temperature drop across the upper thermal boundary layer.

The studies of Forsyth and Uyeda (1975) and Chapple and Tullis (1977) show that the slab pull and resistance to slab penetration are the largest forces acting on the plates. This is consistent with the view of the plates and slabs as upper thermal boundary layers and descending plumes of a vigorous mantle convection system, and as such, the sites of the major bouyancy forces in the system. While only thermal contraction contributes to the negative buoyancy of descending plumes in convection of ordinary Boussinesq fluids, the negative buoyancy of descending slabs is considerably enhanced by the elevation of the olivine–spinel phase change in the slab (Schubert, Turcotte and Oxburgh, 1970; Turcotte and Schubert, 1971; Schubert *et al.* 1975). Since slab pull is largely required to offset the resistance of the mantle to slab penetration, all the remaining forces acting on the plate, although considerably smaller than these main driving and resisting forces, are important in determining the motions of the plates.

While there are many aspects of mantle convection that appear understandable in terms of the convection of constant viscosity, Boussinesq fluids, mantle rocks do not behave so simply, and there are a number of ways in which mantle convection differs fundamentally from the convection of ordinary fluids (Oxburgh and Turcotte, 1978; Schubert, 1979). Most importantly, the subsolidus creep of rocks is highly temperature dependent (Tozer, 1967). One consequence of this is that the upper thermal boundary layer of the mantle convection system also becomes a rheological boundary layer; the plates are nearly rigid on a geological time-scale because the rocks are highly creep resistant at the relatively low temperatures in the upper part of the surface thermal boundary layer. The rigidity of this layer may delay the onset of subduction (Parmentier and Turcotte, 1978) and as a consequence, mantle convection cells may be wider compared to their depths than are convection cells in ordinary viscous fluids.

Another way in which mantle convection may differ from convection in ordinary fluids is the extent of decoupling of the upper boundary layer from the underlying flow. In an ordinary constant viscosity fluid there is little shear in the horizontal velocity profile at the base of the upper thermal boundary layer. In the mantle, however, because of its rheological properties, there may be considerable shear immediately below the lithosphere, i.e. the plates may be effectively decoupled from the underlying mantle and move relative to it (Froidevaux and Schubert, 1975; Schubert, Froidevaux and Yuen, 1976). Hotspot traces on the Earth's surface provide strong evidence of shear between the plates and deeper mantle, but the uncertainty in the source depth of mantle plumes precludes any quantitative determination of the depth dependence and magnitude of the shear. The thermomechanical models of the oceanic upper mantle by Schubert *et al.* (1976, 1978) assume that the oceanic lithosphere is decoupled from the underlying mantle by an asthenosphere, a region of low viscosity and large shear. A low viscosity layer, or asthenosphere, would be a consequence of the temperature and pressure dependences of mantle viscosity. Viscosity tends to decrease with depth as temperature increases, but it tends to increase with depth as pressure increases. These competing tendencies may result in a minimum in the viscosity–depth profile. Peltier (1980) argues that there should be no shear at the base of the upper thermal boundary layer of a mantle convection system since this layer is part of the basic convection. However, there is no essential difficulty in reconciling plate decoupling from the rest of the mantle with the notion that the plates are a part of the total mantle convection system if there exists a low viscosity layer in the upper mantle. The existence of such a layer, especially beneath the old oceans, is uncertain.

We have already noted that the energy for mantle convection derives mainly from the mantle itself, since the heat flow across the core–mantle boundary is only a small fraction of the heat loss at the Earth's surface. In addition, we have also remarked that the heat produced internally in the mantle originates mainly in the decay of the radioactive elements ^{235}U, ^{238}U, ^{232}Th and ^{40}K. Because mantle convection is generally regarded as being vigorous and highly efficient in transporting heat from the deep interior to the surface, it has been widely believed that the present day surface heat loss is in equilibrium with the present day mantle radiogenic heat production (Tozer, 1965; Turcotte and Oxburgh, 1972). This idea permits the determination of the Earth's current total budget of radioactive elements from measurements of heat flow at its surface. Despite the efficiency of mantle convection, however, it is straightforward to show that if the Earth is cooling, and it most probably is at the present day, then there cannot be equality of present day surface heat flow and present day internal heat generation. A substantial fraction, perhaps as much as 50%, of the surface heat flow is due to cooling of the Earth (Schubert, Stevenson and Cassen, 1980). Surface heat flow provides only an upper limit to the present content of radioactive elements in the Earth's interior; the upper limit may exceed the actual content by as much as a factor of 2 (Schubert *et al.*, 1980). The difference between present day surface heat loss and present day radiogenic heat production has recently been studied by Sharpe and Peltier (1978), Schubert (1979), Schubert, Cassen and Young (1979a, b), Stevenson and Turner (1979), Turcotte, Cook and Willeman (1979), Schubert *et al.* (1980), Davies (1980) and Stacey (1980).

The portion of the present day surface heat loss associated with the whole Earth cooling is heat that is partly primordial and partly radiogenic in origin, the latter having been produced at some earlier stage in the Earth's thermal evolution. Present day cooling may be entirely due to past radiogenic heating since primordial heat may have been lost early in the Earth's history by a highly vigorous mantle convection. Nevertheless, although the ultimate source of all the present day surface heat flow may be the radioactive decay of uranium, thorium and potassium, the present day decay rate of these elements cannot account for the present day heat loss of a cooling Earth. To see this it is simply necessary to write a heat balance equation for the mantle (we will neglect the small heat flow from the core)

$$\rho c V\dot{T} = QV - Aq \tag{1}$$

or

$$\frac{qA}{V} = Q + \rho c(-\dot{T}), \tag{2}$$

where ρ is the mantle density, c is the specific heat of the mantle, V is the volume of the mantle, T is the temperature of the mantle (the dot represents a time derivative), Q is the present day volumetric heat production rate in the mantle, A is the surface area of the Earth and q is the present day surface heat flow. If the Earth is cooling $\dot{T}$ is negative and

$$\frac{qA}{V} > Q \tag{3}$$

i.e. the present day rate of decrease of the Earth's internal energy must make a contribution to the present day surface heat loss and qA/V is only an upper bound to the present day radiogenic heat production rate Q. Thus conservation of energy ensures that present day surface heat loss must exceed present day volumetric heat production if the planet is cooling; the only uncertainty is the magnitude of the excess. This is where mantle convection enters the discussion. Can it be so efficient as to make the excess negligible? The answer for the Earth is no, as the following simple calculation will show.

Suppose first that the present day surface heat loss is roughly in balance with the present day mantle radiogenic heat production. Then the surface heat flux is approximately given by

$$q = \frac{QV}{A}. \tag{4}$$

Since the heat sources are decaying in time, the surface heat flow must also be decreasing at the rate

$$\dot{q} = \dot{Q}\frac{V}{A}. \tag{5}$$

The heat sources decay exponentially in time

$$\dot{Q} = -\lambda Q \tag{6}$$

λ is the average decay rate of the assemblage of radioactive elements in the mantle. By combining Eqns (4)–(6) we can write the rate of decay of surface heat loss as

$$\dot{q} = -\lambda q, \tag{7}$$

i.e. the decay of the surface heat flux proceeds at the same rate as the decay of the radiogenic heat sources if there is an approximate "instantaneous" balance between the two quantities.

Equation (7) permits an approximate evaluation of the time rate of change of mantle temperature $\dot{T}$ or internal energy since there is a direct relation between surface heat flow and the interior temperature of a convecting mantle $q = q(T)$, i.e.

$$\dot{T} = \frac{dT}{dq}\dot{q} = -\lambda q \frac{dT}{dq}. \tag{8}$$

The surface heat flow in a vigorously convecting mantle is dependent on temperature approximately according to

$$q \propto T^{1+\beta} \exp\left(-\frac{\beta A'}{T}\right). \tag{9}$$

The exponential arises from the strong dependence of mantle viscosity on temperature; A' is the activation energy of that dependence and β is a constant. From Eqn (9) we find

$$q\frac{dT}{dq} = \left\{\frac{1+\beta}{T} + \frac{\beta A'}{T^2}\right\}^{-1}, \tag{10}$$

so that $\dot{T}$ becomes

$$\dot{T} = -\lambda\left\{\frac{1+\beta}{T} + \frac{\beta A'}{T^2}\right\}^{-1}, \tag{11}$$

or

$$\frac{\dot{T}}{T} = -\lambda\left\{1+\beta+\frac{\beta A'}{T}\right\}^{-1}. \tag{12}$$

The rate of change of mantle temperature can be estimated from Eqn (12) using $T = 2500$ K (Stacey, 1977b; Anderson, 1980), $\lambda = 1{\cdot}4 \times 10^{-17}$/s (corresponding to the time-averaged decay constant for a chondritic composition), $\beta = 0{\cdot}3$ and $A' = 5{\cdot}6 \times 10^4$ K (Schubert *et al.*, 1980). One finds $\dot{T} \approx -100$ K/10^9 yr. If all the present day surface heat loss is attributed to mantle cooling then

$$\dot{T} = -\frac{Aq}{\rho c V}. \tag{13}$$

With $A = 5{\cdot}1 \times 10^{18}$ cm^2, $V = 8{\cdot}9 \times 10^{26}$ cm^3, $\rho = 4{\cdot}5$ g/cm^3, $c = 0{\cdot}25$ cal/g K and $q = 1{\cdot}75 \times 10^{-6}$ cal/cm^2 s, Eqn (13) gives $\dot{T} \approx -300$ K/10^9 yr. The mantle cooling rate inferred from an assumed surface heat loss–mantle heat production balance is a substantial fraction of what the cooling rate would be if the Earth's internal energy supplied all of the surface heat flow. Thus whole Earth cooling must contribute significantly to the present day surface heat flow, a conclusion also reached by Davies (1980) and Stacey (1980).

Acknowledgements

This research was supported by the NSF under grant EAR 77–15198 and by the Planetology Program, Office of Space Science, NASA, under grant NGR 05–007–317 and by NASA under grant NSF 7315.

References

Anderson, O. L., 1980. Temperature profiles in the earth. *Am. Geophys. Union Monograph of Working Group 5 of the Inter-Union Commission of Geodynamics* (in press).

Chapple, W. M. and Tullis, T. E., 1977. Evaluation of the forces that drive the plates. *J. Geophys. Res.* **82**, 1967–1984.

Davies, G. F., 1978. The roles of boundary friction, basal shear stress and deep mantle convection in plate tectonics. *Geophys. Res. Lett.* **5**, 161–164.

Davies, G. F., 1980. Thermal histories of convective earth models and constraints on radiogenic heat production in the earth. *J. Geophys. Res.* **85**, 2517–2530.

Elsasser, W. M., 1969. Convection and stress propagation in the upper mantle. In *The Application of Modern Physics to the Earth and Planetary Interiors* (Ed. Runcorn, S. K.), 223–246, Wiley, London.

Forsyth, D. and Uyeda, S., 1975. On the relative importance of the driving forces of plate motion. *Geophys. J. R. Astr. Soc.* **43**, 163–200.

Froidevaux, C. and Schubert, G., 1975. Plate motion and structure of the continental asthenosphere: a realistic model of the upper mantle. *J. Geophys. Res.* **80**, 2553–2564.

Griggs, D. T., 1972. The sinking lithosphere and the focal mechanism of deep earthquakes. In *The Nature of The Solid Earth* (Ed. Robertson, E. C.), 361–384, McGraw-Hill, New York.

Hager, B. H., 1978. Oceanic plate motions driven by lithospheric thickening and subducted slabs. *Nature* **276**, 156–159.

Hales, A. L., 1969. Gravitational sliding and continental drift. *Earth Planet. Sci. Lett.* **6**, 31–34.

Harper, J. R., 1975. On the driving forces of plate tectonics. *Geophys. J. R. Astr. Soc.* **40**, 465–474.

Jacoby, W. R., 1970. Instability in the upper mantle and global plate movements. *J. Geophys. Res.* **75**, 5671–5680.

Lister, C. R. B., 1975. Gravitational drive on oceanic plates caused by thermal contraction. *Nature* **257**, 663–665.

Lliboutry, L., 1969. Sea-floor spreading, continental drift and lithosphere sinking with an asthenosphere at melting point. *J. Geophys. Res.* **74**, 6525–6540.

McKenzie, D. P., 1969. Speculations on the consequences and causes of plate motions. *Geophys. J. R. Astr. Soc.* **18**, 1–32.

Olson, P. and Corcos, G. M., 1980. A boundary layer model for mantle convection with surface plates. *Geophys. J. R. Astr. Soc.* **62**, 195–219.

Oxburgh, E. R. and Turcotte, D. L., 1978. Mechanisms of continental drift. *Rep. Prog. Phys.* **41**, 1249–1312.

Parmentier, E. M. and Turcotte, D. L., 1978. Two-dimensional mantle flow beneath a rigid, accreting lithosphere. *Phys. Earth Planet. Int.* **17**, 281–289.

Peltier, W. R., 1980. Mantle convection and viscosity. *Proc. Enrico Fermi School of Physics* (Course 78) (in press).

Schubert, G., 1979. Subsolidus convection in the mantles of terrestrial planets. *Ann. Rev. Earth Planet. Sci.* **7**, 289–342.

Schubert, G. and Zebib, A., 1980. Thermal convection of an internally heated infinite Prandtl number fluid in a spherical shell. *Geophys. Astrophys. Fluid Dyn.* **15**, 65–90.

Schubert, G., Turcotte, D. L. and Oxburgh, E. R., 1970. Phase change instability in the mantle. *Science* **169**, 1075–1077.

Schubert, G., Yuen, D. A. and Turcotte, D. L., 1975. Role of phase transitions in a dynamic mantle. *Geophys. J. R. Astr. Soc.* **42**, 705–735.

Schubert, G., Froidevaux, C. and Yuen, D. A., 1976. Oceanic lithosphere and asthenosphere: thermal and mechanical structure. *J. Geophys. Res.* **81**, 3525–3540.

Schubert, G., Yuen, D. A., Froidevaux, C., Fleitout, L. and Souriau, M., 1978. Mantle circulation with partial return flow: effects on stresses in oceanic plates and topography of the sea-floor. *J. Geophys. Res.* **83**, 745–758.

Schubert, G., Cassen, P. and Young, R. E., 1979a. Core cooling by subsolidus mantle convection. *Phys. Earth Planet. Int.* **20**, 194–208.

Schubert, G., Cassen, P. and Young, R. E., 1979b. Subsolidus convective cooling histories of terrestrial planets. *Icarus* **38**, 192–211.

Schubert, G., Stevenson, D. and Cassen, P., 1980. Whole planet cooling and the radiogenic heat source contents of the Earth and Moon. *J. Geophys. Res.* **85**, 2531–2538.

Sharpe, H. N. and Peltier, W. R., 1978. Parameterized mantle convection and the Earth's thermal history. *Geophys. Res. Lett.* **5**, 737–740.

Stacey, F. D., 1977a. *Physics of the Earth*, 198, Wiley, New York.

Stacey, F. D., 1977b. A thermal model of the Earth. *Phys. Earth Planet. Int.* **15**, 341–348.

Stacey, F. D., 1980. The cooling Earth: a reappraisal. *Phys. Earth Planet. Int.* **22**, 89–96.

Stevenson, D. J. and Turner, J. S., 1979. Fluid models of mantle convection. In *The Earth, It's Origin, Evolution and Structure*, (Ed. McElhinny, M.), 277–263, Academic Press, London and New York.

Tozer, D. C., 1965. Heat transfer and convection currents. *Phil. Trans. R. Soc.* **258A**, 252–271.

Tozer, D. C., 1967. Towards a theory of thermal convection in the mantle. In *The Earth's Mantle* (Ed. Gaskell, T. F.), 325–353, Academic Press, London.

Turcotte, D. L. and Oxburgh, E. R., 1967. Finite amplitude convective cells and continental drift. *J. Fluid Mech.* **28**, 29–42.

Turcotte, D. L. and Oxburgh, E. R., 1972. Mantle convection and the new global tectonics. *Ann. Rev. Fluid Mech.* **4**, 33–68.

Turcotte, D. L. and Schubert, G., 1971. Structure of the olivine–spinel phase boundary in the descending lithosphere. *J. Geophys. Res.* **76**, 7980–7987.

Turcotte, D. L., Cook, F. A. and Willeman, R. J., 1979. Parameterized convection within the Moon and the terrestrial planets. *Proc. 10th Lunar Planet, Sci. Conf.* 2375–2392.

Zebib, A., Schubert, G. and Straus, J. M., 1980. Infinite Prandtl number thermal convection in a spherical shell. *J. Fluid Mech.* **97**, 257–277.

Plate Sliding and Sinking in Mantle Convection and the Driving Mechanism

W. R. JACOBY

Institut für Meteorologie und Geophysik der Johann Wolfgang Goethe Universität, Frankfurt, W. Germany

---

Introduction

One of the open questions of Earth dynamics is the role the gravitational forces play in the whole scheme of the driving mechanism of plate tectonics. It can be easily shown that these forces acting on the moving plates are very powerful and that they seem to dominate all conceivable driving forces, equalled only by the forces of resistance. But these forces cannot be responsible for initiating new plates to form and to start moving, or for the whole complex history of plate motions.

It is therefore the purpose of this paper to review and compare various gravitational driving mechanisms and to put them into perspective with the whole problem. Part of the essential mechanism has been called "gravity sliding", referring to the ocean ridges spreading "down-slope"; we shall concentrate on this part. Another part is the sinking of the lithosphere at deep sea trenches into the mantle; this will be reviewed more briefly, as will be the resistive forces to the plate motions.

Initially, let us try to clarify the term "gravity sliding". Since it carries the notion of an isolated block sliding down a slope, it is a misnomer; at ridges, in contrast, material rises and then spreads laterally. Matter is transferred from the rising asthenosphere to the diverging lithospheric plates. Unfortunately, the term "gravity sliding" has become well established, as have other equally misleading terms , for example, "crust" and "spreading centre".

One should perhaps also in the introduction make the comment that any conceivable mechanism of moving the plates must involve gravity. To talk about a gravitational mechanism implies direct action of the force of gravity on the plates. This view is in competition with mechanisms which carry the plates or drive them by tractions from deeper mantle convection (within which gravity, of course, is also the moving agent). On the other hand, if the weight of the plates is important for the driving mechanism and, of course, flow occurs in the mantle to conserve mass, the whole system is also a form of convection (in fact, a form of thermal convection if the density contrasts involved are thermally induced). The question is thus not whether or not gravity is involved, or whether or not it is convection, but where in the system gravity is applied and what form of convection it is. In other words, do the plates move *with* the deeper mantle, carried by it, or do they move against the mantle (return) flow? This notion has been called "passive" versus "active" plates. Again these terms may be misleading since they may carry much more meaning than the direct action of gravity.

Development of the idea of a gravitational driving mechanism

The idea of a direct gravitational mechanism forcing the plates apart at *oceanic ridges* appears to have been first put forward by Hales (1969) who called it "gravitational sliding". He based his arguments on the universal presence of a low-velocity, low-Q and presumably low-viscosity layer and also on the problems which mantle-wide Rayleigh–Bénard convection faced. He argued that if a low-viscosity layer were dipping away from the ridge crests, there would be a gravitational sliding force. His estimate involved the slope and the size of the block. As can be seen from Fig. 1(b) (adapted from Hales' Fig. 3) the block would only slide if the sides were removed. If the whole structure including the surroundings were homogeneous in density there would be no effective sliding force available, no matter what the slope of the low-viscosity layer.

It is essential to have lateral density variations and/or topography to have a lateral "sliding force" or a horizontal pressure gradient. Jacoby (1969) called this "active diapirism" under mid-ocean ridges, but he also used the gravity sliding model to estimate the force. Jacoby (1970) extended the model (within its purely mechanical limitations) by visualizing a force balance on the moving plates between the gravitational driving forces at ridges and trenches and the velocity-proportional resistance at the bottom from the counter flow in the mantle. The trench pull had already been proposed by Elsasser (1967); it had also been discussed by McKenzie (1969) who dismissed its importance since the plate velocities did not seem to be governed by forces at their margins, but rather by areal forces. We shall return to this point later.

Jacoby (1970, 1973a, b) estimated the horizontal "ridge push" H_r from the sliding-block force and obtained, after simplification,

$$H_r \approx h_a^2 \rho g/2$$

with h_a = height of the upward pointing wedge of low density asthenosphere relative to its "normal"depth, ρ = density of lithosphere, g = gravitational

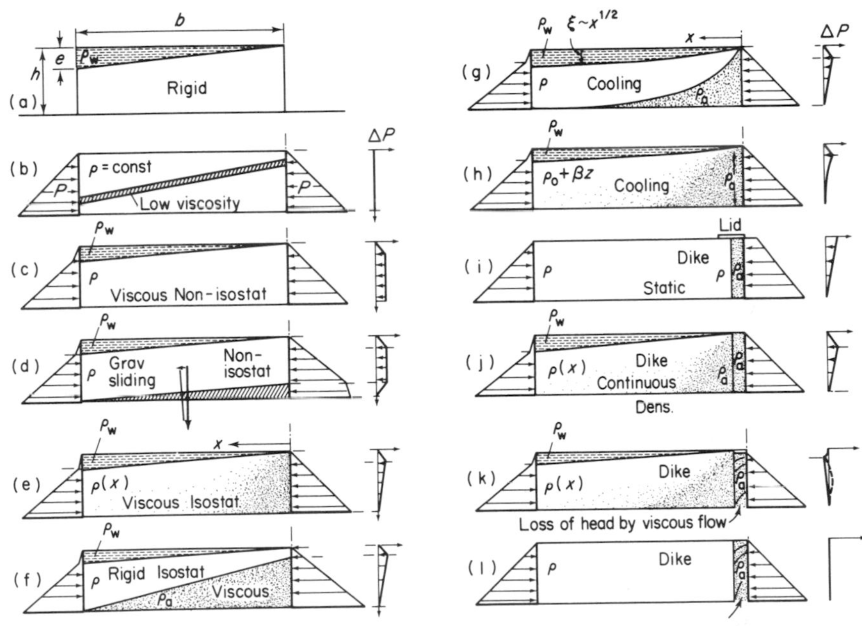

Figure 1. Diagrams of a number of ridge models. Left-hand half-ridges are shown as blocks cut out of the mantle; right-hand boundary at ridge axis. Distribution of normal stresses on both lateral boundaries, and pressure difference versus depth shown on the right. Dimensions and densities indicated by symbols. For discussion, see text.

acceleration. The use of ρ instead of $\rho - \rho_w$ (ρ_w = water density) was in error; the force balance is affected by the water pressure at the sea-floor. Jacoby argued that only a small fraction of the sliding force is available for driving, because lower-density asthenosphere (contrast $\Delta\rho$) has to rise against gravity as the ridge spreads. The ridge push must be scaled down from H_r by a factor of $\Delta\rho/\rho$. This result is correct although the argument is not clear; it is plausible, however, if one considers an analogy of an endless chain hanging from a roller and driven by a density contrast between both sides (which could, for example, be maintained by heating one side).

Order of magnitude estimates of the forces demonstrated that the resistance at the plate bottom can easily be overcome; more than that, it seemed likely that the major resistance to plate motion was encountered where the leading edges plunged into higher-viscosity mantle. Generally, the model was supported by the energy balance, the apparent stress field in the lithosphere, and by gravity over the ridges. An actively pushing role of the ridges (and a pulling role of the slabs) seemed to be plausible, at least in the moving system; the role need not be active in the sense that ridges could explain the initiation of the movements.

Lliboutry's (1969) model of a ridge push from a vertical dike filled with low density magma (Fig. 1(i), (j)) is dynamically similar to that of an upward pointing wedge because it also involves differential pressure. The difference in geometry is of secondary importance. Lliboutry argued that the asthenosphere is at its melting point so that large magma sills would form at the base of the lithosphere. If there is a mechanism for creating cracks, such as tension in the lithosphere, buoyant rise of the magma can be initiated. The difference in lithostatic and magma pressure with depth ($\Delta\rho gh$) across the vertical extent h results in a force driving the magma upward and pushing the walls apart. Lliboutry believed that this force is the most important one in plate dynamics. It may be mentioned in passing that the vertical-dike model was also used to estimate the temperature distribution in the developing lithosphere (e.g. McKenzie, 1967).

Lliboutry (1972) extended his analysis of the driving mechanism to a two-dimensional three-layer model (lithosphere–asthenosphere–mesosphere) in which gravity sliding over broad asthenosperic bumps, the liquid-dike push, slab sinking and neighbouring plate push at trenches are balanced. In spite of its simplicity, this model quite plausibly predicted plate velocities, stress and slopes. Lliboutry concluded that mantle flow should extend to 1000 km depth, that the plates move in response to direct action of gravity upon them, that, in this, the ridges are important, and that the plates drag the asthenosphere. A further spherical extension of the three-layer model (Lliboutry, 1974) strengthened these conclusions.

Frank (1972), using the analogy with glacier flow, demonstrated that it is the surface slope, not that of the bottom, which determines the horizontal pressure gradient $\partial p/\partial x \approx \rho g \partial e/\partial x$ (e = ridge elevation) and hence the spreading force or "ridge push" (Fig. 1(c)–(h)). The notion of the pressure gradient is physically more intelligible than the earlier one of "gravity sliding" or the special one of a "dike push", although it was implicit in them. Frank illustrated with a number of different models that the ridge push is not very sensitive to the particular model assumptions. He also estimated that the continents should spread as vigorously as ocean ridges; the fact that they do not presents a puzzling question, but it may be related to the temperature distribution and the corresponding viscosities at depth.

Artyushkov's (1973, 1974) mathematical analyses of the stresses in the lithosphere caused by density and thickness inhomogeneities (Fig. 1(f)) led to the same conclusions. Ocean ridges with low-density and low-viscosity roots create deviatoric stresses, flow and spreading, quantitatively similar to the earlier estimates. Ultimately, "local" vertical isostasy cannot be maintained by lateral density variations in material capable of flow, although the notion of flow is inherent in the concept of isostasy; but the time constants for the decay of the inhomogeneities depend on the viscosity and may be very long for continents.

Using the same model employed by both Frank and Artyushkov, Jacoby (1978a, b, 1979) demonstrated that lateral density variations under ocean ridges (and elsewhere) not only result in a lateral spreading force, but also in a torque uplifting the crestal zone and depressing the neighbouring low; the local mass balance is thus disturbed and the observed broad positive free air anomaly over spreading ridges (or part of it) may be explained by this mass imbalance.

More complex analytical (e.g. Hager, 1978) and numerical model computations (Andrews, 1972; Neugebauer and Jacoby, 1975) support the simple ridge-push estimates. Gravitational spreading seems also to be important for continental rifting (Neugebauer, 1978; Neugebauer and Braner, 1978).

Another class of ridge models relevant to our problem is concerned with the balance of gravitational forces and pressure, on the one hand, and viscous forces in the conduit between the receding plates, on the other hand (Sleep, 1969; Sleep and Biehler, 1970; Lachenbruch, 1973, 1976; Cann, 1974; Tapponnier and Francheteau, 1978; Sleep and Rosendahl, 1979; Collette, 1980). The viscous drag on the walls of the conduit and the related loss of hydraulic head, evident in topography (rift valley versus horst) and in gravity, reduce the driving pressure. Quantitatively this depends on the viscosity, spreading velocity and conduit geometry, but these are related with one another in a highly non-linear way. Ultimately one would have to solve the whole convection problem with temperature-controlled rheology and thermal conductivity. An estimate of the pressure available for the "ridge-push" can be gained from the actual broad average ridge elevation. The possible reduction of the push by viscous forces and formation of the rift valley is illustrated in Fig. 1(k), (l).

Sinking slabs

We now turn to the sinking slabs. Elsasser (1967) was the first to suggest that they may provide the major driving mechanism if the lithosphere acts as a stress guide. A similar conclusion was reached by Jacoby (1970) who, however, pointed out that the gravitational sinking force may be largely balanced by the viscous resistance, the leading edge meets in the mesosphere. McKenzie (1969) questioned the importance of this driving mechanism for different reasons (see above). The distribution of focal mechanisms within the descending slabs, however, is most plausibly explained by the gravitational pull and viscous forces, usually leaving a net pull acting on the horizontal lithosphere (Isacks, Oliver and Sykes, 1968; Isacks and Molnar, 1969, 1971). The validity of such a mechanism has also been demonstrated by laboratory experiments (Jacoby, 1973a, 1976) and by numerical models (Turcotte and Schubert, 1971; Smith and Toksöz, 1972; Neugebauer and Breitmayer, 1975). Model experiments can, of course, not prove the mechanism for the Earth, but only demonstrate it within a set of assumptions.

The positive density contrast of the sinking slabs is explained as the result of sinking (by delay in heating, i.e. by the low temperatures, thermal contraction and elevated phase boundaries: McKenzie, 1969, 1970; Minear and Toksöz, 1970a, b; Oxburgh and Turcotte, 1970; Toksöz, Minear and Julian, 1971; Toksöz, Sleep and Smith, 1973; Turcotte and Schubert, 1971, 1973; Griggs, 1972; Matheson, 1976). This seems, however, to be a circular argument if we also consider the density contrast to be the cause of sinking; we must therefore assume some kind of initial instability to start the sinking. In the convectively moving system it is, of course, the heating from below or within and the cooling at the surface which maintain the driving density differences (in this connection the latter lose their meaning as ultimate cause and effect, but the sinking force remains, nevertheless, real).

The force F transmitted from the sinking slab to the horizontal lithosphere depends on the coupling between them. Jacoby (1970) assumed a "guiding force" in the bending region and found $F = ld\Delta\rho g \cos\alpha$ (l = thickness of lithosphere, d = depth of penetration into the mantle, $\Delta\rho$ = density contrast, α = dip angle of slab). If, on the other hand, the slab moves as if around a "fixed roller" of immobile asthenosphere, the transmitted force $F = ld\Delta\rho g$ (Jacoby, 1973a, b, 1978a). The acutal coupling should not be imagined as a truly tensile force but rather as a pressure reduction caused by the downward withdrawal of lithosphere; the horizontal part is thus pushed into the potential void (Elsasser, 1971). This would be the case (i.e. $F \neq 0$) even if the lithosphere dropped piecewise into the mantle. In this sense, the sinking force would remain important in the whole system, even if it were largely or wholly spent in overcoming the resistance at the leading edge, and the ridge push or other forces would have to do the rest of the work. Whether sinking as a slab or dropping piecewise, it would set up forced convection by viscous coupling on both sides and add to the forces driving plate convergence (McKenzie, 1969; Richter, 1973a, b, 1977; Andrews and Sleep, 1974; Bodri and Bodri, 1978; Richter and McKenzie, 1978; Dennis and Jacoby, 1979).

The extreme case opposite to the "fixed roller" of immobile asthenosphere would be sinking as a compact slab in the vertical direction with respect to the deeper mantle, such that the horizontal oceanic plate would also remain fixed, but the trench would migrate oceanward, pushing the whole asthenosphere with it (irrespective of slab dip). In this case the force F would be wholly spent in moving the asthenosphere through the channel between the lithosphere and the deeper mantle – obviously an unrealistic model. Neither extreme is likely to be realised in nature. For instance, from the Pacific, plates move out and subduct at about 10 cm/yr maximum velocity while it closes by about 1 cm/yr from east and west (Minster *et al.*, 1974; Solomon, Sleep and Richardson, 1975). The situation in the Pacific may, however, be looked upon as a very special one, in which about 2 cm/yr shrinkage of the upper 700 km would be compensated by about 2 × 7 cm/yr average outflow of 100 km thick lithosphere; in this case no exchange with the deeper mantle and no "return flow" in the upper mantle would have to occur (I owe this point to N. Grohmann, 1979). This view is certainly also too simplistic, but must be kept in mind.

Thus, simple force estimates such as those mentioned above will only give loose guidance toward an understanding of the driving mechanism, because it is not clear how the forces are exerted on or in the system; the simple models are, nevertheless, indispensable on the way to more complex ones.

Velocity-limiting forces

Our discussion of the mechanical aspects of the driving mechanism is concluded by briefly considering the *resistances.* The viscous forces in the asthenospheric conduit below the ridge and the related loss of head as well as the viscous resistance against slab sinking into the mesosphere have already been mentioned. Furthermore, there is dynamic pressure at the leading plate edge (Richter and McKenzie, 1978) of equivalent nature to the above loss of head; friction (viscous or non-viscous) at convergent (trench) and transform plate boundaries; and, of course, viscous coupling at the plate bottom. The latter has been estimated with grossly simplifying assumptions such as one-dimensional models with (1) shear flow between the moving plates and a fixed bottom (Hales, 1969); see also Froidevaux and Schubert, 1975; Schubert *et al.*, 1978); (2) return flow of constant-viscosity asthenosphere between the moving plates and a fixed bottom such that no net mass transport occurs through the vertical section (Lliboutry, 1969; Jacoby, 1970, 1973b, 1978a; Forristal, 1972); (3) the same with variable viscosity (Schubert and Turcotte, 1972; Lliboutry, 1972, 1974; Schubert *et al.*, 1978; Jacoby, 1978b, Jacoby and Ranalli, 1979). The next step is two-dimensional model flow usually in a "box" or harmonically extended, (1) responding to a moving boundary (e.g. Davies, 1977; Richter, 1977; Richter and McKenzie, 1978); (2) driven by body forces distributed inside such that the computed velocities are used for test (Andrews, 1972; Hager, 1978); (3) driven by thermally induced density contrast such that the whole convection problem has to be solved, albeit within stringent restrictions (among many others: Torrance and Turcotte, 1971; Turcotte, Torrance and Hsui, 1973; Richter, 1973a, b, McKenzie, Roberts and Weiss, 1974; Gebrande, 1975; Houston and DeBremaecker, 1975; Parmentier, Turcotte and Torrance, 1976; DeBremaecker, 1977; Schmeling, 1979; Jacoby and Schmeling, 1980). The ultimate step is three-dimensional spherical flow models (1) with moving-plate boundary conditions (Hager and O'Connell, 1978, 1979a; Chase, 1979; Parmentier and Oliver, 1979); (2) with the body forces included (Hager and O'Connell, 1979b); and (3) as the complete convection problem.

By now we have left the realm of simple models and shall not continue in this direction. The simple models, after all, seem to be convincing even if applied to the spherical Earth. This has been demonstrated by dynamic plate models in which the torques from the simple model forces are balanced either on single plates or on all plates (e.g. Solomon and Sleep, 1974; Forsyth and Uyeda, 1975; Harper, 1975; Solomon *et al.*, 1975; Richardson, Solomon and Sleep, 1976,

1979; Braunmühl, 1977; Chapple and Tullis, 1977; Davies, 1978 (whose conclusions differ from the previous ones because he assumes *a priori* large stresses at convergent plate boundaries). It thus appears that the plates move in response to gravitational spreading away from ridges and to gravitational sinking into the mantle, reaching a terminal velocity by the action of velocity-dependent resistances. In view of the observed plate motions, deep mantle convection currents do not appear to govern them mechanically; this suggests a weak coupling, not necessarily a non-existence of deep mantle convection. Evidence for deep convection must therefore be looked for elsewhere, e.g. in the gravity field (Runcorn, 1964, 1967).

Comparison of different "gravity-sliding" or "spreading" models

In summarizing this review, we shall briefly compare the different variants of the "gravity-sliding" model. As the above discussion has shown, the comparison can only show that the sliding or spreading force does not strongly depend on the special model assumptions, with the exception of some unrealistic extremes (e.g. rigidity or no lateral density variation). The different models are sketched in Fig. 1, showing two-dimensional blocks of height h and width b; the ridge elevation is e. Typical values would be $h \approx 50$–100 km; $b \approx$ several hundred kilometres; $e \approx 1$–3 km. Imagine such a block cut out of the mantle by vertical dikes at the ridge crest and b km further down. Water is included in the block because it influences the force balance. Densities are $\rho_w \approx 1$, $\rho \approx 3 \times 10^3$ kg/m^3 and $\Delta\rho = \rho - \rho_a \approx \rho e/h$.

If the block (beside the overlying water) is ideally rigid (Fig. 1(a)), the dikes remain open and no spreading stress or force is exerted onto the neighbourhood. If the block is elastic the situation is similar; any non-hydrostatic stresses would result in elastic deformation, but in no continuous spreading.

All other models (Figs 1(a)–(l)) are at least partly non-rigid or non-elastic and capable of flowing. To keep the above dikes permanently open, we would have to apply normal stresses at the walls. In reality the normal stresses are, of course, simply exerted by the opposite walls. These stresses would have to be close to hydrostatic,

$$g\rho z \quad \text{or} \quad \int_0^z g(\zeta)\,\mathrm{d}\zeta,$$

and we take this as a first approximation to the spreading force estimate. The approximation is plausible as long as the spreading rate is not limited by the internal viscous friction in the spreading block itself, or in other words, as long as the spreading rate is sufficiently slow.

The depth distribution of normal hydrostatic stress on both lateral walls is shown in Figs 1(a)–(l); because of topography and lateral density variation there is a pressure differential between both sides in most models. It is this pressure differential which, integrated over the whole depth range, renders the spreading force. Equal hydrostatic stresses on both sides would, of course, simply cancel each other and lead to zero force.

Figure 1(b) is one of the cases without a pressure differential; it has been adapted from Hales (1969, Fig. 3). We assume here that density is constant and surface topography is negligible although this has probably not been implied by Hales. Without a pressure differential there is no spreading force in spite of the oblique low-viscosity gliding plane. Only with the lateral walls removed would the upper block slide down.

Figure 1(c) represents a block of homogeneously viscous and dense material; the ridge topography is thus not isostatically compensated. The pressure differential shown on the right-hand side leads to the spreading stress:

$$\sigma \approx (\rho - \rho_w) ge \quad \text{or} \quad F \approx \sigma h \tag{1}$$

With the above dimensions and densities, $\sigma \approx 50$ MPa and $F \approx 5 \times 10^{12}$ N/m (force per length of the two-dimensional ridge). The model of Fig. 1(d) is dynamically quite similar; it is that of the sliding block, the typical case of gravity sliding, without isostatic equilibrium. In contrast to Fig. 1(c), the constant thickness lithosphere slides down the slope e/b of the top of the asthenosphere. We can thus compare the force estimate from block sliding with that of spreading, and they turn out to be equal within the approximations. The sliding force is

$$F \approx (\rho - \rho_w) bhge/b = (\rho - \rho_w) heg \tag{2}$$

identical to Eqn (1). The pressure differential is virtually the same as that of Fig. 1(c); only at the bottom of the sliding block, pressure changes and the difference vanishes at the lowest point.

The models of Figs 1(e)–(h) are all isostatically compensated in the sense that the pressure at the bottom is assumed to be constant. In Figs 1(e) and (f) the density variation is linear with distance from the axis, either as a continuous function in space or as a density discontinuity at the lithosphere–asthenosphere boundary. In Fig. 1(g) the lithospheric growth is assumed to be as $t^{1/2}$ or equivalently as $x^{1/2}$. In Figs 1(e)–(g) density is essentially constant vertically at the sides; in such cases the spreading stress is

$$\sigma \approx (\rho - \rho_w) ge/2, \tag{3}$$

i.e. half of that exerted by uncompensated ridges. If density at the outer boundary varies linearly with depth as $\rho_0 + \beta z$ (e.g. Richter and McKenzie, 1978) (Fig. 1(h)),

$$\sigma \approx (\rho - \rho_w) ge/3, \tag{4}$$

The average pressure differential or spreading stress depends only on the density–depth functions on both walls. These functions, because of isostasy are, of course, related to the elevation difference, but details of the lateral density variation between both sides are unimportant.

For the case of lithosphere thickness (h) and water depth (w) growing as the square root of age ($t^{1/2}$) and hence of distance from the axis ($x^{1/2}$), the spreading force F increases linearly with t and x. Denoting various constants with C_1, C_2, etc., we can write $h = C_1 t^{1/2}$; $e = C_2 - w$; $|\mathrm{d}e| = \mathrm{d}w = C_3 t^{-1/2}\mathrm{d}t$; and hence $\mathrm{d}F \approx C_4 \mathrm{d}t$ or $\mathrm{d}F \approx C_5 \mathrm{d}x$, i.e. independent from t or x. If the resistance at the plate bottom is simply constant viscous shear stress, the net force would still vary linearly with x; with spreading and shear stress equal, both would exactly cancel everywhere, but only for a certain velocity.

The last four models are characterized by dikes. Figure 1(i) shows a homogeneous block; a low-density dike has intruded but it is sealed off at the surface. In this case the spreading force is about that of Figs 1(e)–(g). The model is quite hypothetical. In Fig. 1(j) the dike is distinct only in viscosity but not in density; the block cools and contracts as in Figs 1(e)–(g) and the spreading force is also quite similar. The last two models (Figs 1(k) and (l)) are different since the magma flows upward and viscous friction leads to a loss of hydrostatic head. In Fig. 1(l) the vertical pressure gradient in the dike is steeper than hydrostatic; indeed it is exactly the same as that on the higher density side: the additional pressure gradient drives the upward flow. There is no pressure differential in this model, and no driving force.

In Fig. 1(k) the block has a lateral density increase away from the axis, and the vertical pressure gradient in the dike is too steep to allow the magma to reach the top. In this case there may be even a negative pressure differential and the ridge may exert a net resistance to spreading. However, with changes in dike width and viscosity the vertical pressure distribution may be much more complicated than depicted; it may be some combination of models Figs 1(k) and (j), so that the spreading force cannot be predicted so simply.

Conclusions

It is hoped that the merit of very simple models has been demonstrated. It is the help these models render on the way towards an understanding of more complicated models and ultimately of nature itself that makes them useful. If we want to understand how an engine works, we have to take it apart and investigate the transfer of force in detail. It is not sufficient to know where we put in the fuel and where the wheels turn. If we want to know the driving mechanism of plate tectonics, it is not always sufficient to write down the equations of motion of continuum mechanics and to apply them to a complex analytical or numerical model. If the structure is complicated and characterized by strong boundary layers or plates, we may want to know how and where in the system the forces act. For this we first need some simple models. We can then apply them to more and more realistic ones, ultimately of a global system of circulation. Such models are now becoming feasible, and they show that gravitational sliding, or rather spreading and gravitational sinking of the plates, are important aspects of the driving mechanism.

References

Andrews, D. J., 1972. Numerical simulation of sea-floor spreading. *J. Geophys. Res.* **77**, 6470–6481.

Andrews, D. J. and Sleep, N. H., 1974. Numerical modelling of tectonic flow behind island arcs. *Geophys. J. R. Astr. Soc.* **38**, 237–257.

Artyushkov, E. V., 1973. The stresses in the lithosphere caused by crustal thickness inhomogeneities. *J. Geophys. Res.* **78**, 7675–7708.

Artyushkov, E. V., 1974. Can the Earth's crust be in a state of isostasy? *J. Geophys. Res.* **79**, 741–752.

Bodri, L. and Bodri, B., 1978. Numerical investigation of tectonic flow in island arc areas. *Tectonophysics* **50**, 163–175.

Braunmühl, W., 1977. Modelle zur Dynamik kontinentaler und ozeanischer Lithosphärenplatten. Ph.D. Thesis, Univ. Frankfurt.

Cann, J. R., 1974. A model for oceanic crustal structure developed. *Geophys. J. R. Astr. Soc.* **39**, 169–187.

Chapple, W. M. and Tullis, T. E., 1977. Evaluation of the forces that drive the plates. *J. Geophys. Res.* **82**, 1969–1984.

Chase, G. C., 1979. Asthenospheric counterflow: a kinematic model. *Geophys. J. R. Astr. Soc.* **56**, 1–18.

Collette, B. J., Verhoef, J. and deMulder, A. F. J., 1980. Gravity and a model of the median valley. *J. Geophys.* **47**, 91–98.

Davies, G. F., 1977. Whole mantle convection and plate tectonics. *Geophys. J. R. Astr. Soc.* **49**, 459–486.

Davies, G. F., 1978. The roll of boundary friction, basal shear stress, and deep mantle convection in plate tectonics. *Geophys. Res. Lett.* **5**, 161–164.

Dennis, J. G. and Jacoby, W. R., 1980. Geodynamic processes and deformation in orogenic belts. *Tectonophysics* **63**, 261–273.

DeBremaecker, J. C., 1977. Convection in the earth's mantle. *Tectonophysics* **41**, 195–208.

Elsasser, W. M., 1967. Convection and stress propagation in the upper mantle. *Tech. Rep.* **5**, 223–246.

Elsasser, W. M., 1971. Sea-floor spreading as thermal convection. *J. Geophys. Res.* **76**, 1101–1112.

Forristal, G. Z., 1972. Comments on active plate tectonic hypotheses. *J. Geophys. Res.* **77**, 6407–6412.

Forsyth, D. W. and Uyeda, S., 1975. On the relative importance of driving forces of plate motion. *Geophys. J. R. Astr. Soc.* **43**, 163–200.

Frank, F. C., 1972. Plate tectonics, the analogy with glacier flow, and isostasy. In *Flow and Fracture of Rocks* (Eds Heard, H. C., Borg, I. Y. and Carter, N. L.), *Am. Geophys. Union Geophys. Monogr.* **16**, 285–292, Am. Geophys. Union, Washington DC.

Froidevaux, C. and Schubert, G., 1975. Plate motion and structure of the continental asthenosphere: a realistic model of the upper mantle. *J. Geophys. Res.* **80**, 2553–2564.

Froidevaux, C., Schubert, G. and Yuen, D. A., 1977. Thermal and mechanical structure of the upper mantle: a comparison between continental and oceanic models. *Tectonophysics* **37**, 233–246.

Gebrande, H., 1975. Ein Beitrag zur Theorie thermischer Konvektion im Erdmantel mit besonderer Berücksichtigung der Möglichkeit eines Nachweises mit Methoden der Seismologie. Ph.D. Thesis, Univ. München.

Griggs, D. T., 1972. The sinking lithosphere and the focal mechanism of deep earthquakes. In *The Nature of the Solid Earth* (Eds Robertson, E. C., Hays, J. F. and Knopoff, L.),364–384, McGraw-Hill, New York.

Hager, B. H., 1978. Oceanic plate motions driven by lithospheric thickening and subducted slabs. *Nature* **276**, 156–159.

Hager, B. H. and O'Connell, R. J., 1978. Subduction zone dip angles and flow driven by plate motions. *Tectonophysics* **50**, 111–133.

Hager, B. H. and O'Connell, R. J., 1979. Kinematic models of large-scale flow in the Earth's mantle. *J. Geophys. Res.* **84**, 1031–1048.

Hager, B. H. and O'Connell, R. J., 1979b. A simple global model of plate motions and mantle convection. *J. Geophys. Res.* (submitted).

Harper, J. F., 1975. On the driving forces of plate tectonics. *Geophys. J. R. Astr. Soc.* **40**, 465–474.

Houston, M. H. and DeBremaecker, J. C., 1975. Numerical models of convection in the upper mantle. *J. Geophys. Res.* **80**, 742–751.

Isacks, B. and Molnar, P., 1969. Mantle earthquake mechanisms and the sinking of the lithosphere. *Nature* **233**, 1121–1124.

Isacks, B. and Molnar, P., 1971. Distribution of stresses in the descending lithosphere from a global survey of focal mechanism solutions of mantle earthquakes. *Rev. Geophys. Space Phys.* **9**, 103–174.

Isacks, B., Oliver, J. and Sykes, L. R., 1968. Seismology and the New Global Tectonics. *J. Geophys. Res.* **73**, 5855–5899.

Jacoby, W. R., 1969. Active diapirism under mid-oceanic ridges (abstract). *Trans. Am. Geophys. Union* **51** (1970) 204.

Jacoby, W. R., 1970. Instability in the upper mantle and global plate movements. *J. Geophys. Res.* **75**, 5671–5680.

Jacoby, W. R., 1973a. Model experiment of plate movements. *Nature* **242**, 130–134.

Jacoby, W. R., 1973b. Gravitational instability and plate tectonics. In *Gravity and Tectonics* (Eds DeJong, K. A. and Scholten, R.), 17–33, Wiley, New York.

Jacoby, W. R., 1976. Paraffin model of plate tectonics. *Tectonophysics* **35**, 103–113.

Jacoby, W. R., 1978a. Role of gravity in plate tectonics. In *Rockslides and Avalanches* (Ed. Voight, B.), 707–727, Elsevier, Amsterdam.

Jacoby, W. R. 1978b. One-dimensional modelling of mantle flow. *Pure Appl. Geophys.* **116**, 1231–1249.

Jacoby, W. R., 1978c. Lateral density variations and isostasy – or isostasy with balance of torques. *Geoskrifter, Aarhus* **10**, 75–93.

Jacoby, W. R. and Ranalli, G., 1979. Non-linear rheology and return flow in the mantle. *J. Geophys.* **45**, 299–317.

Lachenbruch, A. H., 1973. A simple mechanical model for oceanic spreading centers. *J. Geophys. Res.* **78**, 3395–3417.

Lachenbruch, A. H., 1976. Dynamics of a passive spreading center. *J. Geophys. Res.* **81**, 1883–1902.

Lliboutry, L., 1969. Sea-floor spreading with an asthenosphere at melting point. *J. Geophys. Res.* **74**, 6525–6540.

Lliboutry, L., 1972. The driving mechanism, its source of energy, and its evolution studied with a three-layer model. *J. Geophys. Res.* **77**, 3759–3770.

Lliboutry, L., 1974. Plate movement relative to rigid lower mantle. *Nature* **250**, 298–300.

McKenzie, D. P., 1967. Some remarks on heat flow and gravity anomalies. *J. Geophys. Res.* **72**, 6261–6273.

McKenzie, D. P., 1969. Speculations on the consequences and causes of plate motions. *Geophys. J. R. Astr. Soc.* **18**, 1–32.
McKenzie, D. P., 1970. Temperature and potential temperature beneath island arcs. *Tectonophysics* **10**, 357–366.
McKenzie, D. P., Roberts, J. M. and Weiss, N. O., 1974. Convection in the Earth's mantle: towards a numerical simulation. *J. Fluid Mech.* **62**, 465–538.
Matheson, B., 1976. The thermal regime of a descending lithospheric plate. Ph.D. Thesis, Sheffield.
Minear, J. W. and Toksöz, M. N., 1970a. Thermal regime of a down-going slab and new global tectonics. *J. Geophys. Res.* **75**, 1397–1419.
Minear, J. W. and Toksöz, M. N., 1970b. Thermal regime of a down-going slab. *Tectonophysics* **10**, 367–390.
Minster, J. B., Jordan, T. H., Molnar, P. and Haines, E., 1974. Numerical modelling of instantaneous plate tectonics. *Geophys. J. R. Astr. Soc.* **36**, 541–576.
Neugebauer, H. J., 1978. Crustal doming and the mechanism of rifting. I, Rift formation. *Tectonophysics* **45**, 159–186.
Neugebauer, H. J. and Braner, B., 1978. Crustal doming and the mechanism of rifting. II, Rift development of the upper Rhine graben. *Tectonophysics* **46**, 1–20.
Neugebauer, H. J. and Breitmayer, G., 1975. Dominant creep mechanism and the descending lithosphere. *Geophys. J. R. Astr. Soc.* **43**, 873–895.
O'Connell, R. J., 1977. On the scale of mantle convection. *Tectonophysics* **38**, 119–136.
Oxburgh, E. R. and Turcotte, D. L., 1970. Thermal structure of island arcs. *Bull. Geol. Soc. Am.* **81**, 1665–1688.
Parmentier, E. M. and Oliver, J. E., 1979. A study of shallow global mantle flow due to the accretion and subduction of lithospheric plates. *Geophys. J. R. Astr. Soc.* **57**, 1–21.
Parmentier, E. M., Turcotte, D. L. and Torrance, K. E., 1976. Studies of finite amplitude non-Newtonian thermal convection in the Earth's mantle. *J. Geophys. Res.* **81**, 1839–1846.
Richardson, R. M., Solomon, S. C. and Sleep, N. H., 1976. Intra-plate stress as an indicator of plate tectonic driving forces. *J. Geophys. Res.* **81**, 1847–1856.
Richardson, R. M., Solomon, S. C. and Sleep, N. H., 1979. Tectonic stress in the plates. *Rev. Geophys. Space Phys.* **17**, 981–1019.
Richter, F. M., 1973a. Dynamical models for sea-floor spreading. *Rev. Geophys. Space Phys.* **11**, 223–287.
Richter, F. M., 1973b. Convection and large scale circulation in the mantle. *J. Geophys. Res.* **78**, 8735–8745.
Richter, F. M., 1977. On the driving mechanism of plate tectonics. *Tectonophysics* **38**, 61–88.
Richter, F. M. and McKenzie, D. P., 1978. Simple plate models of mantle convection. *J. Geophys.* **44**, 441–471.
Runcorn, S. K., 1964. Satellite gravity measurements and a laminar viscous flow model of the Earth's mantle. *J. Geophys. Res.* **69**, 4389–4394.
Runcorn, S. K., 1967. Flow in the mantle inferred from the low-degree harmonics of the geopotential. *Geophys. J. R. Astr. Soc.* **14**, 375–384.
Schmeling, H., 1979. Numerische Konvektionsrechnungen unter Annahme verschiedener Viskisitätsverteilungen und Rheologien im Mantel. Diploma Thesis, Univ. Frankfurt.
Schubert, G., Froidevaux, C. and Yuen, D. A., 1976. Oceanic lithosphere and asthenosphere: thermal and mechanical structure. *J. Geophys. Res.* **81**, 3525–3540.
Schubert, G. and Turcotte, D. L., 1972. One-dimensional model of shallow mantle convection. *J. Geophys. Res.* **77**, 945–951.

Schubert, G., Yuen, D. A., Froidevaux, C., Fleitout, L. and Souriau, M., 1978. Mantle circulation with partial shallow return flow: effects on stresses in oceanic plates and topography on the sea-floor. *J. Geophys. Res.* **83**, 745–758.
Sleep, N. H., 1969. Sensitivity of heat flow and gravity to the mechanism of sea-floor spreading. *J. Geophys. Res.* **74**, 542–549.
Sleep, N. H. and Biehler, S., 1970. Topography and tectonics at the intersections of fracture zones with central rifts. *J. Geophys. Res.* **75**, 2748–2752.
Sleep, N. H. and Rosendahl, B. R., 1979. Topography and tectonics of mid-oceanic ridge axes. *J. Geophys. Res.* **84**, 6831–6839.
Smith, A. T. and Toksöz, M. N., 1972. Stress distribution beneath island arcs. *Geophys. J. R. Astr. Soc.* **29**, 289–318.
Solomon, S. C. and Sleep, N. H., 1974. Some simple physical models for absolute plate motions. *J. Geophys. Res.* **79**, 2557–2567.
Solomon, S. C., Sleep, N. H. and Richardson, R. M., 1975. On the forces driving plate tectonics: inferences from absolute plate velocities and intra-plate stress. *Geophys. J. R. Astr. Soc.* **42**, 769–801.
Tapponnier, P. and Francheteau, J., 1978. Necking of the lithosphere and the mechanics of slowly accreting plate boundaries. *J. Geophys. Res.* **83**, 3955–3970.
Toksöz, M. N., Minear, J. W. and Julian, B. R., 1971. Temperature field and geophysical effect of a down-going slab. *J. Geophys. Res.* **76**, 1113–1138.
Toksöz, M. N., Sleep, N. H. and Smith, A. T., 1973. Evolution of the down-going lithosphere and the mechanisms of deep focus earthquakes. *Geophys. J. R. Astr. Soc.* **35**, 285–310.
Torrance, K. E. and Turcotte, D. L., 1971. Structure of convection cells in the mantle. *J. Geophys. Res.* **76**, 1154–1161.
Turcotte, D. L. and Schubert, G., 1971. Structure of the olivine–spinel phase boundary in the descending lithosphere. *J. Geophys. Res.* **76**, 1980–1987.
Turcotte, D. L. and Schubert, G., 1973. Frictional heating in the descending lithosphere. *J. Geophys. Res.* **78**, 5876–5987.
Turcotte, D. L., Torrance, K. E. and Hsui, A. T., 1973. Convection in the Earth's mantle. *Meth. Math. Phys.* **13**, 431–454.

Some Major Questions Concerning Mantle Convection

D. L. TURCOTTE

Department of Geological Sciences, Cornell University, Ithaca, New York, USA

Introduction

The movement of the surface plates of plate tectonics is by definition mantle convection. Thermal convection in the mantle provides a driving mechanism for plate tectonics and continental drift. Buoyancy forces convert heat into directed motion. The fluid behaviour of the mantle is consistent with known mechanisms of thermally activated creep in silicates. If diffusion creep is the appropriate creep mechanism the mantle behaves as a Newtonian fluid; if dislocation creep is applicable then the fluid behaviour is non-Newtonian.

Although the general concept of thermal convection in the mantle is widely accepted, many important aspects of the convection are poorly understood. Some of these are discussed in this chapter.

Mantle Rayleigh number

The primary non-dimensional parameter that determines the type of thermal convection that will occur is the Rayleigh number. Depending upon whether we choose a fluid layer or spherical shell heated from within or below, different Rayleigh numbers can be defined. We choose to consider a fluid layer heated from below and write (Oxburgh and Turcotte, 1978)

$$Ra = \frac{\rho^2 \alpha g H h^5}{k \eta \kappa}. \tag{1}$$

Since other choices for the Rayleigh number give rather similar results the particular choice is not important at this stage. Based on laboratory studies and other available information we take $\rho = 4{\cdot}66\,\text{g/cm}^3$, $\alpha = 3 \times 10^{-5}/\text{K}$, $g = 10^3\,\text{cm/s}^2$, $k = 10^{-2}\,\text{cal/cm s K}$ and $\kappa = 10^{-2}\,\text{cm}^2/\text{s}$. In order to estimate the heat generation H we assume that the mean surface heat flow due to mantle sources is $q = 1{\cdot}45\,\mu\text{cal/cm}^2\,\text{s}$ can be equated to a mean heat production throughout the mantle with the result $H = 1{\cdot}85 \times 10^{-15}$ cal/g s. For the mantle viscosity we take the value obtained from studies of post-glacial rebound $\eta = 10^{22}$ poise (Cathles, 1975). Observational evidence is not sufficient to determine whether convection is occurring in the whole mantle or whether convection is confined to the upper mantle. For whole mantle convection $h = 2880$ km and $Ra = 2{\cdot}4 \times 10^9$; for upper mantle convection $h = 700$ km and $Ra = 2 \times 10^6$.

The critical Rayleigh number for the onset of thermal convection is about 3×10^3; in either case considered above, the mantle Rayleigh number is much larger than the critical value. Studies of thermal convection in a fluid layer with constant properties indicate that three-dimensional convection is likely to occur at about $Ra > 10^6$. At $Ra > 10^9$ unsteady (turbulent) convection is expected. However, the influence of a strongly temperature dependent viscosity on these transitions is not known.

The oceanic lithosphere

At moderate Rayleigh numbers (10^5–10^6) thermal convection in a fluid layer heated from within takes the form of two-dimensional cells. A cross-section of a typical cell is illustrated in Fig. 1(a). A cold thermal boundary layer develops on the cooled upper boundary. The cold dense fluid near the upper boundary becomes gravitationally unstable and separates from the upper boundary to form a cold descending plume. The downward buoyancy force on the plume drives the flow. The large viscosity of the fluid couples the descending plume to the core flow.

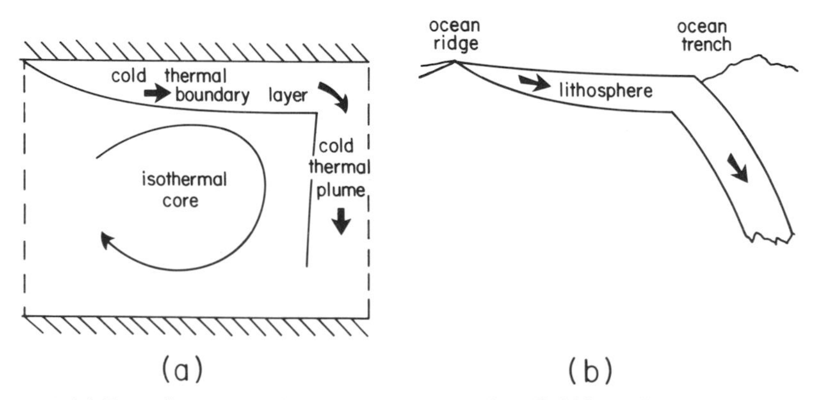

Figure 1. (a) Two-dimensional thermal convection in a fluid layer heated from within and cooled from above. (b) Creation of the oceanic lithosphere as a thermal boundary layer at an ocean ridge and its subduction at an ocean trench.

In Fig. 1(b) the behaviour of the oceanic lithosphere is illustrated. Clearly there are many similarities between the oceanic lithosphere and the cold thermal boundary layer illustrated in Fig. 1(a). The oceanic lithosphere forms due to the loss of heat to the sea-floor. The structure of the lithosphere can be obtained by solving the thermal boundary layer equations (Turcotte and Oxburgh, 1967). The elevation of the ridge is the hydrostatic equivalent of the horizontal pressure in a confined fluid layer.

The oceanic lithosphere becomes gravitationally unstable as it cools and thickens. As a result it founders and sinks into the mantle at an ocean trench. The gravitational body force on the descending lithosphere plays a dominant role in driving the motion of the surface plates. Because of the strong temperature dependence of the rheology the lithosphere is rigid and can act as a stress guide. As a result the gravitational force on the descending plate can be transmitted directly to the surface plate rather than indirectly through the fluid in the core of the convection cell.

The rigidity of the surface plate has other implications. This rigidity inhibits the thermal instability of the upper boundary layer. One result is that the aspect ratio of mantle convection cells may be much larger than the order unity aspect ratios obtained in constant viscosity fluids (Parmentier and Turcotte, 1978). In addition, this rigidity would be expected to suppress the transverse instabilities that would lead to three-dimensional convection.

Differentiation

When the oceanic lithosphere is formed from ascending mantle rock at an ocean ridge considerable partial melting occurs. The basaltic component of undepleted mantle rock melts and rises to the surface to form the oceanic crust with a mean thickness of about seven kilometres. The depleted mantle rock from which the basaltic component has been removed is less dense than undepleted mantle rock. The thickness of the depleted zone of the oceanic lithosphere is approximately 30 kilometres, although the degree of depletion is expected to vary with depth.

This differentiated lithosphere is then subducted into the mantle at an ocean trench. An important and essentially unresolved question is the influence of the compositional layering of the descending lithosphere on mantle convection. At depth the basalt of the oceanic crust is expected to undergo a phase change to eclogite. Eclogite is more dense than undepleted mantle rock. The mean density of an eclogite crust and the depleted mantle is essentially equal to the density of undepleted mantle.

There are essentially two alternative hypotheses for the behaviour of the layered, subducted lithosphere. The first is that the layered lithosphere will mix with the rest of the mantle rock, changing its mean composition and losing its separate identity. The second hypothesis is that the dense eclogite will sink to the base of the lithosphere and that the light depleted mantle will rise to the upper part of the mantle once it has been heated. Oxburgh and Parmentier (1978) suggest that the continental lithosphere is made up of this depleted mantle rock. Significant compositional layering would be expected to inhibit thermal convection in the mantle.

Depth of mantle convection

Seismic studies indicate about a 10% increase in density at a depth of 400 km. This density discontinuity has been shown to be primarily caused by the olivine–spinel phase change. Earthquake epicentres indicate that the descending lithosphere penetrates this phase boundary. Theoretical studies (Schubert and Turcotte, 1971) show that this phase change enhances thermal convection so that thermal convection through the phase change is expected.

Another increase in density of about 10% occurs at a depth of about 650 km. Laboratory studies indicate that this density change can at least partially be attributed to disproportionation of spinel to perovskite plus periclase. Liu (1979), however, has argued that there may also be a change in composition of the mantle at this depth. It is not known whether the disproportionation reaction enhances or suppresses mantle convection. A significant change in a composition could certainly prevent convection in the upper mantle from penetrating the lower mantle.

The complete absence of earthquakes at depths greater than 700 km has been used as an argument for convection restricted to the upper mantle. However, since the mechanism responsible for deep earthquakes is not understood alternative explanations can be given. There is some ambiguous evidence that the descending lithosphere cannot penetrate the 650 km seismic discontinuity. The compressional focal mechanisms found in the deeper parts of the descending lithosphere supports this conclusion.

If there is a significant difference in composition between the upper and lower mantles then convection may be restricted to the upper mantle or two separate layers of convection may occur (McKenzie and Weiss, 1975). Because a significant fraction of the surface heat flow must be attributed to the core in order to drive the dynamo, and since some radioactive heat production is expected in the lower mantle, thermal convection in the lower mantle is certainly expected. If the 650 km seismic discontinuity enhances thermal convection or only weakly inhibits it, then convection throughout the mantle is expected (Davies, 1977).

Parameterized mantle convection

For a fluid layer heated from within it is appropriate to define a non-dimensional temperature, θ, according to

$$\theta = \frac{2k(T_i - T_0)}{\rho H h^2}. \tag{2}$$

In the absence of convection $\theta = 1$ at the lower boundary. For convection at high Rayleigh numbers the internal temperature T_i is nearly constant (T_0 is the temperature of the upper cooled boundary). For the large values of the Prandtl numbers applicable to the mantle the non-dimensional temperature can be parameterized in terms of the Rayleigh number.

Laboratory studies of thermal convection in a layer of uniformly heated water with rigid boundaries have been carried out by Kulacki and Emara (1977). The empirical dependence of the non-dimensional temperature on Rayleigh numbers for large values of the Rayleigh number was found to be

$$\frac{1}{\theta} = 1{\cdot}016\left(\frac{Ra}{Ra_c}\right)^{0{\cdot}227}. \tag{3}$$

The critical Rayleigh number for this problem is $Ra_c = 2772$. For free surface boundary conditions it can be shown that (Turcotte, Cook and Willeman, 1979)

$$\frac{1}{\theta} = 1{\cdot}556\left(\frac{Ra}{Ra_c}\right)^{1/4} \tag{4}$$

with $Ra_c = 1308$.

Validity of the steady-state heat balance

The mean surface heat flow can be related to the mean rate of heat generation in the mantle if a steady-state heat balance is assumed. However, if the Earth is cooling the internal energy of the Earth may be contributing some fraction of the surface heat flow (Sharpe and Peltier, 1978).

If the entire surface heat flow is attributed to the cooling of the Earth we may write

$$4\pi a^2 \bar{q} = M c_p \frac{dT_i}{dt}. \tag{5}$$

Taking $\bar{q} = 1{\cdot}45\,\mu\text{cal/cm}^2\,\text{s}$, $a = 6371$ km, $M = 6 \times 10^{27}$ g and $c_p = 0{\cdot}25$ cal/g we find that $dT_i/dt = 205$ K/Gyr; clearly a small amount of cooling.

Parameterized convection can be used to determine that rate at which the Earth is cooling (McKenzie and Weiss, 1975). Since the use of parameterized convection is basically empirical, many alternative formulations are possible. Using one approach Sharpe and Peltier (1978) have argued that only a small fraction of the surface heat flow is the result of the internal heat generation by radioactive isotopes.

In order to apply parameterized convection to the transient cooling of the Earth we redefine the heat generation, H, according to

$$H = H_0(t) - c_p \frac{dT_i}{dt}, \tag{6}$$

where $H_0(t)$ is the actual time dependent heat production and the transient cooling is also considered part of H_i. The approach has been shown to be in agreement with laboratory experiments by Turcotte *et al.* (1979). Substitution of (6) into (1), (2) and (3), and solving for the time derivative of the temperature yields

$$c_p \frac{dT_i}{dt} = H_0(t) - \frac{k[T_i(t) - T_0]}{\rho h^2} \frac{\rho^2 \alpha g h^5}{2772 k \kappa \eta(T_i)}. \tag{7}$$

A similar substitution into (4) yields

$$c_p \frac{dT_i}{dt} = H_0(t) - \frac{3{\cdot}112 k[T_i(t) - T_0]^{4/3}}{\rho h^2} \frac{\rho^2 \alpha g h^5}{1308 k \kappa \eta(T_i)}. \tag{8}$$

Once the rate of radioactive heat generation with time and the viscosity dependence on temperature have been prescribed, these equations can easily be solved numerically.

For the present concentrations of the radioactive isotopes we take $^{238}U = 3{\cdot}18 \times 10^{-8}$ g/g, $^{235}U = 1{\cdot}93 \times 10^{-10}$ g/g, $^{232}Th = 1{\cdot}28 \times 10^{-7}$ g/g and $^{40}K = 4{\cdot}31 \times 10^{-8}$ g/g. For the viscosity law we take (Oxburgh and Turcotte, 1978)

$$\eta = 4{\cdot}62 \times 10^3 \exp\left[\frac{6{\cdot}2928 \times 10^4}{T_i}\right] \tag{9}$$

with T in K and η in poise. The values of the other parameters have been given above.

If the surface plates play a dominant role in mantle convection as indicated by the boundary layer theory then free surface boundary conditions should be appropriate and Eqn (7) should be applicable. For this case we take the temperature of the upper boundary to be 0 °C. However, if vigorous convection is occurring beneath the surface plates, a fixed surface boundary condition should be appropriate and Eqn (8) should be applicable. For this case the upper boundary temperature is the temperature at the base of the lithospheric plates; we take this to be 1000 °C. For initial internal temperatures we take 1400 °C and 2000 °C.

The internal temperature as a function of time is given in Fig. 2 for the various cases considered. The surface heat flows are given in Fig. 3. For the model including the surface plates 93% of the surface heat flow can be attributed to heating by radioactive isotopes and 7% due to the cooling of the Earth. For the model of convection beneath the surface plates the division is 91% and 9%. A conclusion of these calculations is that the steady-state heat balance is a good approximation.

It should be emphasized that these calculations are based on empirical correlations and many approximations have been made. Whole mantle convection is implicitly assumed and heat loss from the core is neglected.

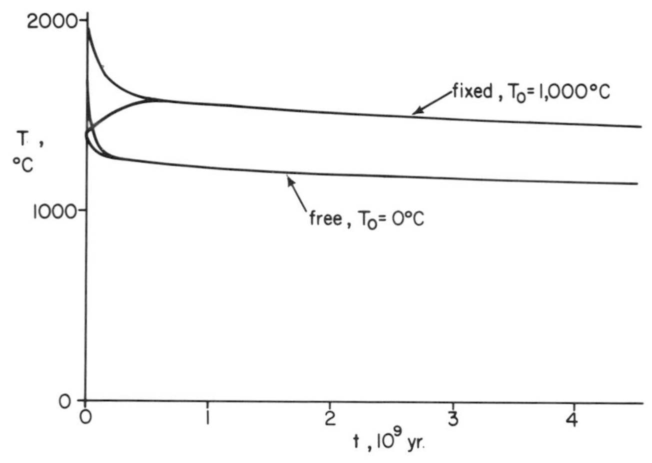

Figure 2. Mantle temperature as a function of time. An upper boundary temperature of 0 °C and free surface boundary conditions model the role of the surface plates in mantle convection. An upper boundary temperature of 1000 °C and fixed surface boundary conditions simulate convection beneath the plates. Initial temperatures of 2000 and 1400 °C are considered.

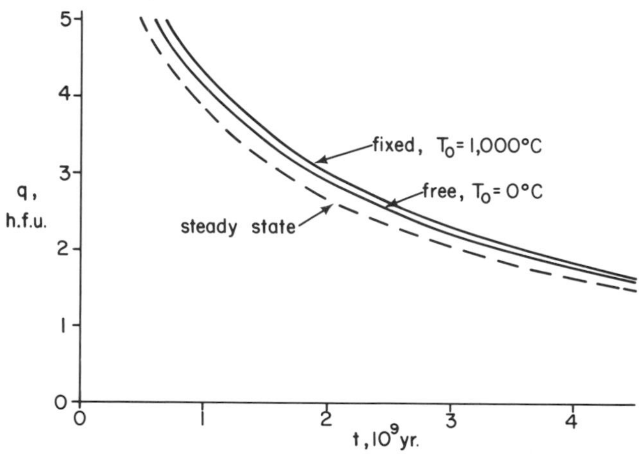

Figure 3. Surface heat flow as a function of time for the two models and the steady-state heat flow.

Multiple scales of convection

The free-surface models including the surface plates implicitly assume that the convection directly associated with the plates is the only convection occurring in the mantle. The fixed-surface models assume that vigorous mantle convection is occurring beneath the relatively slow moving surface plates.

The mantle viscosities as a function of time obtained in our calculations are given in Fig. 4. The present value for the model included the surface plates is 10^{23} poise; about one order of magnitude larger than the value obtained from studies of post-glacial rebound. For the model of convection beneath the surface plates the present viscosity is 4×10^{19} poise; several orders of magnitude smaller. An interpretation of these results is that most heat is transferred from the interior of the earth by the convection of the surface plates but a significant fraction of the heat transport is associated with convection beneath the surface plates. This interpretation is consistent with the observed distribution of surface heat flows. Observations of oceanic heat flow and corresponding studies of bathymetry for oceanic lithosphere older than about 80 Myr indicate heat transport to the base of the lithosphere. There is also conclusive evidence that the continental lithosphere is not continuing to cool; this implies a heat input to the base of the continental lithosphere. This input of heat to the base of the lithosphere can be attributed to convection beneath the lithospheric plates.

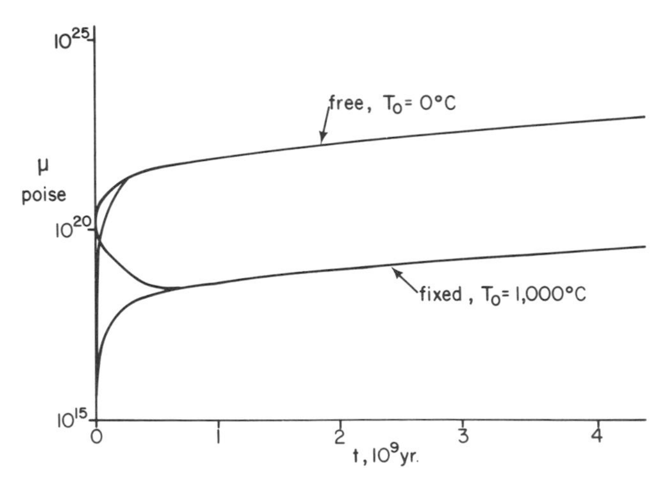

Figure 4. The mantle viscosity as a function of time for the two models.

The Rayleigh numbers as a function of time are given in Fig. 5. The present value for the model including the surface plates is 2×10^8 and for convection beneath the plates is 6×10^{11}. These values are sufficiently high that unsteady or turbulent convection could be expected in the mantle.

It should be noted, however, that there is little surface evidence for the occurrence of convection beneath the surface plates. Some authors attribute centres of intra-plate volcanism to mantle convection beneath the surface plates. Much of this volcanism can be associated with surface tectonic features which would not be expected to influence high Rayleigh number convection in the mantle. An example is the extensive volcanism that occurs in the western United States. Another is the series of volcanic islands adjacent to the west

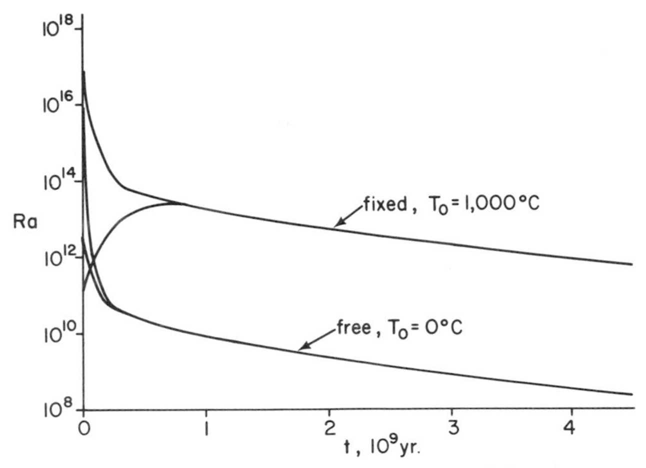

Figure 5. The mantle Rayleigh number as a function of time for the two models.

coast of Africa and the east coast of South America. The presence of these islands seem to be strongly correlated with the continental margin. Studies of seismic attenuation and travel time delays have failed to delineate upper mantle anomalies that would be expected if these volcanic centres were associated with ascending convection beneath the lithospheric plates.

Conclusions

The movement of the surface plates of plate tectonics is by definition mantle convection. The structure of the plates and the forces that drive the plates can be understood in terms of a boundary layer theory for thermal convection in an internally heated fluid. The surface plates are the thermal boundary layers of mantle convection cells.

Although the general role of thermal convection in converting heat to mantle flows is understood, there are many questions that remain unanswered. Some of the major unanswered questions are:

(1) Does significant mantle convection occur beneath the surface plates? The motion of the surface plates represents one scale of mantle convection. Estimates of mantle Rayleigh numbers give very high values. Turbulent, unsteady convection beneath the surface plates would be expected. Some authors attribute intra-plate volcanism to mantle plumes associated with this second scale of mantle. However, there is little direct evidence supporting mantle convection beneath the surface plates.

(2) Is mantle convection restricted to the upper mantle? Earthquake epicentres indicate that mantle convection occurs to depths of 700 km. An increase in mantle density at a depth of about 650 km is attributed primarily to a mantle phase change. There is little direct evidence that mantle convection penetrates this density discontinuity. However, the high Rayleigh number of

the mantle strongly favours convection through this boundary unless this boundary is of a compositional nature.

(3) Is the mantle significantly differentiated? Partial melting at ocean ridges differentiates the oceanic lithosphere. This differentiated lithosphere is subducted into the mantle at ocean trenches. Is the compositionally layered lithosphere rehomogenized by mixing processes or does differentiation occur?

(4) Is the steady-state heat balance a good approximation? What fraction of the surface heat flow is the result of radioactive heat production in the mantle and what fraction can be attributed to the cooling of the mantle? The entire surface heat flow requires only a small decrease in temperature of the mantle. However, empirical calculations using parameterized convection indicate that a large fraction (90%) of the surface heat flow can be attributed to the radioactive decay of isotopes in the mantle.

Acknowledgements

This research has been supported by the Division of Earth Sciences, National Science Foundation, NSF Grant EAR 76–82556. This is contribution 646 of the Department of Geological Sciences, Cornell University.

References

Cathles, L. M., 1975. *The Viscosity of the Earth's Mantle*, Princeton University Press, Princeton, N.J.

Davies, G. F., 1977. Whole-mantle convection and plate tectonics. *Geophys. J. R. Astr. Soc.* **49**, 459–486.

Kulacki, F. A. and Emara, A. A., 1977. Steady and transient thermal convection in a fluid layer with uniform volumetric energy sources. *J. Fluid Mech.* **83**, 375–395.

Liu, L. G., 1979. On the seismic discontinuity at 650 km. *Earth Planet. Sci. Lett.* **42**, 202–208.

McKenzie, D. P. and Weiss, N. O., 1975. Speculations on the thermal and tectonic history of the Earth. *Geophys. J. R. Astr. Soc.* **42**, 131–174.

Oxburgh, E. R. and Parmentier, E. M., 1978. Thermal processes in the formation of continental lithosphere. *Phil. Trans. R. Soc.* **A288**, 415–429.

Oxburgh, E. R. and Turcotte, D. L., 1978. Mechanism of mantle convection. *Rep. Prog. Phys.* **41**, 1249–1312.

Parmentier, E. M. and Turcotte, D. L., 1978. Two-dimensional mantle flow beneath a rigid accreting lithosphere. *Phys. Earth Planet. Int.* **17**, 281–289.

Schubert, G. and Turcotte, D. L., 1971. Phase changes and mantle convection. *J. Geophys. Res.* **76**, 1424–1432.

Sharpe, H. N. and Peltier, W. R., 1978. Parameterized mantle convection and the Earth's thermal history. *Geophys. Res. Lett.* **5**, 737–740.

Turcotte, D. L., Cook, F. A. and Willeman, R. J., 1979. Parameterized convection within the moon and the terrestrial planets. *Proc. 10th Lunar Planet. Sci. Conf.* 2375–2392.

Turcotte, D. L. and Oxburgh, E. R., 1967. Finite amplitude convection cells and continental drift. *J. Fluid Mech.* **28**, 29–42.

The Heat Flow from the Earth: A Review

H. N. POLLACK

Department of Geological Sciences, The University of Michigan, Ann Arbor, Michigan, USA

Introduction

It has been said that plate tectonic activity at the surface of the Earth is the manifestation of a great internal heat engine at work, and that the heat flux through the plates is the exhaust. From a thermodynamic viewpoint, the surface temperature and heat flux are necessary quantities for determining the efficiency of the heat engine, and it is not surprising to find that one of the earliest attempts to measure the Earth's surface heat flux was by the great thermodynamicist Lord Kelvin, more than a century ago. Utilizing both published and personal measurements of temperatures in boreholes and mines throughout Great Britain he was able to estimate a mean geothermal gradient, which when combined with thermal conductivity measurements of representative rocks led to a value of the heat flow in Britain of 70 $mW\,m^{-2}$. Remarkably, this value departs only slightly from the current estimated mean heat flow from the continents of 60 $mW\,m^{-2}$.

Continents and their marine shelves, however, comprise less than 40% of the Earth's surface, and the need for measurements in the ocean basins to enable a proper estimate of the global heat loss went unmet until the 1950s when oceanographic measurement techniques were developed. In the ensuing decades, more than seven thousand heat flow values have been obtained over the surface of the Earth, in both continents and oceans, and which have led to the recognition of rather different thermal regimes and thermal histories for continents and oceans.

Global heat flow data set

Heat flow measurements made prior to 1954 number less than seventy. Subsequently, however, the data set has grown substantially. Table I shows a brief chronology of the growth of the heat flow data collection.

TABLE I *Numbers of heat flow measurements*

Reference	Continental	Oceanic	Total
Birch (1954)	43	20	63
Lee (1963)	73	561	634
Lee and Uyeda (1965)	131	913	1044
Lee (1970)	597	2530	3127
Jessop, Hobart and Sclater (1976)	1699	3718	5417
Chapman and Pollack (1980)	2808	4409	7217

In mid-1979, at the time of preparation of this review, the data set comprised some 7200 measurements, unevenly distributed throughout the oceans and on every continent. These data are located in $5^\circ \times 5^\circ$ geographic grid elements occupying a little more than half the surface area of the Earth. Despite the uneven distribution and incomplete coverage, a number of useful results bearing on the thermal and tectonic history of the continents and oceans have been determined from the studies of the heat flow data; these results are discussed in the following sections.

Heat flow—age relationships

As soon as a modest body of heat flow observations had been accumulated it became apparent that heat flow was correlated with the crustal age of the site of the measurement. In the oceans this is the age of the basaltic crust, magmatically emplaced at and transported away from oceanic ridges. On the continents the significant age is that of the last tectonothermal event to have affected the measurement site. These empirical relationships, as presently known, are shown in Fig. 1. While both oceans and continents show a heat flow in the older segments of about 40–45 mW m^{-2}, oceanic heat flow shows a greater range over a shorter time interval than does continental heat flow.

The oceanic heat flow (q) versus age (t) curve is of the form $q \sim t^{-\frac{1}{2}}$ out to approximately 120 Myr, after which it exponentially approaches an equilibrium value. Such a curve is characteristic of the cooling of a layer from an initial high temperature, with the departure at 120 Myr indicating the presence of the equilibrium flux, likely to be derived from beneath the cooling layer. A cooling model of this type is consistent with the magmatic emplacement of new oceanic crust at mid-ocean ridges and its subsequent cooling. An implication of this model is that the layer should undergo contraction as it cools, an expectation fully supported quantitatively by oceanic bathymetry. The best present estimates (Parsons and Sclater, 1977) for the parameters of the oceanic cooling are 1350 °C for the initial temperature, and 125 km for the cooling layer thickness. The oceanic heat flow versus age relationship is somewhat obscured in the younger oceanic crust because of sea water circulating through the basalt; however measurements at sites where sediments have sealed the crust and prevented sea water penetration verify that the $q \sim t^{-1/2}$ relationship is valid in the younger crust (Anderson, Langseth and Sclater, 1977).

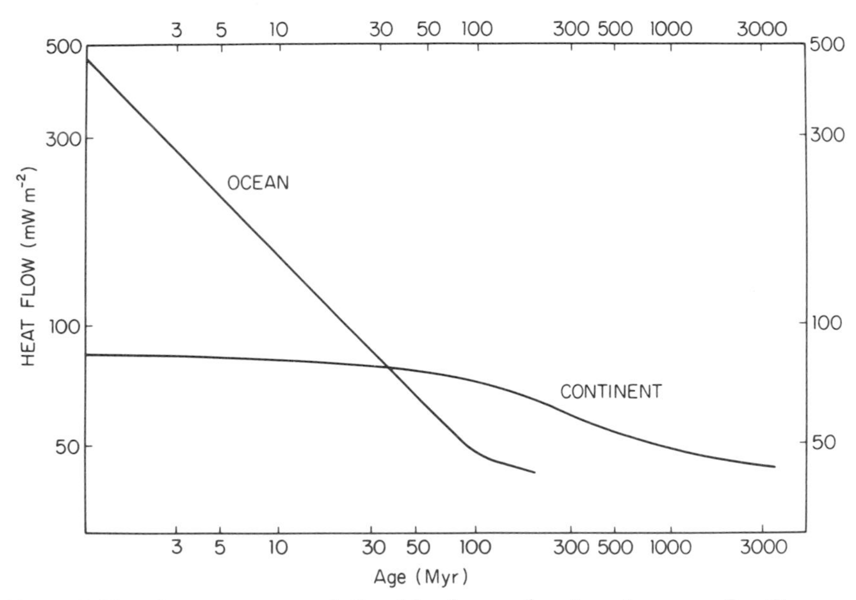

Figure 1. Heat flow versus age relationships for continents and oceans, after Chapman and Furlong (1977, and personal communication, 1979) and Parsons and Sclater (1977). (Reprinted by permission from *Nature* **285**, 393–395. Copyright © 1980 by Macmillan Journals Limited.)

The decay of continental heat flow with age extends over a longer period of time and is more complex than that of the oceans. Vitorello and Pollack (1980) argue that the continental heat flow comprises, in addition to a radiogenic component derived from the enrichment of the continental crust in heat producing radioactive isotopes, a substantial component derived from the cooling of tectonothermally mobilized lithosphere. Two time-scales enter the q versus t relationship, one characterizing the cooling and one characterizing the erosional history of the enriched crust. In addition, a "background" heat flow with no significant time variation and probably of deep origin contributes equally to the heat flow in continental terrains of all ages. The three components of continental heat flow, radiogenic, cooling and background, are shown in Fig. 2. In Cenozoic terrains the three components contribute 36, 27 and 27 mW m^{-2} respectively to the total of 90 mW m^{-2} heat flow characteristic of young regions. The cooling component is largely gone some 500 Myr after the tectonothermal event; thus the heat flow in the older Precambrian regions comprises only the radiogenic and background components. Moreover, the radiogenic component is reduced to only 18 mW m^{-2} by the removal of some of the upper crustal radioisotopes by erosion. The two components together total 45 mW m^{-2} in Archaean terrains.

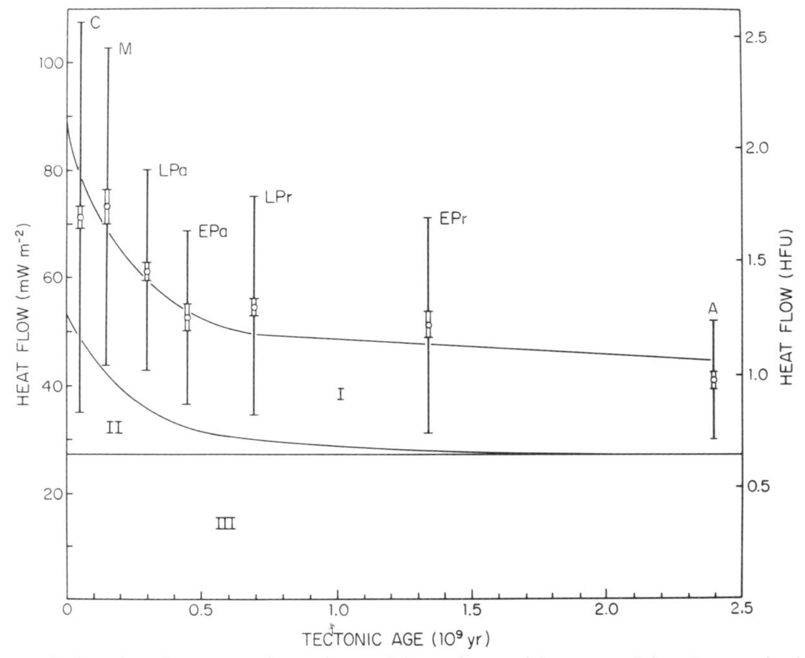

Figure 2. Secular decrease of continental heat flow with age and its three principal components: component I is radiogenic heat from the crust, component II is heat from a transient thermal perturbation associated with tectogenesis and component III is background heat flow from deeper sources (after Vitorello and Pollack, 1980). The data points are from Chapman and Furlong (1977, and personal communication, 1979). C denotes Cenozoic; M. Mesozoic; LPa, late Palaeozoic; EPa, early Palaeozoic; LPr, late Proterozoic; EPr, early Proterozoic; and A, Archaean. Points are plotted at the mean age of the respective age ranges. Double bars represent the standard error of the mean; single bars the standard deviation. The upper curve corresponds to a visually fitted curve through the data points; the middle curve corresponds to 60% of the upper curve; the lower curve represents the background heat flow from deeper sources. (Reprinted by permission from I. Vitorello and H. N. Pollack, *J. Geophys. Res.* **85** (B2), 983–995. Copyright © 1980 by the American Geophysical Union.)

The magnitude of the cooling component following a tectonothermal event has been questioned by England and Richardson (1980). They ascribe that quantity of heat flow to deep and rapid erosion which brings warm rock close to the surface and yields an augmented transient flux of roughly the magnitude and duration of the cooling component of Vitorello and Pollack (1980).

Radiogenic heat in the continental crust

Since the discovery of radioactivity in 1896 and the identification of radioisotopes in rocks in the early part of this century, it has been recognized that the heat production from radioactive decay contributes significantly to the heat flow from the Earth's interior. In fact, the entire surface heat flow on continents could arise within the crust, if the surface concentrations of the important heat-producing isotopes persisted throughout the crust.

The crustal radiogenic component of continental heat flow has in recent years been reasonably well constrained by extensive studies of crustal radioactivity in conjunction with heat flow measurements. A major empiricism, first recognized by Roy, Blackwell and Birch (1968), is that within what has come to be known as a heat flow province, the regional variation in heat flow is linearly related to the heat production of the surface rocks. The relationship is expressed as $q_0 = q_r + bA_0$ where q_0 is the surface heat flow, A_0 is the heat production of the surface rocks, q_r is the reduced heat flow (the heat flow intercept for zero heat production) and b (the slope of the line) is a quantity with dimension of depth which characterizes the vertical distribution of the heat sources. The reduced heat flow is that flux not arising radiogenically from within the crust and therefore it presumably emanates from greater depths within the Earth. A heat flow province is defined as that geographic area in which the heat flow and heat production are linearly related; each heat flow province has a characteristic reduced heat flow and source distribution parameter.

A second empiricism, first proposed by Pollack and Chapman (1977), revealed that the reduced heat flow of a province was approximately 0·6 of the mean surface heat flow of the province, suggesting that the principal source of the variation in surface heat flow between provinces is a variation in the reduced, or deeper, heat flow. The complementary requirement is that only 0·4 of the surface heat flow on continents arises from within the zone of crustal isotopic enrichment, and therefore the surface isotopic concentrations must, in some manner, diminish with depth. The observed values of the heat source distribution parameter b (in the range 4–14 km) confirm that the continental crust is not uniformly endowed with heat sources of the magnitude found in the surface rocks.

The global heat flux

The heat flow versus age relationships shown in Fig. 1 and which provide the critical constraints for thermal history models of the crust and upper mantle can be put to another useful purpose, namely estimating the heat flow in areas where no measurements have been made but for which the age of the crust is known. In a general way, the distribution of ages in the continental crust has been known for some time, and in the last two decades, with the advent of the geomagnetic reversal time-scale, large areas of the oceanic crust have been mapped on the basis of their magnetic signatures. Thus it has become possible to estimate and map the heat flow over the entire surface of the Earth. In this manner Chapman and Pollack (1975) produced the first global heat flow map. A revision by them (1980), incorporating 1800 additional data with adjustments for the more clearly recognized problems of circulation through the

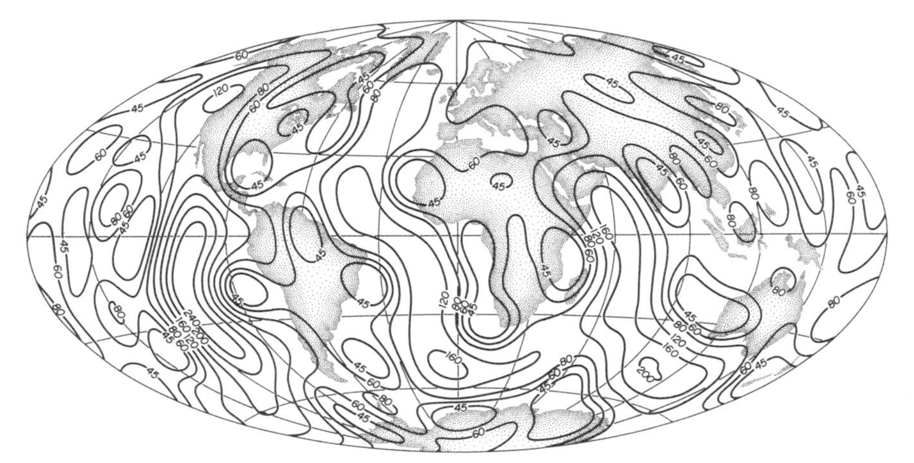

Figure 3. Degree 12 spherical harmonic representation of present-day global heat flow (from Chapman and Pollack, 1980). Contours in $mW\,m^{-2}$.

young oceanic crust, is shown here in Fig. 3. The mean heat flow through continents (including marine shelves) and oceans is 60 and $95\,mW\,m^{-2}$, respectively. The global mean is $81\,mW\,m^{-2}$ and the total heat loss from the Earth is $4{\cdot}1 \times 10^{13}$ W; fully half the Earth's heat loss comes from oceanic lithosphere produced in the Cenozoic, representing only 31% of the Earth's surface. These values are supported by similar calculations of Sclater, Jaupart and Galson (1980) who arrive at $4{\cdot}2 \times 10^{13}$ W; the slight difference arises principally from a moderately different continental heat flow–age curve and somewhat different estimates of the crustal age distribution. It should be noted that these values were anticipated by the brief but cogent analysis of Williams and Von Herzen (1974) which yielded an estimated heat loss of $4{\cdot}3 \times 10^{13}$ W.

Heat flow at past times

Just as the present-day global heat loss can be determined from the heat flow versus age relationships and the present-day distribution of crustal ages, so also can the heat loss at a past time be calculated, provided the age distribution of the crust at that time can be estimated. For the relatively immediate past, i.e. the Cenozoic and Mesozoic, the age distribution of the oceanic crust at a given time can be determined by subtracting from the present-day distribution all oceanic crust produced from that time to the present, and resurrecting and reinstating crust which has been subducted in the interval from that time to the present; the age of the restored lithosphere, however, is uncertain and must be assigned. Sprague and Pollack (1980) present models of crustal age distributions and heat flow for both oceans and continents based on the probability that crust produced or tectonothermally mobilized at an earlier time will have survived to the present day. The probability of existing oceanic crust avoiding subduction or continental crust avoiding a subsequent mobilization is reduced by an increase in oceanic accretion rates. The mean heat flow in oceans and continents, and the global flux are shown in Fig. 4 for the Cenozoic and Mesozoic. The mean global heat flow has varied between 78 and 101 mW m^{-2} compared to the present-day value of 81 mW m^{-2}. Nearly all the variation in global heat flow is due to variation in the oceanic component. Of interest is the fact that throughout most of the period the global heat flow has been within ± 5 mW m^{-2} of the present-day value. The only significant departure was the increase during the rapid sea-floor spreading of the late Cretaceous.

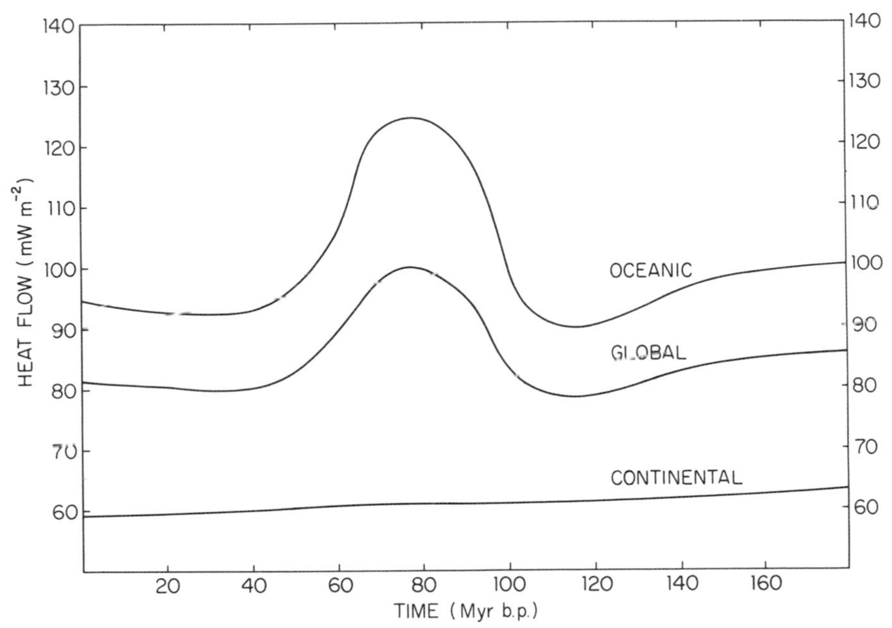

Figure 4. Heat flow in the Mesozoic and Cenozoic (from Sprague and Pollack, 1980), based on probabilistic models of crustal age distributions and the heat flow–age relationships of Fig. 1. (Reprinted by permission from *Nature* **285**, 393–395. Copyright © 1980 by Macmillan Journals Limited.)

For times earlier than the Mesozoic an analysis of the type offered by Sprague and Pollack is unfeasible because there is no ocean floor surviving to the present day from which to assess past accretion rates and survival probabilities. Moreover, as one considers the Earth's heat at times of several hundred million to billions of years before the present, the half-lives and relative abundances of the principal heat producing isotopes must be considered. The significant isotopes include ^{232}Th, ^{238}U, ^{40}K and ^{235}U, with respective half-lives of 14·0, 4·47, 1·25 and $0{\cdot}70 \times 10^9$ years. The isotopic endowment of the bulk of the Earth is unknown, but is commonly thought to be similar to that of the chondritic meteorites, because the present heat flow from the Earth is approximately equal to the heat that would be produced currently if the Earth were of chondritic composition. However, it is becoming apparent from numerical model studies of subsolidus mantle convection (Schubert, 1979) that the heat loss derives not only from radiogenic heat, but also from "initial" heat, that energy residual from the accretion of the planet and settling out of the metallic core. Thus the presumed near-equilibrium between heat production and surface heat flow, called upon as evidence of a chondritic composition, in fact may not exist. Moreover, Wasserburg *et al.* (1964) have pointed out that the rocks of the Earth's crust are potassium-deficient, relative to uranium and thorium, when compared to the chondrites. Because of the relatively short half-life of ^{40}K, the heat production derived from the chondritic and "terrestrial" models (the latter based on K/U ratios of crustal rocks) is very different in earlier times; 3×10^9 years ago the chondritic model would have been yielding more than four times its current heat production, while the "terrestrial" model would yield more than twice its present amount.

In spite of the chemical and dynamical uncertainties, it is probably safe to say that the heat loss from the Earth 3×10^9 years ago was at least twice as great as the present-day heat loss. However, that is not to say that the heat flow was increased uniformly over the surface of the Earth. Bickle (1978) argues persuasively that the increased heat loss is likely to be manifest principally as an increase in the rate of oceanic plate production, not accompanied by a significant augmentation of the heat flow through the continents, the latter assertion being supported by inferred temperature gradients in Archaean metamorphic terrains. If the greater heat loss was principally through the oceans, then the mean age of ocean floor 3×10^9 years ago would be approximately 10–15 Myr, compared to 62 Myr presently.

Summary

Measurements of the amount of heat being conducted to the Earth's surface from the interior, undertaken in earnest only two decades ago, now total more than 7000, unevenly distributed over the globe. One of the early generalities to emerge from the study of these data is that in both oceans and continents, heat flow diminishes with crustal age. A similar quasi-equilibrium heat flow is reached in both the old ocean basins and ancient continental nuclei, but the decay of heat flow with age in oceans and continents is governed by different processes with different time-scales.

The oceanic curve is explained adequately by the cooling of lithosphere, magmatically emplaced at and transported away from oceanic ridges. The continental heat flow derives in part from radiogenic heat produced within the crust, in part from the cooling following a tectonothermal event, and the remainder from deeper sources.

Heat flow provinces have been delineated on the continents, on the basis of a correlation between surface heat flow and radiogenic heat production of surface rocks; each province displays a characteristic heat flow (the "reduced" heat flow) from depths below the zone of crustal isotopic enrichment. The heat flow variation within a heat flow province derives from the regional variability in the enriched zone; the principal difference between heat flow provinces is in the magnitude of the reduced heat flow.

The heat flow–age relationships, in conjunction with the age distribution of the Earth's crust, can be used to estimate the heat flow in areas without measurements and thus yield a determination of the Earth's mean heat flow and total heat loss. These values presently are 81 $mW\,m^{-2}$ and $4{\cdot}1 \times 10^{13}$ W, respectively. Probabilistic models of crustal-age distributions over the past 180 Myr suggest that the Earth's mean heat flow has varied between 78 and 101 $mW\,m^{-2}$. In the Precambrian the heat loss was without doubt greater due to the greater heat production inferred from the half-lives of the significant heat sources. Three billion years (3×10^9) ago the heat loss was probably two to four times that of the present, and likely to have been concentrated principally in the oceans where plate production and consumption proceeded at a much faster pace than at present.

Acknowledgement

The preparation of this paper was supported in part by the Division of Earth Sciences, U.S. National Science Foundation, under NSF Grant EAR 78–09131.

References

Anderson, R. N., Langseth, M. G. and Sclater, J. G., 1977. The mechanisms of heat transfer through the floor of the Indian Ocean. *J. Geophys. Res.* **82**, 3391–3409.

Bickle, M. J., 1978. Heat loss from the Earth: a constraint on Archean tectonics from the relation between geothermal gradients and the rate of plate production. *Earth Planet, Sci. Lett.* **40**, 301–315.

Birch, F., 1954. The present state of geothermal investigations. *Geophysics* **19**, 645–659.

Chapman, D. S. and Furlong, K., 1977. Continental heat flow–age relationships. *EOS Trans. Am. Geophys. Union* **58**, 1240.

Chapman, D. S. and Pollack, H. N., 1975. Global heat flow: a new look. *Earth Planet. Sci. Lett.* **28**, 23–32.

Chapman, D. S. and Pollack, H. N., 1980. Global heat flow: spherical harmonic representation. *EOS Am. Geophys. Union* **61**, 383.

England, P. C. and Richardson, S., 1980. Erosion and the age-dependence of continental heat flow. *Geophys. J. R. Astr. Soc.* **62**, 421–438.

Jessop, A. M., Hobart, M. A., and Sclater, J. G., 1976. The world heat flow data collection – 1975. *Geothermal Series* 5, Energy, Mines and Resources, Earth Physics Branch, Ottawa, Canada.

Lee, W. H. K., 1963, Heat flow data analysis. *Rev. Geophys. Space Phys.* **1**, 449–479.

Lee, W. H. K., 1970. On the global variations of terrestrial heat-flow. *Phys. Earth. Planet. Int.* **2**, 332–341.

Lee, W. H. K. and Uyeda, S., Review of heat flow data. In *Terrestrial Heat Flow* (Ed. Lee, W. H. K.), *Am. Geophys. Union, Geophys. Monogr.* **8**, 87–100, Am. Geophys. Union, Washington, D.C.

Parsons, B. and Sclater, J. G., 1977. An analysis of the variation of ocean floor bathymetry and heat flow with age. *J. Geophys. Res.* **82**, 803–827.

Pollack, H. N. and Chapman, D. S., 1977. On the regional variation of heat flow, geotherms, and the thickness of the lithosphere. *Tectonophysics* **38**, 279–296.

Roy, R. F., Blackwell, D., and Birch, F., 1968. Heat generation of plutonic rocks and continental heat flow provinces. *Earth. Planet. Sci. Lett.* **5**, 1–12.

Schubert, G., 1979. Subsolidus convection in the mantles of terrestrial planets. *Ann. Rev. Earth Planet. Sci.* **7**, 289–342.

Sclater, J. G., Jaupart, C. and Galson, D., 1980. The heat flow through oceanic and continental crust and the heat loss from the Earth. *Rev. Geophys. Space Phys.* **18**, 269–311.

Sprague, D. and Pollack, N. H., 1980. Heat flow in the Mesozoic and Cenozoic. *Nature* **285**, 393–395.

Vitorello, I. and Pollack, H. N., 1980. On the variation of continental heat flow with age and the thermal evolution of continents. *J. Geophys. Res.* **85**, 983–995.

Wasserburg, G. J., MacDonald, G. J. F., Hoyle, F. and Fowler, W. A., 1964. Relative contributions of uranium, thorium, and potassium to heat production in the Earth. *Science* **143**, 465–767.

Williams, D. L. and Von Herzen, R. P., 1974. Heat loss from the Earth: new estimate. *Geology* **2**, 327–328.

Some Comments on the Mechanism of Continental Drift

S. K. RUNCORN

School of Physics, The University, Newcastle upon Tyne, UK

It can hardly be denied that the question of the nature of the forces that move the plates is the most important challenge facing geophysicists today. It is amusing to reflect that the absence of a mechanism was widely held to be a most weighty objection to accepting the geological or palaeomagnetic evidence for continental drift in the 1950s and earlier, whereas the "theory" of plate tectonics is now accepted without any consensus about the geophysical mechanism responsible. Of course the fact that the movement of the plates can be determined in so much detail over the last 10^8 yr from the interpretation of the ocean magnetic anomalies has concentrated so much attention on the properties of plates that their motion has been assumed to be an inherent property. In science a phenomenon or a hypothesis can become so familiar, and its utility in providing an explanation, or consistent description, of a great number of diverse facts so evident, that the underlying mechanism may often be left unstudied. The main geomagnetic field was an example of such a phenomena, for when Chapman and Bartels published their classic work only 10 pages out of 1007 were devoted to its explanation. Evolution is an example of the latter.

This emphasis on the properties of the plates led to the hypothesis of gravity sliding. Hales (1969) argued that the plates move away from the ocean ridge crests because the gravity component on the lithospheric plates on the flanks of the oceanic rises pushes the plates away from them. An alternative likewise appealed to gravity sliding: the seismic evidence that deep earthquakes near the trenches result from the stresses in the edges of the lithospheric plates descending into the mantle. The colder lithosphere, chemically the same as the hotter underlying asthenosphere, can reasonably be supposed to be denser: thus the lithosphere plate edges are sinking under gravity and can be supposed to be pulling the rest of the plate towards the trench and away from the ridge. In these theories, the mantle plays a quite passive role. It must be considered over geological time to behave as a fluid, yielding indefinitely under small stresses and its viscosity is taken to be 10^{21} poise, from the analysis of the uplift of Fennoscandia, and parts of the Canadian Arctic following the removal of ice and Lake Bonneville on the draining of the water from it. I have shown (Runcorn, 1974) that simple calculations show that a force is required of about 10^{15} dyne per cm length of the plate edge to move it over the mantle with the observed speeds, and that either of the two gravity mechanisms seem adequate to provide this. The weakness of these types of theory – it is, I think, a fatal weakness – is that no light is thrown on how the process of continental drift begins: the initiation of the break-up of Pangea is not explained. Supporters of gravity sliding – perhaps the majority of geophysicists – see it as a process readily understood and relying on features, the existence of which is not disputed. Estimates of the forces at trenches and ridges from energy released by earthquakes (Forsyth and Uyeda, 1975) support the quantitative requirements. Many advocates of gravity sliding recognize the fundamental argument in favour of mantle convection, i.e. that it is the general way by which, in Nature, heat is transformed into motion. They meet this argument by saying that the gravity sliding of plates is really a convection process in which the differentiation of magma at the ridge crests and the rising of hot mantle to produce the topography of the oceanic rises and the descending of the cold lithosphere at the trenches, is a kind of convection, though chemical as well as thermal in nature. But to say it again: the fatal flaw in all such theories is that they give no explanation of the process by which the present directions and motions of the plates began during the breakup of Pangea around 100 My ago.

Forsyth and Uyeda (1975) have calculated from earthquake energies the forces at the plate boundaries and argue that further quantitative support for the mechanisms which we have discussed can thus be found. These arguments appear to me to be like taking one element of an electric circuit, say a resistance R and measuring the current I and voltage V across it and demonstrating that $V = IR$. This demonstration does not locate the e.m.f. driving the current through the circuit!

Nevertheless very interesting work has been done, and papers written, stimulated by this philosophy. For example, Hager and O'Connell (1979) have asked what flow pattern would be induced in the mantle given that the plates are moving with the velocities inferred from the ocean-floor data. They show that the drag of the plates produces a flow combined of poloidal and toroidal terms, which has the expected sinking and convergence at the trenches and upwelling and divergence near to the ocean ridges: consistent with observations. But in a fundamental sense they fed this information into their model in making a spherical harmonic analysis of the surface flow vectors so that they seem to have elegantly demonstrated in essence a tautological truth. However, their work is important as it demonstrates clearly how toroidal flow, i.e. one with zero radial velocity component, can be generated in the Earth's mantle. Whereas poloidal flow is generated by the buoyancy forces associated with convection, toroidal flow is produced by movement of the lithosphere over the mantle, such as might be produced, entirely hypothetically, by some external force such as tidal friction. If the driving force of plate motions is, as I believe, convection, then Hager and O'Connell have shown that due to the breaking of the lithosphere into plates, a toroidal flow is also generated in the mantle. However, while the dynamic character of the poloidal flow is caused by density anomalies in the mantle and these have an external expression in the gravity field, the toroidal flow appears not to have any external consequence: it is mainly a secondary not primary phenomenon. Hager and O'Connell (1979) themselves appear to lean to the view that the plate motions are the result of the gravity sliding, discussed above, in which case they can offer no explanation of the low harmonics of the geoid, or would have to seek an explanation in the distortion of the lithospheric plates due to their interaction with each other and the flow which they induce in the mantle. The down warping of the Indian–Australian plates in the region of the central Indian ocean due to its collision with the Asian plate has been suggested as an explanation of the pronounced low in the geoid.

The concept of hot spots has aroused much interest (Morgan, 1971). Relative motion between the plates and these sources of magma seemed to be a simple explanation of the age relationship of the Hawaiian and other Pacific island chains. Searches for an absolute frame of reference in which the hot spots were stationary were undertaken. The development of the hot spot into a mantle plume was the result of assuming that the only absolute frame of reference was the lower mantle, which was assumed to be not convecting. The mantle plume seems utterly implausible. The use of the word plume suggests an analogy with the atmosphere where columns of hot air can rise from a chimney retaining their small horizontal dimensions to great heights. This arises because the viscous term is negligible compared to the inertia term: but in the mantle it is 20 orders of magnitude greater. Nevertheless, the fact that the relative rate of movement of the hot spots is an order of magnitude slower than the plates requires explanation.

The problem with the hypothesis of mantle convection has been the difficulty of determining its pattern by geophysical methods. The reason is that it is driven by extremely small density and temperature differences. If a mantle viscosity of 10^{21} poise is assumed, I showed (Runcorn, 1962, 1964) that density differences $(\Delta\rho) = 10^{-5}\,\mathrm{g/cm^3}$ would produce currents of the order of 10^{-6}–10^{-7} cm/s capable of moving the plates with the observed velocities. I also showed that supposing these density differences arose from thermal convection, the temperature differences are 1 °C, and that such values give a heat transport of the order of that observed by the geothermal gradient. It seems clear that these lateral inhomogeneities are below the level of detection by seismic methods: the travel time anomalies expected are below those observed.

When the existence of long wavelength departures of the geoid from hydrostatic equilibrium was proved by the tracking of Earth satellites, thus confirming Jeffreys' (1959) finding from spherical harmonic analysis of surface determinations of g, it was natural for these celestial mechanics experts to follow Jeffreys also in his interpretation. Indeed the simplest explanation appears to be that the density anomalies were the slight effects to be expected as a result of the origin and differentiation of the Moon and Earth and retained since by the finite strength of the mantle. Such an explanation ignored the phenomena of solid state creep, which had not hitherto played any role in the thinking of geophysicists. The role of solid state creep and its exponential dependence on temperature was recognized by Runcorn (1962) as the fundamental element in any attempt to explain continental drift and existence of a lithosphere which could support loads of relatively short wavelength.

Once the geoid had been recognized as due to thermal convection, it was clear that it should provide the most powerful tool in understanding its pattern. The highs and lows of the geoid must lie above the upgoing and downgoing limbs of the convection. There were however two difficulties which prevented widespread recognition of this. Firstly, at the time the geoid was first determined, it was commonly accepted that convection in the mantle was upgoing under the ocean ridges and descending at the trenches. The second difficulty was that it was variously believed that the uprising limb of a convection pattern was always associated with a positive gravity anomaly – the upward bulge of the surface outweighing the smaller density of the hotter uprising current. The concept that the boundaries of the convection cells were beneath the plate edges soon ran into trouble. The transverse displacements of the mid-Atlantic ridge did not appear to fit the idea that the rift was the locus of the uprising current. Moreover the movement of the mid-Atlantic ridge to the west of the Carlsberg ridge to the east relative to Africa since the break-up of Gondwanaland indicated on the above hypothesis an expanding convection cell and moreover the descending limb under Africa was not associated with compression but rather with extension (in the Rift Valley). These difficulties were cleared away by plate tectonics and the recognition that the volcanism and topography of the ocean ridges were the natural consequence of flow and differentiation of the uppermost mantle as the lithosphere parted. Indeed the success of this idea as the main reason for neglect of mantle convection.

The velocity of convection currents in the mantle must be greater than the plate motions, the greatest of which is about 2×10^{-7} cm/s: the currents might reasonably be supposed to reach 10^{-6} cm/s. Assuming mantle wide convection, cells of 3×10^{-8} cm extent an overturn time of 30 Myr is obtained. It seems therefore likely that the pattern of convection has been constant since at least the break-up of Gondwanaland. The plates therefore must move over the convection pattern as a result of the net viscous stress which the pattern exerts on each plate. The absence of any simple relation between the geoid and the plate boundaries is therefore explained. This stationary convection pattern therefore provides the natural frame of reference for the hot spots. These must be interpreted as being lava brought up in the ascending currents, probably from greater depth in the mantle than the mid-ocean ridge basalts, which must have a more local provenance deriving from the uppermost mantle. The much studied geochemical differences between the mid-ocean ridge basalts and the mid-plate volcanism would be entirely consistent with these ideas. The geochemists have been led to interpret their data by models which, while satisfying the geochemical evidence, do not seem at all adequate geophysically. The plume concept gained support from the geochemical evidence: Schilling (1973) showed that the trace element occurrences in Iceland and in certain other places was distinct from that of normal ridge volcanism and thought it natural to suppose that the "plumes" were tapping mantle material in a less differentiated state than that of the upper mantle, presumably the asthenosphere, tapped by the ocean ridges: Wasserburg and de Paulo (1979) also require for their geochemical evidence two mantle reservoirs, one from which the ocean ridge basalts are derived and depleted in certain elements and one for the "hot spots" characterized by more nearly chrondritic abundances of trace elements. They propose naturally that the upper mantle is the former and for the latter they suggest the lower mantle which they assume is still undifferentiated primeval material. O'Nions, Hamilton and Evensen (1977) come to a similar model. A general point about all models based on a body of data from one discipline, however important, must be made: geoscientists should have learnt from experience that such models are not likely to be true and are highly misleading. Examples abound: why have so many geoscientists believed that the lower mantle is rigid or possesses a very high viscosity (the value of 10^{26} poise is listed in very many textbooks as if it had been actually measured). In the first determination of the Earth's gravitational field by the tracking of satellites, the actually ellipticity was shown to be greater than that calculated on the hydrostatic theory. An "instant" explanation was given in terms of the slowing down of the Earth's rotation by tidal friction: a high viscosity of the lower mantle enables it to remember the angular rotation rate of some 10 Myr ago. Yet a non-convecting lower mantle was not compatible with the requirement that heat must be transported away from the core much more effectively than molecular conduction could do. It seems characteristic that a model which serves only one group of data can be deceptive: it provides perfect satisfaction for one goal. Just because the model can be so well specified, it fails in its broader objective. It appears so precise just because it is so limited.

References

Forsyth, D. W. and Uyeda, S., 1975. On the relative importance of driving forces of plate motion. *Geophys. J. R. Astr. Soc.* **43**, 163–200.

Hager, B. H. and O'Connell, R. J., 1979. *J. Geophys. Res.* **84**, (B3), 1031–1048.

Hales, A. L., 1969. Gravitational sliding and continental drift. *Earth Planet. Sci. Lett.* **6**, 31–34.

Jeffreys, H., 1959. *The Earth*, Cambridge University Press, Cambridge.

Morgan, W. J., 1971. Convection plumes in the lower mantle. *Nature* **230**, 42–43.
O'Nions, R. K., Hamilton, P. J. and Evensen, N. M., 1977. *Earth Planet. Sci. Lett.* **34**, 13–22.
Runcorn, S. K., 1962. Convection currents in the Earth's mantle. *Nature* **195**, 1248–1249.
Runcorn, S. K., 1964. Satellite gravity measurements and a laminar viscous flow model of the Earth's mantle. *J. Geophys. Res.* **69**, 4389–4394.
Runcorn, S. K., 1974. On the forces not moving lithospheric plates. *Tectonophysics* **21**, 197–202.
Schilling, J. G., 1973. *Nature Phys. Sci.* **242**, 2–5.
Wasserburg, G. J. and de Paulo, D. J., 1979. Models of Earth structure inferred from neodymium and strontium isotopic abundances. *Proc. Nat. Acad. Sci. USA* **76**, 3594–3598.

Rheology, Plate Motions and Mantle Convection

B. H. HAGER

Seismology Laboratory, California Institute of Technology, Pasadena, California, USA

R. J. O'CONNELL

Department of Geological Sciences, Harvard University, Cambridge, Massachusetts, USA

Introduction

Since its serious proposal at the beginning of the century, the hypothesis of continental drift has been progressively refined, culminating in the last decade as the theory of plate tectonics. Over fifty years ago, Holmes (1928) proposed that thermal convection powered by the heat liberated by radioactive decay was responsible for the relative motions of the continents. For a time, the hypothesis of thermal convection in the mantle, along with that of continental drift, was highly controversial. We now know, however, that thermal convection in the mantle is even more vigorous than the rather slow overturning proposed by Holmes. Rather than being confined beneath a strong shell at the surface of the Earth, the material in the deep interior actually reaches the surface at the axes of the world-encircling system of mid-oceanic ridges. Rather than being rafted around like the buoyant scum on a jam pot, the plates at the Earth's surface actually form an important part of the convecting system.

Our present understanding of plate motions and plate structure explains far more than the observation that continents move. Sea-floor topography, heat flow, the undulations of the geoid over oceanic ridges, and the pattern of magnetic anomalies observed over the sea-floor can all be understood in terms of the cooling of the material which has come up from the deep interior of the mantle to form the oceanic plates at the spreading centres.

The motions of the plates have a dominant influence on the chemical and thermal state of the Earth. Most of the heat flowing from the Earth results from the cooling of the plates as they move along the surface. The chemical differentiation of the Earth is mainly a result of tectonic activity, which to a large extent is governed by plate motions. To understand the thermal and chemical evolution of the Earth, as well as its present state, it is necessary to understand the dynamics of plate motions. This involves understanding both what determines the positions of plate boundaries and, given the distribution of plate boundaries, what determines the velocities of the plates.

Any proposed driving mechanism must provide the energy associated with plate motions. The average elastic energy release from earthquakes associated with plate motions is $1{\cdot}4 \times 10^{10}$ W, while the total strain energy associated with earthquakes is at least 10 and may be up to 100 times this great (Kanamori, 1978). Holmes (1965) estimated the energy dissipated in volcanism as 10^{11} W, while the dissipation associated with metamorphism is probably twice this amount (Oxburgh and Turcotte, 1978). Thus plate motion and the associated tectonic activity require at least 5×10^{11} W.

Thermal convection is the most plausible agent causing motion of the plates. The present-day heat flux of about 4×10^{13} W is much greater than the dissipation associated with plate motions (O'Connell and Hager, 1980). Three-quarters of this heat flux results from the cooling of the lithosphere as it moves away from spreading centres. The cooling of the lithosphere leads to large-scale horizontal density contrasts, which must tend to drive flow in the mantle (Lister, 1975; Hager, 1978). Thus plate motion itself is necessarily a form of thermal convection.

Convection has been extensively studied theoretically, in the laboratory and by numerical simulation, usually for materials of constant properties. A boundary layer theory developed by Turcotte and Oxburgh (1967) shows that thermal convection may be capable of generating motions in the Earth's mantle with velocities comparable to the observed plate velocities. According to the boundary layer theory the horizontal velocity of a convecting fluid, u, is given by

$$u = 0{\cdot}142 Ra^{2/3} \kappa / d, \tag{1}$$

where κ is the thermal diffusivity of the fluid, d is the depth and Ra is the Rayleigh number, defined as

$$Ra = g\alpha \Delta T d^3 / \kappa \nu. \tag{2}$$

Taking as representative values: coefficient of thermal expansivity, α, 3×10^{-5}/K; gravitational acceleration, g, 10^3 cm/s^2; superadiabatic temperature change, ΔT, 2000 K; thermal diffusivity, κ, 10^{-2} cm^2/s; kinematic viscosity, ν, 10^{22} cm^2/s; gives horizontal velocities of 10 cm/yr if convection extends throughout the mantle. Observed plate velocities are of this order, so it seems reasonable that thermal convection in some form is responsible for plate motions.

Whenever possible, we should test geophysical theories using other planetary bodies. Applying the above analysis to the Moon predicts that lunar surface velocities should be on the order of 10 cm/yr, yet the surface of the Moon has not undergone extensive motion for at least the last 4 Gyr. Also, in laboratory experiments at Rayleigh numbers appropriate for the parameter values given above for the Earth, the surface velocity typically varies over horizontal distances of the order of the thickness of the thermal boundary layer, which for the Earth is of the order of the plate thickness, about 100 km (cf. Elder, 1970). Surface velocities on the Earth are constant over distances of thousands of kilometres. It is apparent that at least some results of laboratory convection experiments cannot be applied directly to the Earth.

The reason that extrapolation from laboratory experiments to the Earth fails to reproduce such first-order observations as the coherent motion of plates over large distances is that the rheology of mantle material is highly dependent upon temperature and stress. Figure 1 summarizes theoretical estimates of the rheology of olivine, the mineral believed to be the major constituent of the uppermost mantle. This map is similar to the deformation maps of Ashby and Verrall (1977) but uses contours of effective viscosity (stress/strain rate) rather than strain rate. The effective viscosity, defined as the ratio of stress to strain rate, is plotted as a function of shear stress and depth. The temperature profile used (believed to be appropriate to the oceanic upper mantle at an age of 80 Myr) is shown at the top of the figure.

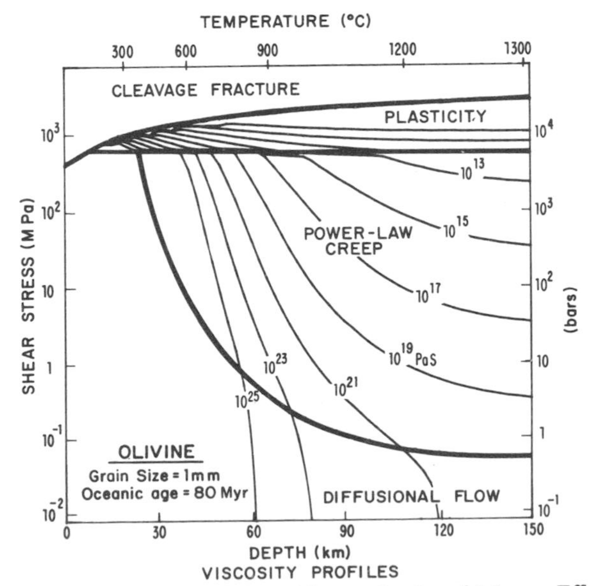

Figure 1. A deformation map for olivine with a grain size of 1·0 mm. Effective viscosity contours (stress/strain rate) are plotted as a function of depth for the upper 150 km of the Earth. The geotherm used is from the boundary layer theory for oceanic plates and is appropriate to a lithospheric age of 80 Myr.

In the lower section of the map, viscosity is independent of stress, and diffusional flow is the prevailing deformation mechanism. At somewhat higher stresses, dislocation motions become important, and the effective viscosity depends inversely upon the square of the stress. At still higher stresses, plastic deformation and fracture become the dominant mechanisms of deformation.

For a given stress, in the upper few tens of kilometres, the effective viscosity is very high. As the depth increases, there is a very rapid decrease in effective viscosity until an approximately constant value is reached. The thermal boundary layer at the surface (the lithosphere) is quite clearly mechanically distinct; because of the strong constrast in rheology between the lithosphere and the mantle below, it is meaningful to consider the lithosphere as a mechanically separate unit. Nonetheless, it is always necessary to bear in mind that motion of the lithosphere is strongly coupled to the more fluid mantle below.

No significant deformation will occur over the lifetime of a plate (50 Gyr) in the lithosphere, where the effective viscosity is greater than 10^{27} poise. For stresses greater than 5 kbar, the lithosphere deforms rapidly, and essentially fails catastrophically. Thus the lithosphere behaves more like an elastic–plastic body than like a viscous fluid; for this reason, the results of fluid mechanical models cannot be extended to the plates in a straightforward way, and convection in the mantle cannot be parameterized in terms of the Rayleigh number alone.

Although the lithosphere itself does not behave as a fluid, the sub-lithospheric mantle can usefully be regarded as fluid, although it may not behave in a Newtonian manner. The stress, τ, generated by convection in this fluid can be estimated as

$$\tau \sim \eta u/d, \tag{3}$$

where η is the viscosity of the fluid. Using our previous estimates of u and assuming η is 10^{22} poise, stresses on the order of 10 bars are likely. This is far below the expected yield stress of the lithosphere, even if the stress is coherent over a distance of order d. Fracture of the lithosphere and the initiation of new plate boundaries may be indirectly the result of fluid instability and convection in the mantle below, but they are primarily determined by the mechanical properties of the lithosphere. Otherwise it is hard to understand why the young ($\sim$25 Myr) lithosphere of the Cocos plate is vigorously subducting, while the very old ($\sim$150 Myr) lithosphere off the east coast of North America is stable, even though it is loaded by a thick sediment prism and already torn at the Puerto Rico trench.

The determination of what conditions are necessary to initiate new plate boundaries is one of the most important, but also one of the more difficult, problems in Earth sciences. We will focus here on the problem of what is the distribution of forces which drives the present plates with their observed velocities, taking the geometries as given. Although it is not possible at this time to give a complete description of plate dynamics, by understanding the forces acting on the plates at present we can place constraints on Earth structure and gain insight into the thermal and chemical evolution of the Earth.

The forces responsible for plate motions ultimately result from density variations within the convecting mantle. The plates, of course, form an important part of this convecting system, both because the strength of the plates suppresses small-scale variations in surface velocity, averaging motions over areas corresponding to the lithospheric plates, and because the temperature gradients in the lithosphere are large in comparison to other temperature gradients in the mantle. In addition to being large, the horizontal density contrasts in the plates are well constrained in comparison to other density contrasts in the mantle. It is possible to quantify the "driving forces" within the lithosphere since we know reasonably well the age of the sea-floor and geometries of subducting slabs and understand the thermal evolution of the lithosphere, at least in broad terms.

Although the density contrasts in the lithosphere are the only ones which are fairly well constrained, there are undoubtedly density contrasts in the mantle which are not confined to the lithosphere. For example, the magnetic field of the Earth is almost certainly a result of convection within the core, so heat must flow from the core into the base of the mantle. Gubbins (1977) estimated that between 10^{11} and 10^{13} watts, or up to a quarter of the surface heat flow, may originate in the core. Since the area of the core–mantle boundary is about a quarter of that of the surface of the Earth, the heat flux across these two boundaries may be comparable.

If all things were equal, the thermal boundary layer at the core–mantle boundary would be similar to that at the surface, leading to significant large-scale horizontal density contrasts. However, the effect of the temperature dependence of viscosity is to stabilize the relatively cold surface boundary layer (lithosphere) and to destabilize the hot boundary layer at the core–mantle boundary. We might expect, then, that the boundary layer at the core–mantle boundary would exhibit behaviour like that observed for laboratory experiments at high Rayleigh numbers. In these experiments, the boundary layer is highly unstable and breaks apart in the form of plumes (Elder, 1970). The spacing of these plumes is governed by the thickness of the boundary layer (~ 100 km) rather than by the depth of the convecting system (~ 3000 km). It seems likely that convection driven by heat flowing from the core would lead to small-scale motions, rather than to motions coherent on the scale of plates.

The geometry of the plates also suggests that ridges are passive features controlled by the lithosphere, rather than being the focus of coherent flow upwelling from the core–mantle boundary. Transform faults offset ridges in a discontinuous manner and ridges have shown discontinuous jumps. Ridges have also been subducted, as when the Aleutian arc overran the Kula ridge (Grow and Atwater, 1970). Both the Antarctic and the African plates are nearly surrounded by ridges, again an indication that there is probably not a one-to-one correspondence between ridges and thermals upwelling from the core–mantle boundary.

The model of plate motions and mantle convection which appears most plausible in the light of the temperature dependence of rheology and the observed large-scale coherence in the motion of the plates is one in which the large-scale flow in the mantle which must accompany the observed plate motions, as well as the motions of the plates themselves, is governed by the strength of the lithosphere and the density contrasts within the lithosphere. Smaller scale density contrasts probably do not have an important influence on plate motions, although they may result in motions on a smaller scale.

Although this model is qualitatively plausible, it remains to demonstrate that it works quantitatively. Because the density contrasts within the lithosphere are fairly well understood, it is possible to test this model. In this chapter we outline a simple numerical model of mantle flow and plate motion that takes into account current knowledge of the thermal evolution of the plates and thermal effects of subduction, as well as the interaction between plate motion and flow in the underlying mantle. The model, which is described in more detail elsewhere (Hager and O'Connell, 1980), attempts to evaluate explicitly the forces acting on the plates and to determine the extent to which the known density inhomogeneities associated with the creation of plates at ridges and their subsequent subduction can account quantitatively for the observed plate motion and the accompanying global circulation in the interior. Such a simple model can adequately account for much of the plate motion. Although our results do not rule out more complex models, they do determine the important contribution of known and well-constrained processes such as lithospheric creation and subduction and viscous coupling of the plates to the mantle flow that would be included in any realistic model of mantle convection.

Plate characteristics

The most obvious observation of mantle convection is the pattern of surface velocities. Figure 2 shows recent estimates of the velocities of the major plates (Minster and Jordan, 1978). The absolute reference frame shown here assumes no net rotation of the lithosphere, but other reference frames give nearly the same absolute velocities (Solomon, Sleep and Richardson, 1975; Minster and Jordan, 1978).

Several generalizations can be made about plate motions (cf. Forsyth and Uyeda, 1975). Absolute plate velocities form a bimodal distribution. Plates with significant fractions of their boundaries made up of subducted slabs

(Pacific, Indian, Nazca and Cocos) move at greater than 6 cm/yr. Those plates without major subducted slabs (North American, South American, African, Eurasian and Antarctic) have velocities under 3 cm/yr. Plate velocity is independent of area and shape – the tiny triangular Cocos, the sickle-shaped Indian and the large Pacific plate have essentially identical velocities. Slowly moving plates tend to have more continental area than rapidly moving ones, although the rapidly moving Indian plate shows that continental plates are capable of moving as fast as oceanic plates. There is no apparent correlation between transform fault length and plate velocity.

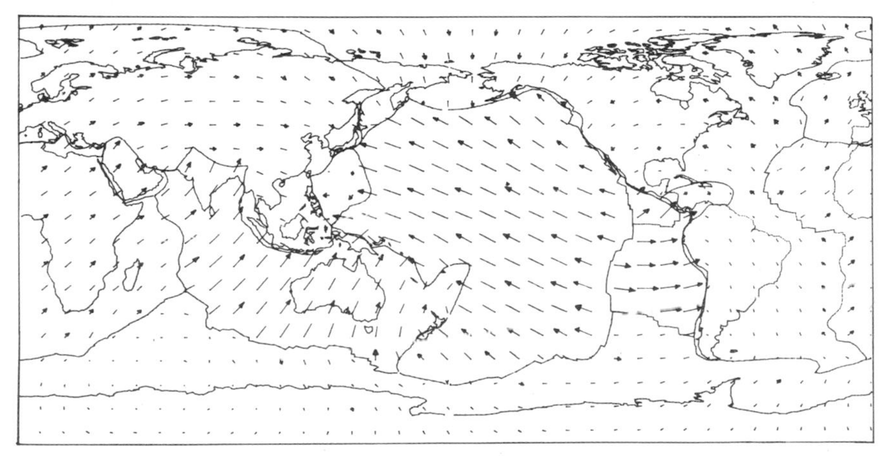

Figure 2. Spherical harmonic expansion through degree and order 40 of the observed plate velocities, from Minister and Jordan (1978), in the no net rotation frame. Plotted displacements are extrapolations of instantaneous plate velocities for 10 Myr.

These observations suggest that subducted slabs are important in determining plate velocities. Since the density contrast between the lithosphere and the interior of the Earth depends on the age of the lithosphere, it is of interest to determine the ages of the subducted slabs. From magnetic isochrons we see that the age of the Cocos plate presently subducting is from 0 to 40 Myr. The age of the subducting Nazca plate is for the most part between 35 and 55 Myr. The age of the Indian plate is poorly constrained, but for the most part is 80 Myr or greater, while the age of the Pacific plate in places exceeds 130 Myr (Pitman, Larson and Herron, 1974).

If the thickness of the lithosphere is determined by the depth to which cooling penetrates into the mantle, its thickness should be proportional to the square root of its age, as is explained in more detail in the following section. In analogy to Stokes' law for a sinking body we might expect the velocity of a slab sinking into the mantle to depend on its cross-sectional area which increases with the age of the slab. That plate velocity is independent of slab age is another observation which must be explained by any quantitative model of mantle dynamics.

Density contrasts in the surface plates

The thermal structure of oceanic lithosphere has been explained in terms of models involving cooling of a boundary layer (Turcotte and Oxburgh, 1967; Davis and Lister, 1974), freezing of partial melt at the lithosphere–asthenosphere boundary (Parker and Oldenburgh, 1973), and cooling of a plate of finite thickness (Langseth, LePichon and Ewing, 1966; McKenzie, 1967; Sclater and Francheteau, 1970). Near the ridge crest these models give equivalent solutions (Parsons and Sclater, 1977) for the temperature T as a function of depth z and lithospheric age t:

$$T(z) = T_0 \operatorname{erf}[t/2\sqrt{(\kappa t)}]. \tag{4}$$

Here T_0 is the initial uniform temperature of the lithosphere. If the base of the lithosphere is defined as the depth at temperature T_l, then the lithosphere thickness h is

$$h = \sqrt{(\kappa t)} \operatorname{erf}^{-1}(T_l/T_0). \tag{5}$$

It is a property of the error function solution that the average temperature of the lithosphere, and thus its average density (considering only thermal effects), is independent of age.

Since gravity anomalies over ocean ridges and basins are small (cf. Kaula, 1972) the lithosphere is close to isostatic equilibrium and the excess mass of the lithosphere as it thickens while moving away from the ridge is nearly exactly compensated by the subsidence of the lithosphere into the hotter and less dense mantle below. The increase in depth, d, of the ocean floor as the plate moves away from the ridge is directly proportional to the thickness of the lithosphere:

$$(\rho_l - \rho_w)\, d = (\rho_l - \rho_a)\, h. \tag{6}$$

Here ρ_l, ρ_w and ρ_a are the densities of the lithosphere, of seawater and of the asthenosphere.

The increase in ocean depth away from spreading centres varies linearly with the square root of the age of the lithosphere (Davis and Lister, 1974) (as predicted by the boundary layer model) up to ages 70–80 Myr (Parsons and Sclater, 1977). Beyond this age, the increase in ocean depth with age is less rapid, suggesting that heat is supplied to the base of the lithosphere, as in the plate model, by small-scale convection (Richter, 1973; Parsons and McKenzie, 1978) by viscous heating (Schubert, Froidevaux and Yuen, 1976), by variations of thermal parameters with depth or by radioactive heating (Forsyth, 1977). Alternatively, the boundary layer thermal model may be correct, with isostatic compensation prevented by dynamic effects (Oxburgh and Turcotte, 1978;

Schubert *et al.*, 1978). In the dynamic model summarized here, we assumed that density contrasts are confined to the lithosphere and that the thickness of the lithosphere, taken to be of constant density, follows the observed increase in ocean depth with age.

As a result of the subsidence of the cooling lithosphere into the mantle there is a pressure difference between the region beneath the ridge and that beneath the older oceanic lithosphere. The slope of the base of the lithosphere is small, so the stress at any point is close to the hydrostatic pressure given simply by the weight of the overburden. We are interested here in the difference in pressure beneath the ridge and far from the ridge. It is convenient to subtract the pressure resulting from the seawater (fictive below the ocean floor) and define a reduced pressure:

$$p(z) = \int_0^z g(\rho(\zeta) - \rho_w)\,d\zeta. \tag{7}$$

In Fig. 3(b), the reduced pressure is plotted as a function of depth beneath a ridge crest R and beneath a point B far away from the ridge crest. Beneath the lithosphere–asthenosphere boundary at L, isostasy implies equal pressures,

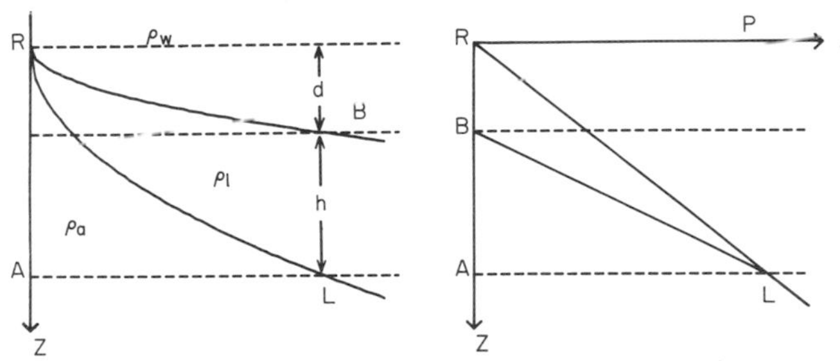

Figure 3. (a) Boundary layer model of a cooling plate; (b) reduced pressure gradient.

but above this depth, the pressure is higher beneath the ridge. The difference in horizontal force is given by the integral of the pressure difference over depth:

$$\Delta F = \int_0^z (p_R - p_B)\,d\zeta \tag{8}$$

and is equal to the area of triangle RBL:

$$\Delta F = (\rho_w - \rho_a)gd(h+d)/2. \tag{9}$$

For old ocean basins, d is 3·7 km, h is 128 km and ΔF has a maximum value of 550 bars. Other models of the thermal structure of the lithosphere give slightly different values (Forsyth and Uyeda, 1975; Richardson, Solomon and Sleep, 1976) but it is clear that the excess elevation of the ridge results in a driving force of several hundred bars for mature oceanic lithosphere. This driving force is one of the best constrained forces acting on the plates and has been used to calibrate the magnitude of other driving forces in simple parameterizations of the forces driving the plates (Forsyth and Uyeda, 1975; Solomon *et al.*, 1975; Richardson *et al.*, 1976).

Most parameterizations of plate driving forces have treated this pressure difference resulting from the subsidence of the lithosphere as it cools and thickens while moving away from the ridge as a boundary force ("ridge push"). Lister (1975) pointed out that this parameterization is not correct, however. From Eqn (9) we see that the total force depends on the product of the increase in ocean depth, d, and the thickness of the lithosphere, h. Both d and h depend initially on the square root of plate age, so the total driving force from lithospheric thickening is proportional to the age of the plate and is distributed over the area of the plate, rather than being concentrated at plate boundaries (Hales, 1969; Lister, 1975; Hager, 1978). Since plate velocity is independent of area, a straightforward interpretation is that there is significant drag at the base of the plates providing a resisting force that also is proportional to area. The driving force per unit length away from the ridge is inversely proportional to spreading velocity. The faster a ridge spreads, the less the driving force per unit area becomes. The velocity of oceanic plates may be determined by an upper limit at which ridges can spread, rather than by a terminal velocity for sinking slabs (Hager, 1978).

The above argument is for a two-dimensional model. The surface of the Earth is not covered with oceanic plates moving directly away from ridges. Complications include the existence of marginal basins and continental crust.

Marginal basins are thought to be formed by a process of injection of melt into the lithosphere along a narrow axis (Karig, 1971a, b) suggesting that the crustal structure and thermal evolution of marginal basins is similar to that of mid-ocean ridges. We assume that the plate model used for the oceanic lithosphere is also applicable for marginal basins, and assign density contrasts accordingly. Mauk (1977) tabulated the ages of marginal basins and oceanic lithosphere using data from the Deep Sea Drilling Program (DSDP), compilations of magnetic lineations, extrapolation and interpolation. We use this data file as a representative interpretation of lithospheric ages in assigning the density contrasts given by the cooling plate model.

Even when a continent is in isostatic equilibrium it is not in hydrostatic equilibrium, and has a tendency to spread (Frank, 1972; Artyushkov, 1973). The stresses resulting from the density contrast between continental and oceanic structures are on the order of those resulting from the cooling and subsiding of the oceanic lithosphere (Frank, 1972), and must be considered in any complete model of global dynamics.

Continental structure is not uniform, but depends on tectonic province. Unfortunately, simple models of continental evolution comparable to the models of oceanic plate evolution have not yet been established. We use the same density model for all continental regions, regardless of tectonic province, recognizing that our model could be refined by considering continental structure as a function of tectonic province.

Subducted slabs

As the cold lithosphere plunges into the hot mantle at a subduction zone, it begins to reheat as a result of conduction (McKenzie, 1969) viscous heating from the shear between the slab and the mantle, heat generated by radioactive decay within the slab, latent heat from phase transitions and adiabatic compression (Minear and Toksoz, 1970). In general, the density contrast between the slab and the mantle decreases as a function of depth, except for localized effects resulting from the elevation of phase boundaries within the slab (Turcotte and Schubert, 1971), although we note that conduction of heat into the slab cools the surrounding material and merely redistributes the temperature anomaly. The cool part of the slab is presumably marked by seismicity, and the cessation of seismicity at depth may be due to heating of the slab (McKenzie, 1969; Molnar, Freedman and Shih, 1979), a change in phase such that failure no longer occurs catastrophically (O'Connell, 1977) or a reduction in stress.

We assigned the density contrasts to the sinking lithosphere on the basis of observed seismicity. Rcgions of the Earth's interior with seismic activity recorded in the catalogue of the Bulletin of the International Seismological Centre (1967–1977) were assigned density contrasts appropriate for a 128 km thick slab with an average density 0·0665 g/cm^3 greater than the mean density of the mantle. (This choice of density results from the thermal model of the lithosphere discussed previously.)

It is likely that in some parts of subducted slabs earthquakes did not occur during the time interval covered by the catalogue either because the stresses in the slabs were not high enough to cause fracture, or because the slabs did not deform in a brittle manner. Thus the lengths of the slabs used are probably underestimated. On the other hand, deep slabs should be closer to thermal equilibrium with the mantle than shallow slabs, so assigning a density contrast of 0·0665 g/cm^2, appropriate to lithosphere before any heating occurs, should overestimate the density contrast between a deep slab and the surrounding mantle. The two effects oppose one another. No additional density contrasts from elevations of the phase boundaries at 350 km and 670 km were included.

Model calculations

The most obvious way to determine if the relatively well-constrained density contrasts associated with the lithosphere are sufficient to drive the plates with their observed velocities would be to solve for the flow driven by these density contrasts in a viscous spherical shell. Free-slip boundary conditions would be applied at the surface and at the core–mantle boundary. If the surface velocities matched the observed plate velocities, the model might be considered successful.

Unfortunately, this obvious approach neglects the fact that the lithosphere behaves more like an ideal plastic material than a viscous material. The plates move over large distances with little internal deformation; deformation is concentrated at plate boundaries, where velocities change discontinuously. While it is possible to simulate the behaviour of plate interiors using a high viscosity fluid to suppress deformation, it is not possible to reproduce the discontinuous velocities at plate boundaries using a viscous rheology for the lithosphere without accepting unrealistically large stresses in convergence zones, which must then be artificially excluded from any force calculations (such as was done by Richter and McKenzie, 1978).

A more appropriate description of the rheology of the plates is plastic behaviour: the plates are strong enough to distribute stresses over their interiors, but fail at some fixed stress at their convergent boundaries. The magnitude of the yield stress is related to the average stress across faults at convergent boundaries, which is not well known and therefore cannot be prescribed *a priori*. Nevertheless, we find that an acceptable force balance on the plates can be obtained using a value of the yield stress of 100 bars, which is compatible with seismic estimates of stress drops on faults at converging plate boundaries.

There are important components of the global mantle flow that may be a direct consequence of the non-linear rheology of the lithosphere. A spherical harmonic expansion of plate velocities reveals that toroidal components of flow are nearly as large as spheroidal components (Hager and O'Connell, 1978). Inspection of the linearized equations of motions for a viscous fluid reveals that only spheroidal components of flow would be excited by density contrasts; thus the presence of toroidal components of flow can be ascribed to non-linear interaction betweeen various flow components and material properties. Since the toroidal terms are needed to satisfy the boundary conditions represented by the plates, it could be said that it is the non-linear temperature dependent constitutive relation of the plates that results in the non-linear excitation of toroidal flow from gravity related body forces. The results of the model considered here bear out this interpretation.

Beneath the plates, the asthenosphere is treated as a linearly viscous fluid with constant viscosity within concentric spherical shells.

Since the model contains both linear (viscous) elements and non-linear (plastic) elements, a solution is obtained by treating each part separately and then combining the appropriate results. In essence, the body forces and viscous forces are evaluated for a linearly viscous model, and the resulting tractions on the plates are then applied to a force balance model for rigid plates with a fixed collisional yield stress. The resulting tractions depend on the value of viscosity chosen for the asthenosphere. An acceptable force balance is obtained with values of viscosity consistent with those inferred from studies of postglacial rebound and current estimates of the creep properties of mantle material.

The energy equation, describing the transport of heat, has not been included explicitly in the solution. Nevertheless, the conductive cooling of the lithosphere, and the implied convective heat transfer to the lithosphere, is implicitly included in the solution through the specification of body forces resulting from density differences associated with the cooling lithosphere. In so far as these accurately describe the large-scale density variations associated with the generation, cooling and subduction of plates, we expect the model to yield sensible results. The consistency of the model provides a check on these assumptions.

Our method of solving the equations governing flow is applicable only for linear viscous fluids, whereas our conceptual model consists of a plastic lithosphere overlying a viscous asthenosphere. In order to obtain results applicable to such a model, we have resorted to the following scheme.

(1) We calculate flow, in the absence of body forces, with the observed plate velocities imposed as boundary conditions; the model consists of a viscous lithosphere with a viscosity higher than that of the underlying asthenosphere to ensure that the effects of the mass flux from the lithosphere are included.

Since there are no body forces included in this preliminary calculation, the flow is driven by the tractions exerted on the surface of the lithosphere that are required to satisfy the velocity boundary conditions. These tractions partly do work to deform the high viscosity "lithospheric" layer included in the model to provide for the mass flux associated with plate motion. The amount of work depends on lithospheric viscosity; the magnitude of this work is an artifact of the formulation of the model and has no direct physical significance. The remainder of the tractions are transmitted to the base of the lithosphere, where they drive the accompanying flow in the sub-lithospheric mantle. These tractions at the base of the lithosphere are quite insensitive to the value chosen for the viscosity of the lithospheric layer, and are presumably then relatively independent of the rheology of the lithosphere.

(2) We then repeat the calculation, but with the body forces from density contrasts in the slabs included. The presence of body forces reduces the work supplied by the surface tractions required to satisfy the boundary conditions, indicating that the body forces are such to help drive the observed plate motions. As before, the total work supplied by the boundary conditions depends on the value chosen for the lithospheric viscosity. But the reduction of work resulting from the inclusion of body forces is essentially independent of the properties of the lithosphere. The change in tractions resulting from the inclusion of body forces is taken as the traction on the plates resulting from the presence of the body forces.

The tractions on the lithosphere resulting from the density contrasts in the slabs alone are shown in Fig. 4. As expected, the oceanic plates are pulled toward their subducted slabs; the longer the slab, the greater the tractions. Note that continental plates are also pulled toward the slabs subducting beneath them, owing to the symmetric distribution of tractions on either side of sinking slab.

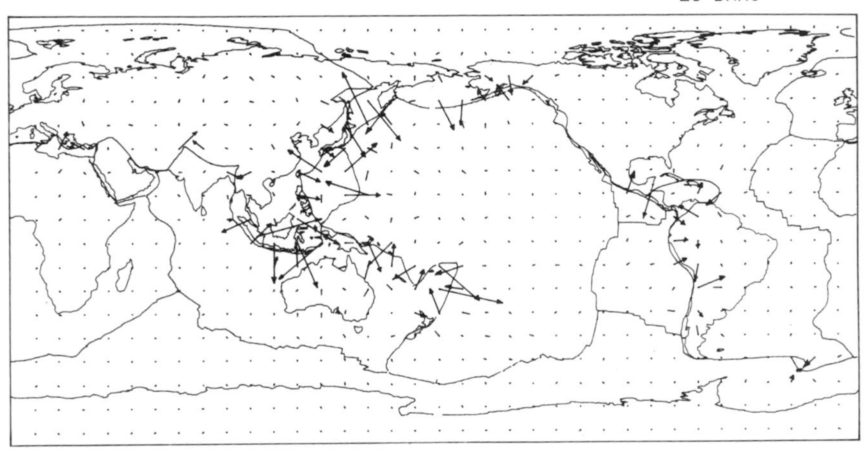

Figure 4. Shear tractions on the lithosphere from the density contrasts in the subducted slabs.

Density contrasts from the cooling of the lithosphere and continent–ocean differences give rise to tractions distributed over much of the plates; these are calculated as were those from subducted slabs and are plotted in Fig. 5. In areas of simple lithospheric structure, such as near the Mid-Atlantic Ridge or Carlsberg Ridge, the tractions are oriented in a fairly simple pattern, directed away from the ridge. Tractions on the Nazca and Cocos plates are also oriented away from the ridges, but are distributed over the plate areas and are smaller in magnitude than those for the steeper Mid-Atlantic Ridge, as would be expected from the relative spreading velocities. Tractions on the Pacific plates are in general directed toward the region of ocean floor of Mesozoic age

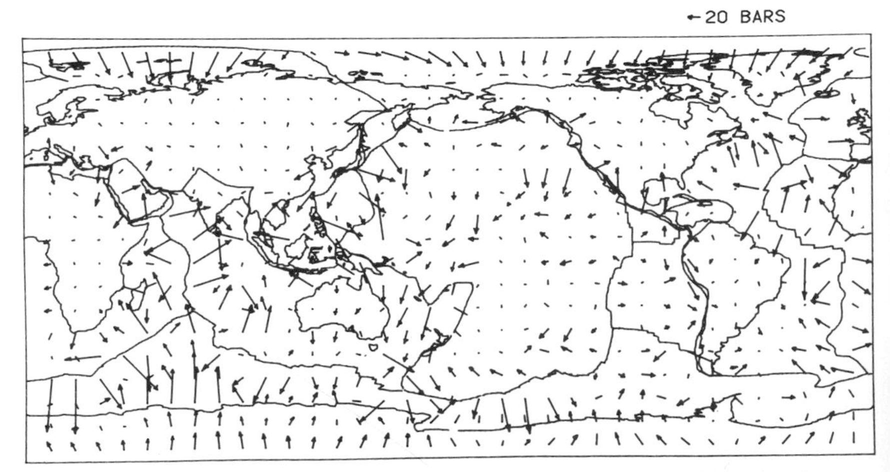

Figure 5. Shear tractions resulting from the density contrasts within the plates.

in the Western Pacific, with perturbations resulting from the young lithosphere created by Miocene spreading north of New Guinea. The tractions are not directed everywhere in the direction of plate motion. The tractions generated by the tendency for continental crust to spread are apparent around Africa and Antarctica.

Since plate motion involves rigid rotation on a sphere, the appropriate measure for the driving force is the total torque:

$$\boldsymbol{\tau} = \int_A \mathbf{r} \times \boldsymbol{\sigma}_s \, \mathrm{d}A, \tag{10}$$

where $\boldsymbol{\sigma}_s$ is the shear traction at position $\mathbf{r}$ on a plate with area A.

The components of the driving torques exerted on each plate by the subducted slabs and lithospheric density contrasts are given in Table I. Also included are the torque components from "ridge push" as given by Forsyth and Uyeda (1975) assuming a "push" of 300 bars on a 100 km thick plate. For most plates, the assumption that the torques exerted by the density contrasts within the lithosphere given by the plate model are equivalent to "ridge push" is a substantial oversimplification.

(3) The tractions exerted on the plate by the asthenosphere are evaluated from the shear stress at the top of the asthenosphere; these also are essentially independent of the lithospheric viscosity, but do depend on the viscosity distribution in the asthenosphere. Results for several models indicate that the magnitude of the tractions depends upon the viscosity model used, but the direction is relatively insensitive to viscosity model. Figure 6 shows the shear tractions (drag) at the base of the lithosphere for a mantle with a 64 km thick low viscosity layer of 4×10^{20} poise overlying a mantle of 10^{22} poise. These tractions are those exerted by the lithosphere on the underlying mantle; the tractions exerted on the lithosphere by the mantle are equal in magnitude, but opposite in sign, and act to retard plate motion. Tractions are relatively higher for the small Cocos plate than for the large Pacific plate even though both have about the same velocity. The Cocos plate is smaller and moves counter to the large-scale flow set up by the Pacific plate. Tractions under North America are at right angles to the absolute motion of that plate (Hager and O'Connell, 1979).

(4) The tractions on convergent plate boundaries are evaluated assuming a fixed collisional resisting stress, and using the geometrical parameters tabulated by Forsyth and Uyeda (1975). This calculation is independent of those for the viscous flow models used to evaluate the other tractions. The resisting stress corresponds to a plastic yield stress, or a critical stress for slip.

According to this model, stress increases along the fault separating two converging plates until a critical stress is reached, at which failure occurs, returning the stress to its original level. The time needed to complete this cycle decreases with increasing converging velocity, but the average stress is independent of converging velocity.

TABLE I *Torques on the plates (10^{33} dyne cm)*

	Ridge push	Lithospheric thickening	Slab	Basal drag	Collision resistance	Residual
Pacific	−0·13	0·03	−0·40	−0·39	−0·24	0·26
	1·15	0·96	0·35	0·89	0·26	0·16
	−2·47	−1·11	−1·65	−1·69	−1·28	0·21
	2·72	1·47	1·74	1·95	1·33	0·37
Indian	1·85	0·60	0·81	0·77	0·67	−0·03
	0·62	−0·01	0·91	0·51	0·20	0·19
	1·44	1·17	0·92	0·58	1·00	0·51
	2·42	1·31	1·53	1·09	1·21	0·55
S. Amer.	−0·59	−0·06	−0·10	−0·11	0·04	−0·09
	0·47	0·48	0·06	−0·02	0·28	0·29
	−1·56	−1·05	−0·34	−0·19	−0·79	−0·41
	1·73	1·16	0·36	0·22	0·84	0·51
Euras.	−1·42	−0·26	−0·78	−0·20	−1·28	0·44
	1·29	0·54	−0·73	−0·37	0·37	−0·20
	0·48	0·04	0·67	0·40	0·04	0·27
	1·97	0·60	1·26	0·58	1·33	0·55
N. Amer.	1·58	0·53	0·31	−0·14	0·41	0·57
	−0·85	−0·38	−0·53	−0·33	−0·22	−0·36
	−1·55	−0·21	0·10	−0·09	−0·10	0·08
	2·37	0·69	0·62	0·37	0·48	0·67
Nazca	0·45	−0·07	0·01	−0·04	−0·06	0·04
	−0·49	−0·21	−0·07	−0·16	−0·19	0·08
	1·19	0·59	0·33	0·44	0·56	−0·10
	1·86	0·63	0·31	0·48	0·59	0·13
Africa	−0·20	−0·01	0·04	0·12	0·13	−0·22
	−0·86	−0·67	−0·11	−0·36	−0·59	0·17
	0·48	0·06	−0·03	0·36	−0·11	−0·22
	1·00	0·67	0·12	0·52	0·62	0·36
Antar.	−1·90	−0·58	−0·01	0·05	0·02	−0·66
	−0·51	−0·29	−0·09	−0·10	−0·10	−0·18
	−0·36	0·06	0·01	0·07	0·10	−0·10
	2·00	0·65	0·09	0·13	0·14	0·69
Cocos	−0·52	−0·16	−0·23	−0·07	−0·24	−0·10
	0·11	0·03	0·03	0·01	0·05	0·00
	0·45	0·11	0·09	0·07	0·18	−0·04
	0·70	0·20	0·25	0·09	0·30	0·10
Arab.	−1·56	0·21	0·00	0·03	0·20	−0·02
	0·38	−0·23	0·00	−0·05	−0·22	−0·05
	0·44	0·06	0·01	0·05	0·07	−0·06
	1·66	0·31	0·01	0·08	0·31	0·08
Phil.	0·00	−0·10	0·23	0·00	0·32	−0·19
	0·00	−0·19	0·15	−0·01	0·17	−0·20
	0·00	0·23	0·00	0·00	0·15	0·07
	0·00	0·31	0·27	0·01	0·38	0·28
Carib.	0·00	−0·11	0·13	0·00	0·06	−0·04
	0·00	−0·01	0·03	0·00	0·02	0·01
	0·00	0·06	−0·07	0·01	0·19	−0·22
	0·00	0·13	0·15	0·01	0·20	0·22

For each plate, the four rows give the components about 0°N, 0°E; 0°N, 90°E; 90°N; and the magnitude of the torque in each column. The column labelled "Ridge push" is from the study by Forsyth and Uyeda (1975) and is to be contrasted with the torques due to lithospheric thickening.

The magnitude of this resisting stress is not well determined. At present the average stress across faults is a matter of controversy. High values, on the order of kilobars, would be expected if laboratory experiments on fracturing of rocks were directly applicable to the faulting process (Hanks, 1977). The absence of large heat flow anomalies along faults suggests that average stresses are smaller – of the order to tens or hundreds of bars (Brune, Henley and Roy, 1969).

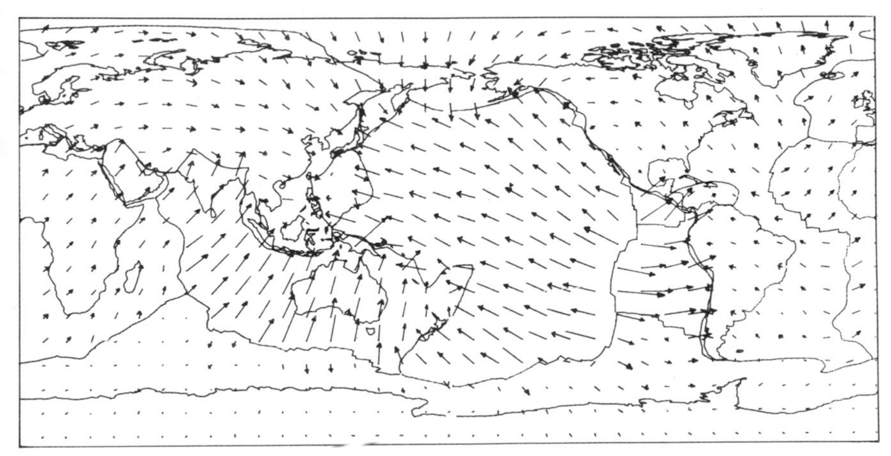

Figure 6. Shear tractions exerted on the asthenosphere by the base of the lithosphere for the global flow model. The magnitudes of the tractions depend on the viscosity distribution, but the directions are relatively insensitive to the viscosity distribution. These tractions are for a viscosity distribution with a 64 km thick layer of 4×10^{20} poise overlying a 10^{22} poise mantle.

(5) The resulting tractions on the plates resulting from all the effects noted are combined to give the net torque acting on each plate.

The driving torques from lithospheric density contrasts are very well constrained. The driving torques from the slabs are not as well constrained, but are probably reasonably correct. The directions but not the magnitudes of the resisting torques from drag at the base of the plates and resistance to collision are also well determined. These are driving and resisting torques that must exist if our understanding of the thermal evolution of the lithosphere and determinations of plate motions are reasonably correct.

For plates in dynamic equilibrium the net torque applied to each plate must vanish. If the driving torques and the directions of the resisting torques are taken as known, the magnitudes of the resisting torques which minimize, in a least squares sense, the residual torques on the plates can be determined. The resisting torques determined in this manner are given in Table I.

The tabulated magnitudes of the resisting torques from drag at the base of the lithosphere would be obtained if the viscosity in the low viscosity layer in the model shown in Fig. 6 were reduced by a factor of 2·5, or alternatively, if its

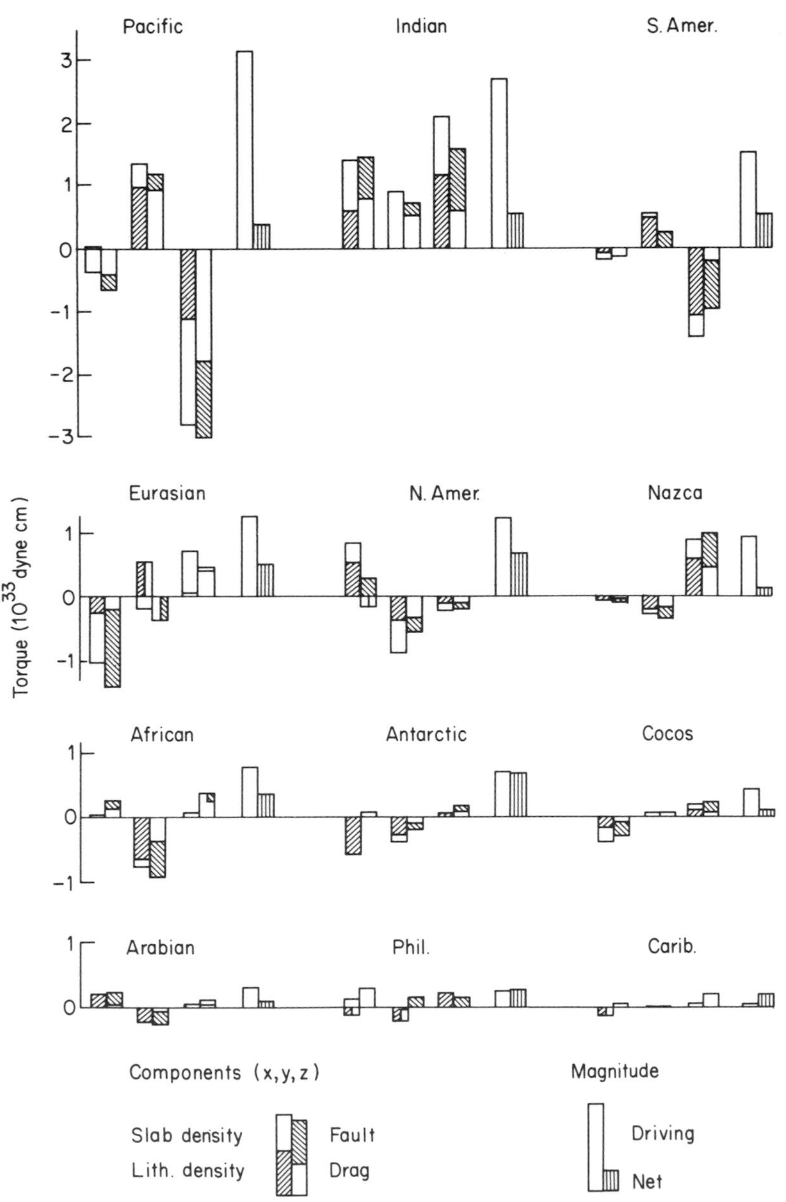

Figure 7. For each plate, the x, y and z (0°N, 0°E; 0°N, 90°E; 90°N) components of the driving torques from density contrasts in the plates and slabs are compared with the components of the resisting torques from basal drag and resistance at the faults separating colliding plates. The magnitudes of the driving torques are also shown, along with the residual torques not accounted for by the simple model. If the plates were in dynamic equilibrium, the residual torques would vanish.

thickness were increased by a factor of 2·5. Such a small change in viscosity structure would have little effect on the global flow pattern generated by the plate motions (Hager and O'Connell, 1979) or on postglacial rebound (Parsons, 1972).

The magnitudes of the torques from collision resistence given in Table I would result from average normal stresses of 100 bars distributed over 100 km deep faults at colliding plate boundaries. Peak stresses could be higher as a result of stress concentrations, narrower fault zones or transient effects.

The components of the driving and resisting torques on each plate are shown in Fig. 7. Also shown is the magnitude of the driving torque and the magnitude of the residual torque for each plate. If the simple model accounted for all the torques exerted on the plates, the residual torques would be zero. The root mean square of the residuals is only 30% of the root mean square of the driving torques. The largest residual component is 24% of the largest driving component.

The model used here should be most appropriate for oceanic plates. If the Pacific, Indian, Nazca and Cocos plates are considered, the r.m.s. residual is only 17% of the r.m.s. driving torque, while the largest residual component is 19% of the largest driving component.

Discussion

The model of plate motions and global dynamics outlined here can be viewed as a simple boundary layer model of mantle convection that takes into account the mechanical strength of the boundary layer and the observed geometry and thermal structure of the plates. Motion is driven by thermally generated density contrasts that are concentrated in the cooling and sinking boundary layer, which is strong enough to move as a rigid plate. Density contrasts in the interior are not taken into account; if the interior were well stirred their net effect on a given plate would be small. The forces on the plates result from gravity acting on the horizontal density contrasts within the plates, from the net traction exerted by the fluid in the interior, and from resistance at convergent zones; the model implicitly contains and evaluates quantitatively the forces parameterized as ridge push, slab pull, collisional resistance and viscous drag in models confined to force balance of the plates alone (Forsyth and Uyeda, 1975; Richardson *et al.*, 1975; Solomon *et al.*, 1975).

Even though the model is simple, based primarily on our knowledge of motions and thermal structure of plates and ignoring horizontal density contrasts from temperature contrasts within the Earth's deeper interior, we expect that it provides considerable insight into some important aspects of mantle dynamics. This expectation is based on the observations that

(a) the plates themselves include (or constitute) large density contrasts that can drive convective flow;

(b) the plate motions and geometries prescribe a global circulation that must constitute a major component of mantle flow (Richter, 1973; Hager and O'Connell, 1978, 1979);

(c) the observed plate velocities require substantial toroidal components of flow that are probably a consequence of the properties of the plates since they would not be expected to be strongly excited by buoyancy forces in the interior (Hager and O'Connell, 1978).

If the plate model of the thermal evolution of the lithosphere is essentially correct, the torques resulting from lithospheric thickening are fairly well constrained. They are insensitive to the rheology of the lithosphere, and are the best constrained forces driving plate motions and the accompanying mantle flow.

The driving torques from the subducted slabs are more model dependent. We assumed a radially symmetric viscosity model in the interior of our flow model, so did not consider any increased strength which the slabs may have. In these models, stress is transmitted to the surface through suction created by the sinking slab and secondarily by viscous shear from the flow established by the sinking slab.

The viscous formulation of the models may not be a serious limitation. If Kanamori's (1971) interpretation that great earthquakes periodically cut through the entire thickness of the elastic oceanic lithosphere is correct, subducted slabs may not pull directly on the oceanic plates. The observation that subducted slabs do not curl up, but rather appear to unbend after subduction, may indicate that the slabs are weak (Lliboutry, 1969). The dips of the subducted slabs are predicted by the large-scale flow entrained by the moving plates, further indication that slabs follow the flow passively (Hager and O'Connell, 1978). For two-dimensional models of flow driven by subducted slabs, Hager (1978) found that detached slabs sink at terminal velocities of the order of 8 cm/yr; although detached slabs can sink at reasonable velocities, relatively little additional force may be available to drive the motions of the surface plates. This conclusion is supported by the numerical models of Richter (1977; Richter and McKenzie, 1978) which show that the buoyancy of the slab is resisted locally primarily by excess pressure at the tip of the slab. If our characterization of the driving force exerted on the surface plates is approximately correct, then, as is apparent from examination of Table I and Fig. 7, the forces from lithospheric thickness variations and subducted slabs are of comparable importance in driving the surface plates.

In our models, drag at the base of the plates and resistance at colliding plate boundaries are of about equal importance in resisting plate motions. The resistance at colliding boundaries would be provided by an average stress normal to convergence of about 100 bars distributed over a 100 km deep fault plane. Higher stresses could occur locally because of stress concentration, but our results are incompatible with average residual stresses of the order of kilobars across plate boundaries.

The appropriate drag at the base of the plates would be provided by a 65 km thick low viscosity layer of 10^{20} poise or a 200 km thick layer of 10^{21} poise overlying a 10^{22} poise mantle. The continental ice sheets are in general too large in area for the viscosity structure of the low viscosity zone to be resolved from postglacial uplift data (Parsons, 1972).

If the viscosity of the low viscosity layer is as high as is indicated by these models, there is significant drag at the base of the plates. Even a viscosity as low as 10^{19} poise is insufficient to decouple completely the plate motions from the mantle below (Hager and O'Connell, 1979). The motion of the plates should exert a strong organizing influence on the large-scale flow in the mantle. Consideration of the flow driven by the plate motions should give a reasonable zero-order model of the large-scale mantle flow (Hager and O'Connell, 1978, 1979; Chase, 1979; Parmentier and Oliver, 1979).

One of the major differences between our models, using body forces from the observed distribution of plate ages and geometries, and previous studies of plate driving mechanisms, using parameterizations such as "ridge push" concentrated at plate boundaries, is that our models indicate that the drag at the base of the lithosphere is significant. This then implies that resistance to plate motion depends on the area of a plate. The observations that plate velocities are independent of area has previously been used to suggest that drag at the base of the lithosphere is negligible (McKenzie and Weiss, 1975; Richter, 1977; Richter and McKenzie, 1978). But the driving force from lithospheric thickening depends on the square of the thickness of the lithosphere, so the net driving force from lithospheric thickening depends directly on the age of the plate and, hence, its area (Lister, 1975; Hager, 1978). Because the depth of penetration of a subducted slab increases linearly with age (Molnar *et al.*, 1979), the total driving force exerted on a plate from its subducted slab is also proportional to plate area. A simple interpretation of the observed independence of velocity and area is that the forces from thickening of the oceanic lithosphere with age and from slabs, both effects which are proportional to area, are balanced by viscous drag.

Conclusions

The horizontal density contrasts in the lithosphere are likely to be the largest horizontal density contrasts in the Earth since the thermal gradient across the lithosphere is large. We have described a simple, self-consistent model of instantaneous mantle dynamics which incorporates the horizontal density contrasts predicted by models of the thermal evolution of lithospheric plates and slabs and applied it to the Earth using the observed plate ages, geometries and locations of slabs and continents. Resistance to plate motions is assumed to be provided by viscous drag from the large-scale flow and frictional resistance on faults at collisional plate boundaries. Density contrasts outside the lithosphere with dimensions on the order of the plate sizes are assumed negligible.

The model is of course non-unique – we may have neglected other factors which have major influence on mantle dynamics. Nonetheless, using this simple model, the plates are found to be close to dynamic equilibrium, suggesting that the plates and accompanying large-scale mantle flow may be driven primarily by the horizontal temperature gradients in the surface thermal boundary layer, the lithosphere, and that distributed density contrasts in the Earth's interior may be of secondary importance in determining the large-scale flow in the mantle.

Although these simple models do not address the more complex question of the initiation of plate motions, they do provide insight into some important aspects of the structure and dynamics of the mantle. From these models we can conclude that the viscosity of the asthenosphere is fairly high – of the order of 10^{20}–10^{21} poise for the asthenospheric thicknesses of from 50 to 500 km. This implies that there is significant coupling between the plates and the underlying mantle. The plate motions should determine the large-scale flow pattern in the mantle.

The driving forces from the well-constrained body forces lead to average stresses at colliding plate boundaries of the order of 100 bars if distributed over fault planes of 100 km width. Local concentration could lead to higher stresses, but average stresses on the order of kilobars are not compatible with these models.

The main limitation of the model is the neglect of forces arising from temperature and density variations not associated with plates. At present we have no simple way of determining these, and their importance is difficult to assess. As mentioned before, the net effect of small-scale density heterogeneites may be small when averaged over the size of a plate. The large-scale flow, however, will introduce large-scale convected temperature heterogeneities that might tend to augment the forces moving the plates; the magnitude and significance of such forces is as yet unknown. The magnitudes of the forces calculated from our present model are such, though, that their effect must certainly be included in any further investigations of mantle flow and plate driving mechanisms, since we have shown that by themselves they can reasonably account for the major fraction of forces causing plate motions.

Acknowledgements

This research was carried out while Bradford H. Hager was at Harvard University. It was supported by NSF grant EAR78–15184. Bradford H. Hager thanks NSF for a Graduate Fellowship and the Chaim Weizmann Foundation for a Postdoctorate Fellowship. We thank R. A. Verrall for providing the deformation map.

References

Artyushkov, E. V., 1973. Stresses in the lithosphere caused by crustal thickness inhomogeneities. *J. Geophys. Res.* **78**, 7675–7708.

Ashby, M. F. and Verrall, R. A., 1977. Micromechanisms of flow and fracture, and their relevance to the rheology of the mantle. *Phil. Trans. R. Soc.* **A288**, 59–95.

Brune, J. N., Henley, T. L. and Roy, R. F., 1969. Heat flow, stress, and rate of slip along the San Andreas fault, California. *J. Geophys. Res.* **74**, 3821–3827.

Bulletin of the International Seismological Centre: Catalogue of Events and Associated Observations (Years 1964-1975), Vol. 1–10, International Seismological Centre, Edinburgh, Scotland, 1967–1977.

Chase, C. G., 1979. Asthenospheric counterflow: a kinematic model. *Geophys. J. R. Astr. Soc.* **56**, 1–18.

Dàvis, E. E. and Lister, C. R. B., 1974. Fundamentals of ridge crest topography. *Earth Planet. Sci. Lett.* **21**, 405–413.

Elder, J. W., 1970. Quantitative laboratory studies of dynamical models of igneous intrusions. In *Mechanism of Igneous Intrusion* (Eds. Newall, G. and Rast, N.), 245–260, Gallery Press, Liverpool.

Forsyth, D. W., 1977. The evolution of the upper mantle beneath mid-ocean ridges. *Tectonophysics* **38**, 98–118.

Forsyth, D. W. and Uyeda, S., 1975. On the relative importance of the driving forces of plate motion. *Geophys. J. R. Astr. Soc.* **43**, 163–200.

Frank, F. C., 1972. Plate tectonics, the analogy with glacial flow, and isostacy. In *Flow and Fracture of Rocks, Am. Geophys. Union Geophys. Monogr. Ser., Vol. 16* (Eds. Heard, H. C., Borg, I. Y., Carter, N. L., and Raleigh, C. B.), 285–292, Am. Geophys. Union, Washington, D.C.

Grow, J. A. and Atwater, T., 1970. Mid-Tertiary tectonic transition in the Aleutian arc. *Bull. Geol. Soc. Am.* **81**, 3715–3722.

Gubbins, D., 1977. Energetics of the Earth's core. *J. Geophys. Res.* **43**, 453–464.

Hager, B. H., 1978. Oceanic plate motions driven by lithospheric thickening and subducted slabs. *Nature* **276**, 156–159.

Hager, B. H. and O'Connell, R. J., 1978. Subduction zone dip angles and flow driven by plate motion. *Tectonophysics* **50**, 111–133.

Hager, B. H. and O'Connell, R. J., 1979. Kinematic models of large-scale flow in the Earth's mantle. *J. Geophys. Res.* **84**, 1031–1048.

Hager, B. H. and O'Connell, R. J., 1980. A simple global model of plate motions and mantle convection. (submitted.)

Hales, A. L., 1969. Gravitational sliding and continental drift. *Earth Planet. Sci. Lett.* **6**, 31–34.

Hanks, T. C., 1977. Earthquake stress drops, ambient tectonic stresses and stresses that drive plate motions. *Pure Appl. Geophys.* **115**, 441–448.

Holmes, A., 1928. Radioactivity and Earth movements. *Trans. Geol. Soc. Glasgow* **18**, 559–606.

Holmes, A., 1965, *Principles of Physical Geology*, 463, London.

Langseth, M. G. Jr, LePichon, X. and Ewing, M., 1966. Crustal structure of the mid-ocean ridges. 5 Heat flow through the Atlantic Ocean floor and convection currents. *J. Geophys. Res.* **71**, 5321–5355.

Lister, C. R. B., 1975. Gravational drive on oceanic plates caused by thermal contraction. *Nature* **257**, 663–665.

Kanamori, H., 1971. Great earthquakes at island arcs and the lithosphere. *Tectonophysics* **12**, 187–198.

Kanamori, H., 1978. Quantification of earthquakes. *Nature* **271**, 411–414.

Karig, D. E., 1971a. Structural history of the Mariana island arc system. *Bull. Geol. Soc. Am.* **82**, 323–344.
Karig, D. E., 1971b. Structural history of the Mariana island arc system. *Bull. Geol. Soc. Am.* **82**, 323–344.
Kaula, W. M., 1972. Global gravity and tectonics. In *The Nature of the Solid Earth* (Ed. Robertson, E. C.), 385–405, McGraw-Hill, New York.
Lliboutry, L., 1969. Sea-floor spreading, continental drift and lithosphere sinking with an asthenosphere at melting point. *J. Geophys. Res.* **74**, 6525–6540.
McKenzie, D. P., 1967. Some remarks on heat flow and gravity anomalies. *J. Geophys. Res.* **72**, 6261–6273.
McKenzie, D. P., 1969. Speculations on the consequences and causes of plate motions. *Geophys. J. R. Astr. Soc.* **18**, 1–32.
McKenzie, D. P. and Weiss, N., 1975. Speculations on the thermal and tectonic history of the Earth. *Geophys. J. R. Astr. Soc.* **42**, 131–174.
Mauk, F. J., 1977. A tectonic-based Rayleigh wave group velocity model for prediction of dispersion character through ocean basins. Ph.D. thesis, University of Michigan, Ann Arbor, Michigan.
Minear, J. W. and Toksoz, M. N., 1970. Thermal regime of a downgoing slab and new global tectonics. *J. Geophys. Res.* **75**, 1397–1419.
Minster, J. B. and Jordan, T. H., 1978. Present-day plate motions. *J. Geophys. Res.* **83**, 5331–5354.
Molnar, P., Freedman, D. and Shih, J. S. F., Lengths of intermediate and deep seismic zones and temperatures in downgoing slabs of lithosphere. *Geophys. J. R. Astr. Soc.* **56**, 41–54.
O'Connell, R. J., 1977. On the scale of mantle convection. *Tectonophysics* **38**, 119–136.
O'Connell, R. J. and Hager, B. H., 1980. On the thermal state of the Earth. In *Physics of the Earth's Interior,* Enrico Fermi International School of Physics, Course 78 (Eds. Dziewonski, A. and Boschi, E.), North-Holland-Elsevier, Amsterdam (in press).
Oxburgh, E. R. and Turcotte, D. L., 1978. Mechanisms of continental drift. *Rep. Prog. Phys.* **41**, 1249–1312.
Parker, R. L. and Oldenburg, D. W., 1973. Thermal model of ocean ridges. *Nature Phys. Sci.* **242**, 137–139.
Parmentier, E. M. and Oliver, J. E., 1979. A study of shallow mantle flow due to the accretion and subduction of lithospheric plates. *Geophys. J. R. Astr. Soc.* **57**, 1–22.
Parsons, B., 1972. Changes in the Earth's shape. Ph.D. thesis, Downing College, Cambridge.
Parsons, B. and McKenzie, D. P., 1978. Mantle convection and the thermal structure of the plates. *J. Geophys. Res.* **83**, 4485–4496.
Parsons, B. and Sclater, J. G., 1977. An analysis of the variation of ocean floor bathymetry and heat flow with age. *J. Geophys. Res.* **82**, 803–827.
Pitman, W. C., III, Larson, R. L. and Herron, E. M., 1974. Age of the ocean basins determined from magnetic anomaly lineations. *Geol. Soc. Am. Charts.*
Richardson, R. M., Solomon, S. C. and Sleep, N. H., 1976, Intra-plate stress as an indicator of plate tectonic driving forces. *J. Geophys. Res.* **81**, 1847–1856.
Richter, F. M., 1973. Dynamical models for sea-floor spreading. *Rev. Geophys. Space Phys.* **11**, 223–287.
Richter, F. M., 1977. On the driving mechanism of plate tectonics. *Tectonophysics* **38**, 61–88.
Richter, F. M. and McKenzie, D. P., 1978. Simple plate models of mantle convection. *J. Geophys.* **44**, 441–471.
Schubert, G., Froidevaux C., and Yuen, D. A., 1976. Oceanic lithosphere and asthenosphere: thermal and mechanical structure. *J. Geophys. Res.* **81**, 3525–3540.

Schubert, G., Yuen, D. A., Froidevaux, C., Fleitout, L. and Souriau, M., 1978. Mantle circulation with partial shallow return flow: effects on stresses in oceanic plates and topography of the sea-floor. *J. Geophys. Res.* **83**, 745–758.

Sclater, J. G. and Francheteau, J., 1970. The implications of terrestrial heat flow observations on current tectonic and geochemical models of the crust and upper mantle of the Earth. *Geophys. J. R. Astr. Soc.* **20**, 509–542.

Solomon, S. C., Sleep, N. H. and Richardson, R. M., 1975. On the forces driving plate tectonics: inferences from absolute plate velocities and intraplate stress. *Geophys. J. R. Astr. Soc.* **42**, 769–801.

Turcotte, D. L., and Oxburgh, E. R., 1967. Finite amplitude convective cells and continental drift. *J. Fluid Mech.* **28**, 29–42.

Turcotte, D. L. and Schubert, G., 1971. Structure of the olivine–spinel phase boundary in the descending lithosphere. *J. Geophys. Res.* **76**, 7980–7987.

Laboratory Modelling of Mantle Flows

P. A. DAVIES*

School of Physics, The University, Newcastle upon Tyne, UK

1 Introduction

The problem of thermal convection in a fluid region subjected to differential heating is one which has been regarded as a classical problem (see, for example, Chandrasekhar, 1961) in the field of fluid dynamics. The number of analytical and numerical contributions on the subject is immense (for an up to date bibliography see, for example, the review articles by Palm (1975) and Busse (1978)) but the problem is one which is also particularly well suited to laboratory experimentation. There is a wealth of laboratory data available from studies of Bénard convection (e.g. Chen and Whitehead, 1968; Krishnamurti, 1970a, b, 1973; Whitehead and Chan, 1976; Dubois and Bergé, 1978; Whitehead and Parsons, 1978) – the case in which heat is supplied continuously to the lower boundary of a fluid layer (the top surface of which may be free or bounded but held at a lower temperature than the lower); convection with internal heat generation (Tritton and Zarraga, 1967; Schwiderski and Schwab, 1971; Kulacki and Goldstein, 1972; De la Cruz-Reyna, 1976), where the fluid is warmed uniformly throughout its volume; and convection from inclined heated surfaces (see, for example, Turner, 1973b; Tritton, 1977, Ch. 14). Many studies from the first two of these groups are evidently particularly relevant to mantle motions, and in the following sections of this chapter an attempt will be made to introduce the reader to a selection of such studies. The chapter makes no claim to be a comprehensive critical review of laboratory simulations of thermally driven motions in the interiors of the Earth and terrestrial planets, since not only are more complete and detailed reviews already available (Koschmieder, 1974; Elder, 1976, 1977; Whitehead, 1976) but also many authors in this volume make extensive reference to related laboratory work performed by themselves and/or their colleagues. The intention here is to scan the range of laboratory models which are in the geophysical literature and to concentrate attention, rather

* Present address: Department of Civil Engineering, The University, Dundee, UK.

arbitrarily, on several which are of particular interest. In addition, in this multidisciplinary volume, this chapter attempts to familiarize the non-fluid dynamicist with some of the ideas and concepts in fluid dynamics which are discussed in following articles and which underlie the subject of "realistic" modelling of motions in the Earth's interior. The limitations of physical modelling of large-scale planetary interior flows have been emphasized elsewhere ("a thorny problem" (Tozer, 1977a)). It is a general principle of fluid dynamics that if information from a laboratory model is to be applied to, say, a geophysical flow (and vice versa) the two systems should be geometrically and dynamically similar (see, for example, Tritton, 1977). For a Boussinesq fluid in which the flow is dominated by viscous and buoyancy forces (as in the Earth's interior), conditions for dynamical similarity can be quantified in terms of a set of non-dimensional parameters, the most important of which, for steady flows, are the Rayleigh number, Ra, and the Prandtl number, Pr. The latter, defined as

$$Pr = \nu/\kappa,$$

where ν and κ are the kinematic viscosity and thermal diffusivity of the fluid respectively, is itself a property of the fluid and for the Earth assumes a value many orders of magnitude greater than unity (typically 10^{22}–10^{23}). The Rayleigh number for Bénard convection is defined by

$$Ra = \frac{g\alpha(\Delta T)\,L^3}{\nu\kappa},$$

where g is the acceleration due to gravity, α is the thermal expansion coefficient of the fluid and ΔT is the destabilizing temperature difference imposed across the fluid layer of depth L. For the case in which there is power generation H per unit volume throughout the fluid, the appropriate Rayleigh number Ra_H (Roberts, 1967; Turcotte and Oxburgh, 1972) is

$$Ra_H = \frac{g\alpha H L^5}{\nu\kappa k},$$

where k is the thermal conductivity of the fluid. Estimates of Ra and Ra_H for the mantle vary, primarily because of uncertainties in the values of many of the mantle's individual physical properties and their variation with depth (Oxburgh and Turcotte, 1978) and the length scale(s) over which convection overturning is thought to extend. Typical values of Ra, Ra_H for the Earth are thought to be in the range 10^6–10^9 (Tozer, 1967; McKenzie and Richter, 1976; Elder, 1977; Oxburgh and Turcotte, 1978). Most significantly, the range above ensures that both Ra and Ra_H exceed, by several orders of magnitude, the critical values necessary for convection to occur (Chandrasekhar, 1961; Krishnamurti, 1970a). From the preceding discussion, it is therefore clear that in order to model steady Boussinesq mantle convection, laboratory studies should be made with high Prandtl number fluids and be conducted at Rayleigh numbers three to five orders of magnitude greater than their critical values. Some of the consequences regarding the compatibility of steady convection with the latter constraints upon Ra, Ra_H and Pr are considered elsewhere in this volume. It is clear from laboratory experiments that, even at high values of Pr, high Rayleigh number flows develop time-dependent and eventually turbulent characteristics. The form of the flow for a given high Pr–high Ra combination depends upon many factors, as shall be discussed later, but is primarily determined by the upper and lower boundary conditions, the variations with temperature of the physical properties of the fluid (Booker, 1976; Richter, 1978), the orientation of the fluid layer with respect to the gravitational acceleration vector (Palm, 1975), the geometry of the container (Koschmieder, 1974) and the initial conditions (Whitehead and Chan, 1976; Whitehead and Parsons, 1978).

It is well known, of course (see, for example, Elder, 1977), that the velocity scale of any convective motion which occurs under the above circumstance is not imposed, but is determined internally. This is significant when discussing mantle motions in terms of two other relevant non-dimensional parameters, Re and Pe (the Reynolds number and Peclet number respectively), since the value of the former often gives an indication of the nature (laminar, turbulent) of the flow and the latter indicates the relative importance of convection to conductive heat flow. (High values of Re and Pe have been regarded as corresponding to "momentum turbulence" and "thermal turbulence" respectively (Elder, 1976, Ch. 7).) Both parameters contain a velocity scale U as

$$Re = \frac{UL}{\nu}$$

and

$$Pe = \frac{UL}{\kappa}$$

and have global values for the Earth based upon, say, observed sea-floor spreading rates of typically 10^{-17} and at least 10^3, respectively. (As explained earlier, it is important to realize that Re and Pe are not independent of Ra and/or Pr – their definition and inclusion here merely serves to complete the set of fluid dynamical parameters usually referred to in the geophysical literature, in the context of mantle convection.) The other important non-dimensional parameter of direct relevance here is the Nusselt number, Nu, the ratio of total heat transfer through a fluid region to the heat transfer which would be transferred through the same region by conduction alone. Experiments indicate that for $Ra > 10^5, Pr \gg 1$ the following relationship applies (Elder, 1976):

$$Nu \propto Ra^{1/3}.$$

The significance of this relationship to the general problem of mantle convection has been recognized particularly by Tozer (1967, 1977b).

The limitations of physical models (and numerical and analytical models) of the Earth's convecting interior are not only determined by a failure to explore fully the high *Ra*–high *Pr* flow regimes (with appropriate boundary conditions) in the laboratory, but also by fundamental difficulties in modelling non-Boussinesq effects and/or non-Newtonian flow behaviour. Most laboratory investigations, with few exceptions (Liang and Arcrivos, 1970), have made use of Newtonian fluids, for example, though several computational studies* have been made of convection in fluids having non-linear rheology (e.g. Parmentier, Turcotte and Torrance, 1976; Schubert, Froidevaux and Yuen, 1976).

Conditions under which the Boussinesq approximation is valid (Spiegel and Veronis, 1960) impose restrictions upon laboratory modelling of geophysical flows in a subtle way. Non-Boussinesq effects associated with the physical properties of the fluid varying significantly with temperature are present in both laboratory and geophysical situations, but the importance of other non-Boussinesq effects is quite different in the two cases. The two leading first-order non-Boussinesq parameters in this respect are the independent thermodynamic parameters B and C defined by

$$B = g\rho\beta L$$

and

$$C = g\alpha L/C_p,$$

where ρ is the reference density of the fluid, C_p is the specific heat at constant pressure and β is the isothermal compressibility. As Tritton (1977) points out in his book, though B and C are independent parameters they are linked by the ratio, B/C (the Grüneisen parameter), the value of which seldom differs from unity for liquids and solids. (The reader is referred to the appendix to Chapter 14 of Tritton's book for a full and detailed specification of the conditions under which the Boussinesq approximation is valid.)

So far as laboratory experiments are concerned, it is impossible to achieve values of C which are comparable with those in the mantle (Tozer, 1977b).† Since C can be regarded as the ratio of heat generation by viscous dissipation to primary heat production (Tozer, 1977b; Tritton, 1977), the above incompatibility may be one of the most serious limitations to laboratory models. Viscous dissipation will not be negligible for convection in the interior of the Earth (Houston and De Bremaecker, 1975) though for constant fluid properties the Boussinesq approximation may be valid in geophysical

* In order to obtain solutions many computational studies, whilst retaining non-Newtonian and viscous dissipation effects, are themselves subject to limiting factors not present in laboratory studies; for example, the employment of periodic boundary conditions which impose a cell aspect ratio on the flow and a restriction to two-dimensional flows.

† In the laboratory C takes typical values of 10^{-7}–10^{-8}, whilst in the mantle it is of order 10^{-1}–10^{0}.

applications if "temperature" is replaced by "potential temperature" (Busse, 1978).

Before proceeding to discuss particular laboratory studies it is of interest to note that most of them have been made with plane horizontal layers of fluid. There have been attempts to simulate in the laboratory the radial gravitational field of the Earth but these have been few. They have been of two types; those which have used strong radial electric field gradients (both steady (Gross and Porter, 1966; Gross, 1967) and alternating (Smylie, 1966; Chandra and Smylie, 1972)) in dielectric liquids, and those in which a cylinder containing fluid is rotated rapidly about an axis perpendicular to the gravitational acceleration vector so that the centrifugal force dominates over the gravitational force (Busse, 1978). Though both simulations produce novel results, we will restrict ourselves here to convection in fluids contained between plane horizontal boundaries.

2 Convection with internal heat generation

Though for historical and other reasons global mantle convection has been (and is sometimes still) treated in terms of the Bénard problem, it seems more probable (Tozer, 1965; Turcotte and Oxburgh, 1972) that the dominant heat source presently driving mantle convection is that derived from radiogenic decay of ^{40}K, ^{238}U, ^{235}U and ^{232}Th within the mantle. The ratio of heat being generated in this way to that entering the lower mantle from the core is, of course, open to conjecture – many numerical models retain both sources and compute isotherm and streamline patterns for a range of values between the extremes of zero and ninety respectively (e.g. Houston and De Bremaecker, 1975). Even though radiogenic heating may be proposed as the driving force for mantle convection, the distribution of radioactive sources with depth remains uncertain (Whitehead, 1972). It has been argued that in fact the sources are not distributed uniformly in the mantle but are confined to relatively shallow depths (Whitehead, 1972). Indeed, because continental crustal rocks are relatively rich in radiogenic heat sources compared with oceanic crustal rocks, it has been proposed that horizontal temperature variations may play a significant role in either driving the plates or modifying flow patterns initiated by vertical temperature gradients (Richter, 1973).

Laboratory experiments have been performed (Whitehead, 1972) to demonstrate the effects of various types of mobile floating heat sources in a viscous fluid, although several authors (Tozer, 1967; McKenzie, Roberts and Weiss, 1974) have argued that, at least for the Earth's mantle, convection driven by horizontal temperature differences constitutes a secondary effect. Experiments which have attempted to model convective flows driven by total volumetric heating have themselves produced unexpected results, and the problem is still not clearly solved. There is a degree of inconsistency between experiments conducted by different investigators (Tritton and Zarraga, 1967;

De la Cruz-Reyna, 1970; Schwiderski and Schwab, 1971; Schwiderski, 1972) and there are discrepancies between experimental results and theoretical predictions.

The first qualitative experiments of this type were performed by Tritton and Zarraga (1967) (though these experiments were later repeated in a more quantitative manner by Hooper (unpublished)), using a plane layer of electrolyte internally heated by an alternating electric current. The thermal boundary conditions were such that the lower boundary was insulated and the top was kept at a constant temperature. Heat was therefore produced throughout the volume of the fluid but removed at the top surface. The studies revealed that the internally heated layer did indeed convect, but the overall features of the motion differed from those observed in the appropriate Bénard problem in two important respects; firstly, in the range of Ra/Ra_{crit}* from 4 to 40 the preferred plan form of the convection is down-going hexagons, whilst at higher values of Ra/Ra_{crit} the hexagon pattern breaks down and rolls are preferred. It is interesting to note that the preferred plan form for Bénard convection throughout this range is two-dimensional rolls (for a more detailed discussion see the following chapter) unless the fluid properties vary with temperature: if the viscosity of the Bénard configuration decreases with temperature up-going hexagons are preferred, whilst if the viscosity increases with temperature down-going hexagons are preferred (Palm, 1978). The result compares favourably with theoretical treatments in several respects. The observed sense of circulation is predicted by linear stability theory (Roberts, 1967) and numerical computation (Tveitereid and Palm, 1976), though the above theoretical studies disagree over the range of stability of the hexagonal plan form (see also Thirlby, 1970).

The second and perhaps more controversial observation from the Tritton and Zarraga experiments was that as the values of Ra/Ra_{crit} increased, the length-to-depth aspect ratio, l/h, of the cells also increased. The elongation of the cells reached a value of about 5 for the maximum Rayleigh numbers investigated. This result has possible general significance for mantle convection theories, in that if taken at face value, it would seem to give some support to shallow mantle convection. However, there is a clear disagreement between the observations and theoretical treatments (either analytical or numerical) of this problem; the latter do not predict elongation of the cells. This disagreement has been highlighted by the suggestions of Schwiderski and Shwab (1971) that the elongations are caused by uneven electrical heating in the experiments because of the temperature-dependence of the electrical conductivity of the fluid, although the observations of elongated cells by Hooper (unpublished) and De la Cruz-Reyna (1976) – see Fig. 1 – seem to indicate that the uneven heating effect is not of first-order importance. Hooper,

* Here Ra and Ra_{crit} are defined, according to the earlier definition appropriate to the internally heated case – the subscript "crit" referring to the critical Rayleigh number for this case below which convection will not occur (Roberts, 1967; Tveitereid and Palm, 1976).

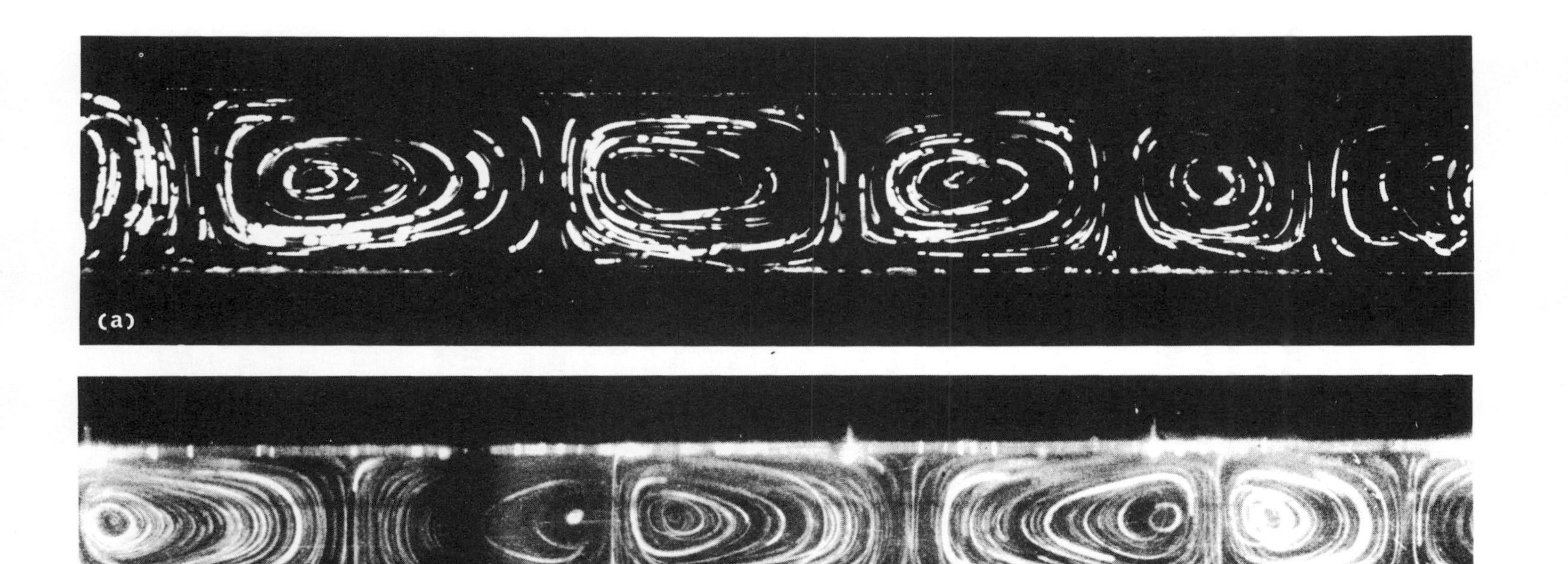

Figure 1. Cross-sectional views of convection cells in a fluid layer heated internally, showing elongation as the Rayleigh number increases from (a) 10^4 to (b) 2×10^5 (from De la Cruz-Reyna, 1976).

for example, observed the same elongation for the same Rayleigh number in layers of different depth, whilst De la Cruz-Reyna still observed elongation in a system in which the electric currents were flowing in a direction perpendicular to that in the experiments of Tritton and Zarraga. Elongation seems to be a real effect in internally-heated flows having the same boundary conditions, and more quantitative work of the standard achieved by Kulacki* and his colleagues (Kulacki and Goldstein, 1972, 1975; Kulacki and Nagle, 1975; Kulacki and Emara, 1977) is needed for this question to be solved satisfactorily.

3 The effects of moving boundaries

Most of the early studies of mantle convection ascribe to the crust (later "lithosphere", later "plate"!) a rather passive role. The possibility of the motion of the plate itself imposing a restraint and control upon the primary driving convection pattern has been considered recently by several authors (Richter, 1973; Richter and Parsons, 1975; McKenzie and Richter, 1976). The basic idea, as originally proposed, is the following; the moving plates do not constitute the cold boundary layer of a Bénard cell but are slabs distinct from the underlying convecting fluid – they are in some sense "decoupled" from the mantle (see also Tozer, 1973, on this terminology!). In order to satisfy continuity, according to the proposition, there must exist a large-scale circulation in the mantle – the mass flux produced by the moving plates must be matched by a return flow (of unspecified profile) at depth. This large-scale flow will then be expected to interact with the thermally-driven convective circulation, which is presumably responsible for the initial motion of the plates. The two circulations may have different length scales and numerical experiments (Richter, 1973) suggest that this is the case. As a result, an interesting series of laboratory experiments have been conducted to investigate the effects of shear upon Bénard convection (Richter and Parsons, 1975; McKenzie and Richter, 1976) and these have been compared with linear stability theory (Richter and Parsons, 1975). From the latter, which was a two-dimensional model using stress-free boundaries and a shear of the form

$$u = U \cos \pi z,$$

where u is the horizontal velocity of the large-scale fluid circulation at depth z and U is the imposed horizontal velocity of the plate (at $z = 0$), the amplitudes of transverse and longitudinal rolls were computed for various values of U. It was found that above a certain value of U, the amplitudes of the transverse and longitudinal rolls fell and grew respectively, such that the Bénard flow aligns with the shear (Tritton and Davies, 1980).

* See the contribution by this author in a later chapter of this volume.

The experiments performed to verify these predictions were of a special type: in general, as described in the next chapter, a supercritical Bénard fluid will choose a wavenumber from an array of possibilities. However, the technique developed by Chen and Whitehead (1968), Busse and Whitehead (1971, 1974) and Whitehead and Chan (1976) allows a chosen wavenumber scale to be imposed on the experimental system, with the result that the various transitions between different states of flow as *Ra* increases can be rather beautifully illustrated. The notation for these states, some of which are illustrated in Fig. 2, is clarified in the next chapter. Figure 3, taken from the

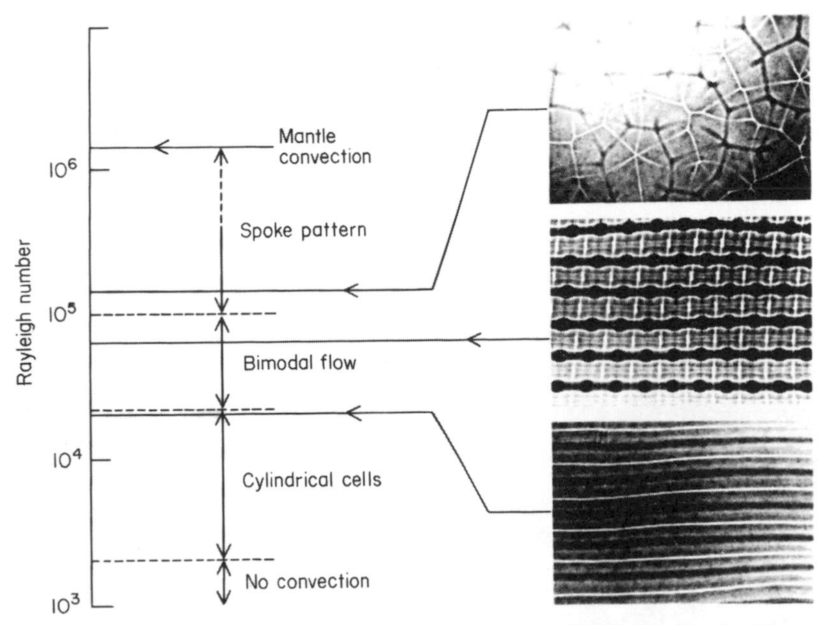

Figure 2. Schematic representation of the types of flow observed using the Chen and Whitehead technique as *Ra* is increased (from McKenzie and Richter, 1976).

experiments of Richter and Parsons (1975), shows the effects of shear in aligning both transverse roles and bimodal convection in the direction of shear.

The notion of a multiplicity of length scales over which mantle convection is occurring is now attracting considerable attention (Richter and Daly, 1978; O'Nions, Hamilton and Evenson, 1980)*. There is rudimentary laboratory evidence (Parsons and McKenzie, 1978) exhibiting the co-existence of large- and small-scale circulations in a particular heat source–sink configuration. Experiments are now being conducted by the author to investigate Bénard convection in a two-layer fluid. The depths of the

* See also the chapter in this volume by Turcotte.

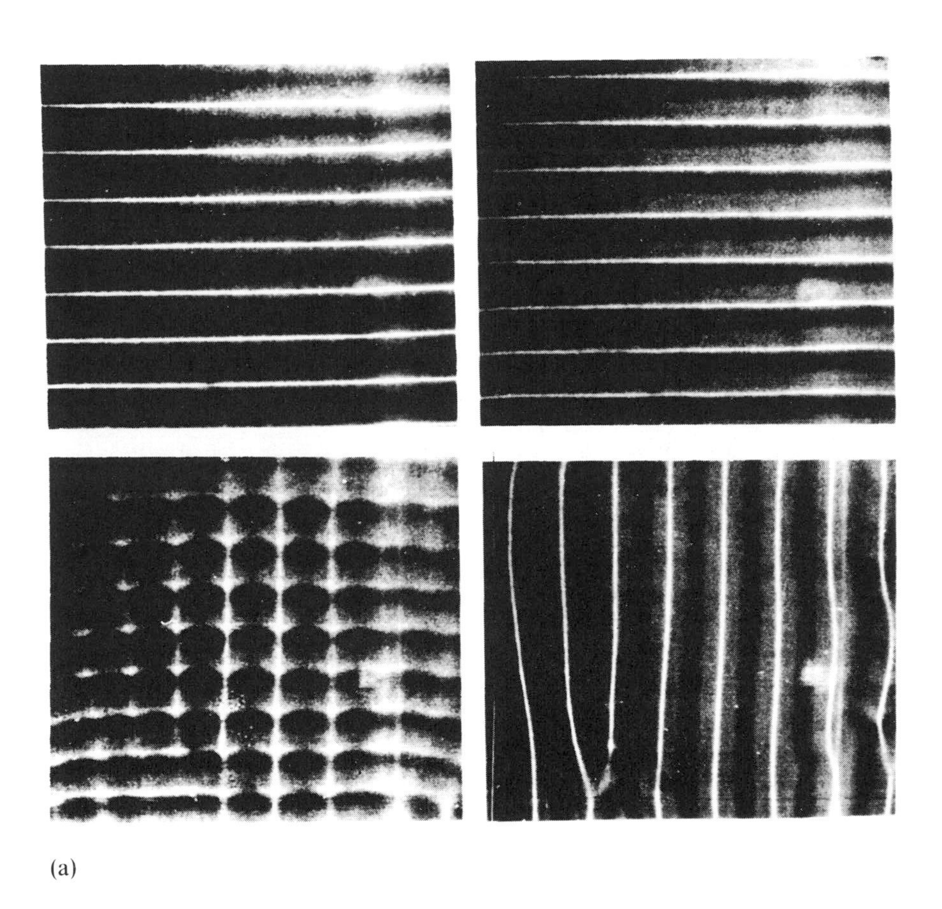

(a)

Figure 3. Shadowgraphs illustrating the development with time of (a) transverse rolls and (b) bimodal convection as a shear is applied to the top surface of the convecting fluid. In (a) $Ra = 3{\cdot}3 \times 10^3$, $Pe = 31$, and in (b) $Ra = 5{\cdot}3 \times 10^4$, $Pe = 298$. In both cases the sequence of events proceeds from top left to top right to bottom left to bottom right (from Richter and Parsons, 1975).

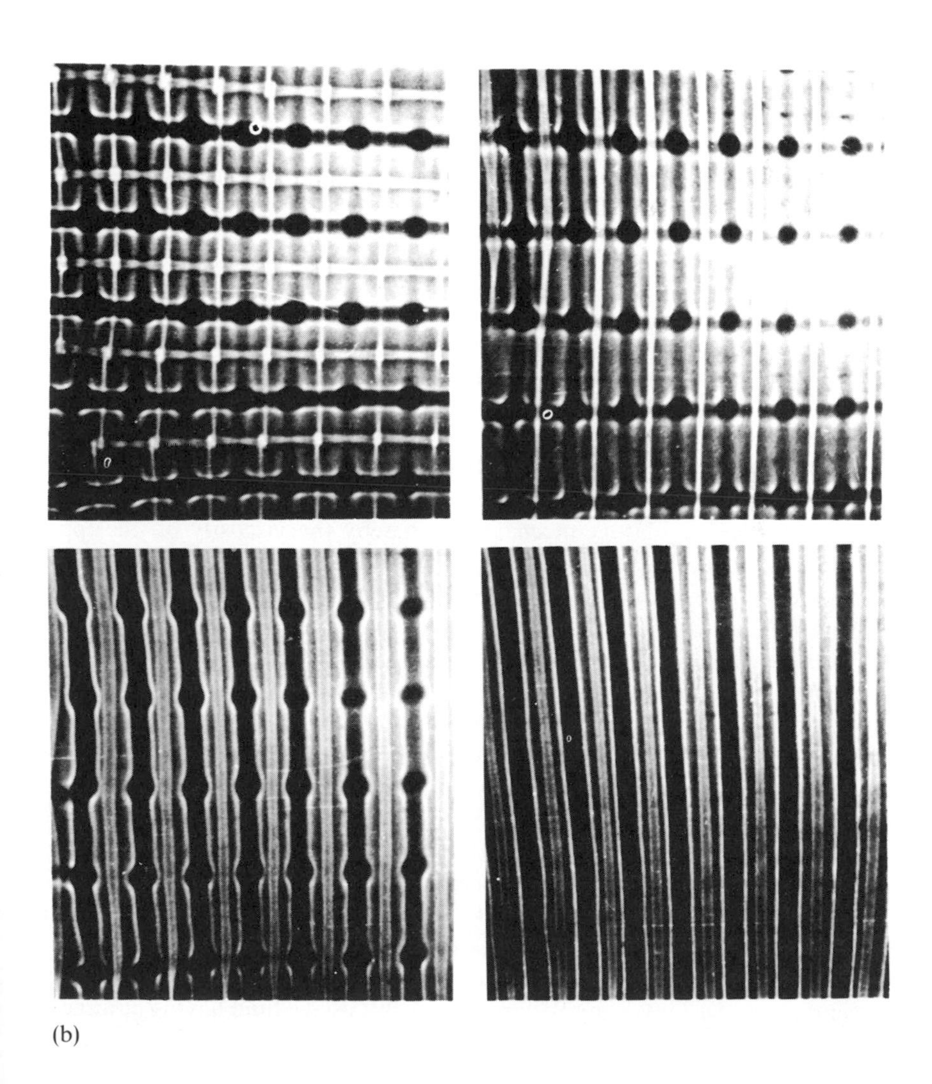

(b)

Figure 3. (cont.)

(immiscible) fluids are chosen such that the Rayleigh number is initially supercritical for one layer but subcritical for the other. As the convection in the former proceeds distortion of the interface occurs, but the enhanced heat transfer through the convecting fluid appears to stimulate motion in the initially non-convecting fluid. It is of some interest, in the light of the remarks made earlier, to determine whether the length scale of the induced motion is comparable with that of the driving cell, or whether it chooses its own aspect ratio. If the latter, it may well be the case that this length scale feeds back into the other layer, and so on, until eventually a steady optimum heat transfer state is reached. Preliminary experimental work in this problem is in progress in the author's laboratory.

4 Variation in fluid properties with temperature

We have already remarked upon the geophysical importance of the variation of fluid viscosity, with temperature and pressure. Great efforts have been made to deduce viscosity–depth relationships for the Earth (for a summary of this work the reader is referred to Turcotte and Oxburgh, 1972) though the significance of these have been questioned by Tozer (1967, 1973, 1977) who has advocated a self-regulatory mechanism whereby the viscosity of the upper mantle, at least (Tozer, 1967), retains a fairly uniform value of 10^{20}–10^{21} poise. Experiments have been performed on convection in fluids having viscosities varying rapidly with temperature, but these have been of different types.

The first group of experiments (Booker, 1976; Richter, 1978) have been concerned with exploring the Bénard problem with fluids having the properties outlined above, though the methods used and the observables considered were different in the two cases. Booker (1976) made careful heat flow measurements, in addition to visualizing the velocity field cross-section, using a fluid whose viscosity varied by up to a factor of 300 over the range of temperatures imposed. He found that the flow's overall character was determined by the upper and lower mechanical boundary conditions and that *Nu* decreased with increasing viscosity variation. *Nu*, normalized by a mean viscosity of the fluid, was essentially independent of *Ra*.

Richter (1978) employed the technique outlined earlier for imposing a wavenumber on a convecting Bénard system but used a fluid having a range of viscosity an order of magnitude less than that employed by Booker. Photographs of up-going hexagons (see Section 2) obtained by Richter are shown on Fig. 4. He found that with his highest values of ν_{max}/ν_{min} this was the preferred plan form near Ra_{crit}, but the hexagons broke down to rolls as *Ra* increased. Thereafter, the behaviour of the rolls (transitions through bimodal, cross-roll and zig-zag instability – see Fig. 5) was almost identical to the behaviour of constant viscosity fluids, under the same initial boundary conditions. There were indications that the Rayleigh number at which the

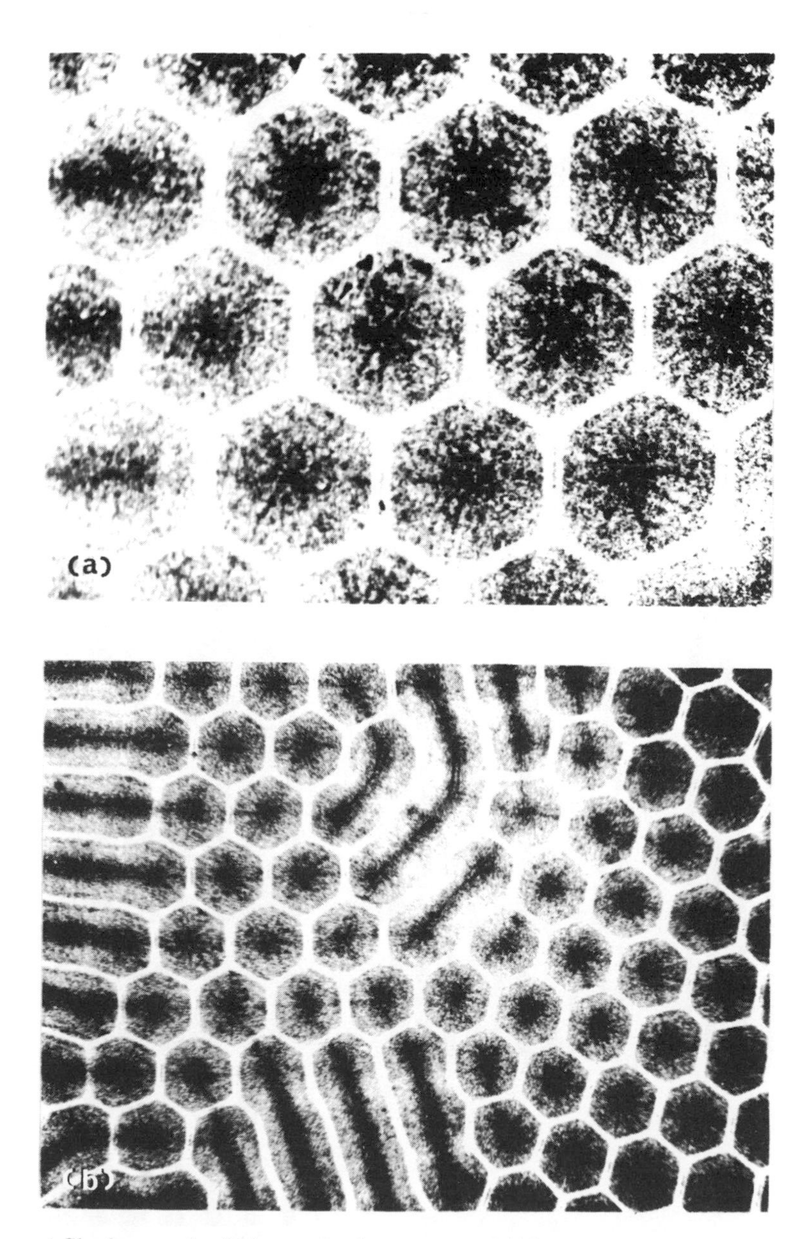

Figure 4. Shadowgraph of (a) up-going hexagons and (b) hexagons transforming to rolls in a fluid in which the viscosity changes significantly with temperature (from Richter, 1978).

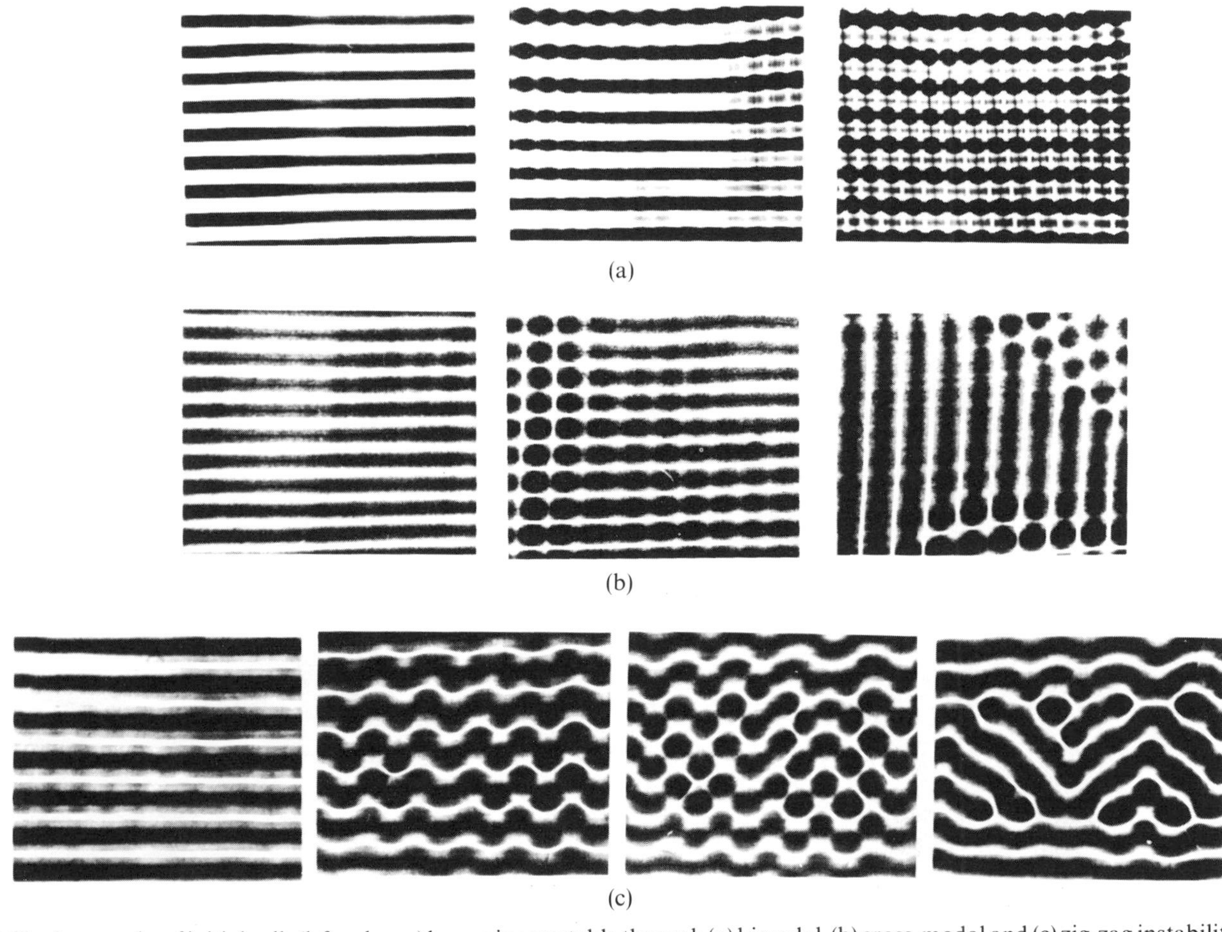

Figure 5. Shadowgraphs of initial rolls (left column) becoming unstable through (a) bimodal, (b) cross-modal and (c) zig-zag instability in fluids in which viscosity varies significantly with temperature (from Richter, 1978).

transition occurred between stable rolls and bimodal flow was lower than that for a constant viscosity fluid as ν_{max}/ν_{min} increased.

Finally we mention an experiment which is different from those already described but which still investigates the consequences of flow in a fluid whose viscosity is strongly dependent upon temperature. The approach (Turner, 1973a) was to study with such a fluid the interaction of vigorous convection in a relatively high temperature region of the fluid with surface cooling. The convection was driven not by basal heating, as in the Bénard problem, but by stirring set up by bubbles emitted from pellets of dry ice dropped into the fluid. The same bubbles produced surface cooling of the fluid (glycerine) such that the resulting viscosity increase caused the surface skin to spread out and plunge down into the interior of the fluid. Examples of such flows (which are similar in many respects to those investigated by Jacoby, 1976) are shown on Fig. 6.

5 Final remarks

To reiterate, this article has made no claim to serve as a critical and complete review of laboratory models of mantle convection. I have included points and references which I find interesting and relevant but my selection has been, of necessity, subjective. For reasons of space limitations I have concentrated upon global models of mantle flows and I have avoided discussing important and controversial smaller-scale processes such as mantle plumes and diapirs. Laboratory studies of these processes are in the literature (Whitehead and Luther, 1975) though relatively few investigations of these problems have been attempted. (Perhaps as a result of stimulating articles such as those by Tozer (1973) and Holden and Vogt (1977).)

It is clear from the range of studies indicated in this article that laboratory and numerical modelling can play important roles in understanding geophysical processes. Nevertheless, it is important to be aware of the limitations and ranges of validity of each of these approaches to geotectonic problems.

Acknowledgements

I wish to thank NERC for their financial support and I wish to thank, in particular, Dr D. C. Tozer and Dr D. J. Tritton for stimulating discussions of planetary convection problems. I am personally grateful to Louise Tarn and Denise Orton for their secretarial help both during and after the NATO meeting and in the preparation of this manuscipt. This article is for Ann.

Figure 6. Shadowgraphs showing the interaction of a convective flow with surface cooling in a fluid whose viscosity varies strongly with density. The circulation and surface cooling are produced by bubbles emitted from dry ice pellets on the floor of the tank and examples are shown of convection from (a) a single and (b) a double source respectively (from Turner, 1973).

(b)

References

Booker, J. R., 1976. Thermal convection with strongly temperature-dependent viscosity. *J. Fluid Mech.* **76**, 741–754.

Busse, F. H., 1978. Non-linear properties of thermal convection. *Rep. Prog. Phys.* **41**, 1929–1967.

Busse, F. H. and Whitehead, J. A., 1971. Instabilities of convection rolls in high Prandtl number fluid. *J. Fluid Mech.* **47**, 305–320.

Busse, F. H. and Whitehead, J. A., 1974. Oscillatory and collective instabilities in large Prandtl number convection. *J. Fluid Mech.* **66**, 67–80.

Chandra, B. and Smylie, D. E., 1972. A laboratory model of thermal convection under a central force field. *Geophys. Fluid Dyn.* **3**, 211–224.

Chandrasekhar, S., 1961. *Hydrodynamic and Hydrodynamic Stability*, Oxford University Press.

Chen, M. M. and Whitehead, J. A., 1968. Evolution of two-dimensional periodic Rayleigh convection cells of arbitrary wavenumbers. *J. Fluid Mech.* **31**, 1–15.

De la Cruz-Reyna, S., 1970. Asymmetric convection in the upper mantle. *Geofis. Int.* **10**, 49–56.

De la Cruz-Reyna, S., 1976. The thermal boundary layer and seismic focal mechanisms in mantle convection. *Tectonophysics* **35**, 149–160.

Dubois, M. and Bergé, P., 1978. Experimental study of the velocity field in Rayleigh–Bénard convection. *J. Fluid Mech.* **85**, 641–653.

Elder, J., 1976. *The Bowels of the Earth*, Oxford University Press.

Elder, J., 1977 Thermal convection. *J. Geol. Soc. Lond.* **133**, 293–309

Gross, M. J., 1967. Laboratory analogies for convection problems. In *Mantles of the Earth and Terrestrial Planets* (Ed. Runcorn, S. K.), 499–503, Interscience Publishers, New York.

Gross, M. J. and Porter, J. E., 1966. Electrically induced convection in dielectric liquids. *Nature* **212**, 1343–1346.

Holden, J. C. and Vogt, P. R., 1977. Graphic solutions to problems of plumacy. *EOS, Trans. Am. Geophys. Un.*, 573–580.

Houston, M. M. and De Bremaecker, J. C., 1975. Numerical models of convection in the upper mantle. *J. Geophys. Res.* **80**, 742–751.

Jacoby, W. R., 1976. Paraffin model experiment of plate tectonics. *Tectonophysics* **35**, 103–113.

Koschmieder, E. L., 1974. Bénard convection. In *Advances in Chemical Physics* (Eds. Prigogine, I. and Rice, S. A.), Vol. 26, 177–212, John Wiley, New York.

Krishnamurti, R., 1970a. On the transition to turbulent convection, 1. The transition from two- to three-dimensional flow. *J. Fluid Mech.* **42**, 295–307.

Krishnamurti, R., 1970b. On the transition to turbulent convection, 2. The transition to time-dependent flow. *J. Fluid Mech.* **42**, 309–320.

Krishnamurti, R., 1973. Some further studies on the transition to turbulent convection. *J. Fluid Mech.* **60**, 285–303.

Kulacki, F. A. and Emara, A. A., 1977. Steady and transient thermal convection in a fluid layer with uniform volumetric energy sources. *J. Fluid Mech.* **83**, 375–395.

Kulacki, F. A. and Goldstein, R. J., 1972. Thermal convection in a horizontal fluid layer with uniform volumetric energy sources. *J. Fluid Mech.* **55**, 271–287.

Kulacki, F. A. and Goldstein, R. J., 1975. Hydrodynamic instability in fluid layers with uniform volumetric energy sources. *Appl. Sci. Res.* **31**, 81–97.

Kulacki, F. A. and Nagle, M. E., 1975. Natural convection in a horizontal fluid layer with volumetric energy sources. *J. Heat Transfer.* **91**, 204–215.

Liang, S. F. and Acrivos, A., 1970. Experiments on buoyancy driven convection in non-Newtonian fluid. *Rheol. Acta* **9**, 447–455.

McKenzie, D. P. and Richter, F., 1976. Convection currents in the Earth's mantle. *Scientific American* **235**, 72–89.

McKenzie, D. P., Roberts, J. M. and Weiss, N. O., 1974. Convection in the Earth's mantle. Towards a numerical simulation. *J. Fluid Mech.* **62**, 465–538.

O'Nions, R. K., Hamilton, P. J. and Evensen, N. M., 1980. The chemical evolution of the Earth's mantle. *Scientific American* **242**, 120–133.

Oxburgh, E. R. and Turcotte, D. L., 1978. Mechanisms of continental drift. *Rep. Prog. Phys.* **41**, 1249–1312.

Palm, E., 1975. Non-linear thermal convection. *Ann. Rev. Fluid Mech.* **7**, 39–61.

Palm, E., 1978. Thermal convection (invited review paper delivered at the European Geophysical Society Meeting, Strasbourg, 1978).

Parmentier, E. M., Turcotte, D. L. and Torrance, K. E., 1976. Studies of finite amplitude non-Newtonian thermal convection with application to convection in the Earth's mantle. *J. Geophys. Res.* **81**, 1839–1846.

Parsons, B. and McKenzie, D. P., 1978. Mantle convection and the thermal structure of the plates. *J. Geophys. Res.*, **83**, 4485–4496.

Richter, F. M. 1973. Dynamical models for sea-floor spreading. *Rev. Geophys. Space Phys.* **11**, 223–287.

Richter, F. M., 1978. Experiments on the stability of convection rolls in fluids whose viscosity depends on temperature. *J. Fluid Mech.* **89**, 553–560.

Richter, F. M. and Daly, S. F., 1978. Convection models having a multiplicity of large horizontal scales. *J. Geophys. Res.* **83**, 4951–4956.

Richter, F. M. and Parsons B., 1975. On the interaction of two scales of convection in the mantle. *J. Geophys. Res.* **80**, 2529–2541.

Roberts, P. H., 1967. Convection in horizontal layers with internal heat generation: theory. *J. Fluid Mech.* **30**, 33–49.

Schubert, G., Froidevaux, C. and Yuen, D. A., 1976. Oceanic lithosphere and asthenosphere: thermal and mechanical structure. *J. Geophys. Res.* **81**, 3525–3540.

Schwiderski, E. W., 1972. Current dependence of convection in electrolytically heated fluid layers. *Phys. Fluids.* **15**, 1189–1196.

Schwiderski, E. W. and Schwab, H. J. A., 1971. Convection experiments with electrolytically heated fluid layers. *J. Fluid Mech.* **48**, 703–719.

Smylie, D. E., 1966. Thermal convection in dielectric liquids and modelling in geophysical fluid dynamics. *Earth Planet. Sci. Lett.* **1**, 339–342.

Spiegel, E. A. and Veronis, G., 1960. On the Boussinesq approximation for a compressible fluid. *Astrophys. J.* **131**, 442–447.

Thirlby, R., 1970. Convection in an internally heated fluid layer. *J. Fluid Mech.* **44**, 673–686.

Tozer, D. C., 1965. Heat transfer and convection currents. *Phil. Trans. R. Soc. Lond.* **A258**, 252–271.

Tozer, D. C., 1967. Some aspects of thermal convection theory for the Earth's mantle. *Geophys. J. R. Astr. Soc.* **14**, 395–402.

Tozer, D. C., 1973. Thermal plumes in the Earth's mantle. *Nature* **244**, 398–400.

Tozer, D. C., 1977a. The thermal state and evolution of the Earth and terrestrial planets. *Sci. Prog. Oxf.* **64**, 1–28.

Tozer, D. C., 1977b. Terrestrial planet evolution and the observational consequences of their formation. In *The Origin of the Solar System* (Ed. Dermott, S. F.), 433–462, John Wiley, London.

Tritton, D. J., 1977. *Physical Fluid Dynamics*, Van Nostrand Reinhold, Wokingham, England.

Tritton, D. J. and Davies, P. A., 1980. Instabilities in geophysical fluid dynamics. In *Hydrodynamic Instabilities and the Transition to Turbulence* (Eds Swinney, H. L. and Gollub, J. P.), Topics in Applied Physics, Springer-Verlag, Berlin (in press).

Tritton, D. J. and Zarraga, M. N., 1967. Convection in horizontal layers with internal heat generation: experiments. *J. Fluid Mech.* **30**, 21–31.

Turcotte, D. L. and Oxburgh, E. R., 1972. Mantle convection and the new global tectonics. *Ann. Rev. Fluid Mech.* **4**, 33–68.

Turner, J. S., 1973a. Convection in the mantle: a laboratory model with temperature dependent viscosity. *Earth. Planet. Sci. Lett.* **17**, 369–374.

Turner, J. S., 1973b. *Bouyancy Effects in Fluids*, Cambridge University Press.

Tveitereid, M. and Palm, E., 1976. Convection due to internal heat sources. *J. Fluid Mech.* **76**, 481–499.

Whitehead, J. A., 1972. Moving heaters as a model of continental drift. *Phys. Earth Planet. Int.* **5**, 199–212.

Whitehead, J. A., 1976. Convection models: laboratory versus mantle. *Tectonophysics* **35**, 215–228.

Whitehead, J. A. and Chan, G. L., 1976. Stability of Rayleigh–Bénard convection rolls and bimodal flow at moderate Prandtl number. *Dyn. Atmos. Oceans* **1**, 33–49.

Whitehead, J. A. and Luther, D. S., 1975. Dynamics of laboratory diapir and plume models. *J. Geophys. Res.* **80**, 705–717.

Whitehead, J. A. and Parsons, B., 1978. Observations of convection at Rayleigh numbers up to 760,000 in a field with large Prandtl number. *Geophys. Astrophys. Fluid Dyn.* **9**, 201–217.

Theory and Experiment in Cellular Convection

R. KRISHNAMURTI

Department of Oceanography and Geophysical Fluid Dynamics Institute, Florida State University, Florida, USA

I wish to discuss the problem of convection in a very simple, idealized situation. In his chapter (p. 225) Dr Peter Davies discusses possible application to the Earth's mantle. Let us consider a horizontal layer of fluid uniformly and steadily heated below and cooled above. Even in such a simple configuration many different kinds of flow are observed (see Fig. 1). We will start the discussion at low Rayleigh number, where theory and experiment agree, and then proceed to higher Rayleigh numbers where I will describe some new laboratory observations (at $Ra = 10^6$–$10^7, Pr = 7$).

The governing equations in the Boussinesq approximation may be written as follows:

$$\nabla^2 u_i + \lambda_i \theta - \frac{\partial P}{\partial x_i} = \frac{1}{Pr}\left(\frac{\partial u_i}{\partial_t} + u_j \frac{\partial u_i}{\partial x_j}\right), \tag{1}$$

$$\frac{\partial u_j}{\partial x_j} = 0, \tag{2}$$

$$\nabla^2 \theta + R u_j \lambda_j = \frac{\partial \theta}{\partial t} + u_j \frac{\partial \theta}{\partial x_j}, \tag{3}$$

where $i = 1, 2, 3; \lambda_i = (0, 0, 1); u_i = i$th velocity component; $\theta =$ perturbation temperature; $= T - T_s$; $T =$ actual temperature field; $T_s =$ temperature field in the static state; $P =$ modified pressure; $R =$ Rayleigh number $= (g\alpha/\kappa\nu)\Delta T d^3$; $Pr =$ Prandtl number $= \nu/\kappa$; $g =$ acceleration of gravity; $\alpha =$ thermal expansion coefficient; $\kappa =$ thermal diffusivity; $\nu =$ kinematic viscosity; $d =$ layer depth; and $\Delta T =$ temperature difference between bottom and top of the layer.

At R less than a critical value R_c (which is 1708 for slip-free, perfectly conducting boundaries) the state of no motion and heat conduction down the gradient is a stable state. At $R = R_c$ an infinity of possible cellular flows,

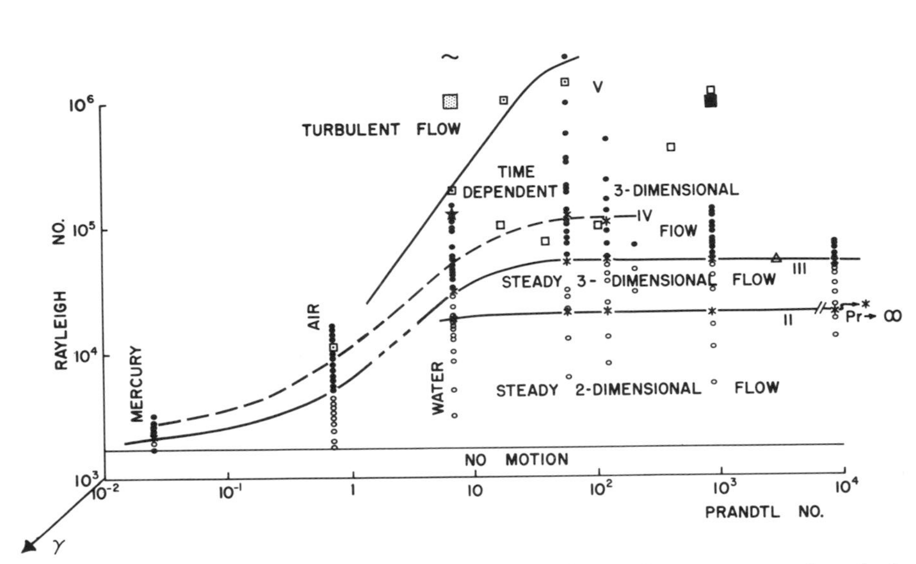

Figure 1. Regime diagram for the $\gamma = 0$ plane, where γ is a measure of vertical asymmetry (see text). The circles represent steady flow, the circular dots time-dependent flow. The stars represent transition points.

■ $Ra = 10^6$, $Pr = 10^3$; cells plus hot spots.
★ $Ra = 10^5$, $Pr = 7$; cells, some bubbles (transitional).
▩ $Ra = 10^6$, $Pr = 7$; no cells, only bubbles.
~ $Ra = 2 \times 10^6$, $Pr = 7$; no cells, only bubbles, occasional large-scale flow.
~ $Ra = 10^7$, $Pr = 7$; large-scale flow.
~ $Ra = 2 \times 10^7$, $Pr = 7$; large-scale flow.

including rolls, rectangles, squares, hexagons, become marginally stable if these disturbances have an overall size which is very nearly equal to the depth of the fluid. This degeneracy is resolved in the following manner.

Steady finite amplitude solutions to the Eqns (1)–(3) are obtained by formally expanding

$$u_i = \sum_{n=1}^{\infty} \varepsilon^n u_i^{(n)},$$

$$\theta = \sum_{n=1}^{\infty} \varepsilon^n \theta^{(n)},$$

$$P = \sum_{n=1}^{\infty} \varepsilon^n P^{(n)},$$

$$R = \sum_{n=0}^{\infty} \varepsilon^n R^{(n)}.$$

The as yet undetermined amplitude ε is used as an ordering parameter to generate an infinite set of linear inhomogeneous partial differential equations. The lowest order in ε yields the linear marginal stability problem if $R^{(0)}$ is identified with R_c. The higher order equations are solved sequentially, with $R^{(n)}$ determined in such a way as to satisfy the solubility condition on the inhomogeneous equations governing $u^{(n+1)}$, $\theta^{(n+1)}$. The amplitude ε is then determined in terms of the Rayleigh number by inverting the expansion of R.

The result of such a procedure still yields an infinity of solutions, corresponding to finite amplitude rolls, hexagons, rectangles, etc. The stability of these non-linear solutions is tested by perturbing each of them by a disturbance field $\tilde{u}_i\,\tilde{\theta}, \tilde{P}$. Linearizing about the steady non-linear solutions $u_i\,\theta, P$ yields the following equations governing the perturbations:

$$\nabla^2 \tilde{u}_i + \lambda_i\, \tilde{\theta} - \frac{\partial \tilde{P}}{\partial x_i} = \frac{1}{Pr}\left(\frac{\partial}{\partial t}\tilde{u}_i + u_j \frac{\partial \tilde{u}_i}{\partial x_j} + \tilde{u}_j \frac{\partial u_i}{\partial x_j}\right), \tag{4}$$

$$\nabla^2 \tilde{\theta} + R\tilde{u}_j\,\lambda_j = \frac{\partial \tilde{\theta}}{\partial t} + u_j \frac{\partial \tilde{\theta}}{\partial x_j} + \tilde{u}_j \frac{\partial \theta}{\partial x_j}, \tag{5}$$

$$\frac{\partial \tilde{u}_j}{\partial x_j} = 0. \tag{6}$$

Again expanding the perturbation fields in powers of ε, we have

$$\tilde{u}_i = \sum_{n=0}^{\infty} \varepsilon^n\, \tilde{u}_i^{(n)},$$

$$\tilde{\theta} = \sum_{n=0}^{\infty} \varepsilon^n \tilde{\theta}^{(n)},$$

$$\tilde{P} = \sum_{n=0}^{\infty} \varepsilon^n\, \tilde{P}^{(n)},$$

$$\sigma = \sum_{n=0}^{\infty} \varepsilon^n\, \sigma^{(n)},$$

where σ is the growth rate of the disturbance.

The result of the analysis to this stage leads to the conclusion (Schlüter, Lortz and Busse, 1965) that for R slightly in excess of R_c only the roll (i.e. two-dimensional) type of cellular flow is stable to infinitesimal disturbances and hence realizable (see Fig. 2(a)).

Hexagonal cells which are certainly realized in nature (see Fig. 2(b)), do not occur in the Rayleigh number–Prandtl number plane of Fig. 1. There is another dimensionless parameter γ, which we may call an asymmetry parameter, which must be sufficiently large, for a given ε, before hexagonal cells can be realized. This parameter γ may be due to the effect of viscosity variation with mean temperature (Segel and Stuart, 1962; Palm, Ellingsen and Gjevik, 1967; Busse, 1967), it may be due to curvature of the conduction

Figure 2. (a) Rolls; (b) hexagons (see text).

temperature profile (Krishnamurti, 1968), or it may be due to asymmetries produced by an imposed vertical mass flux either upwards or downwards through the layer (Krishnamurti, 1975). In this last case γ is the Peclet number defined as

$$\gamma = \frac{w_0 d}{\kappa},$$

where w_0 is the imposed, uniform, vertical velocity.

In this case the governing equations are

$$\nabla^2 u_i + \lambda_i \theta - \frac{\partial P}{\partial x_i} = \frac{1}{Pr}\left(\frac{\partial}{\partial t} u_i + u_j \frac{\partial}{\partial x_j} u_i\right) + \frac{\gamma}{Pr}\frac{\partial u_i}{\partial z}, \tag{7}$$

$$\nabla^2 \theta - R u_j \lambda_j \frac{\partial}{\partial z} T_s = \frac{\partial \theta}{\partial t} + u_j \frac{\partial \theta}{\partial x_j} + \gamma \frac{\partial \theta}{\partial z}, \tag{8}$$

$$\frac{\partial u_j}{\partial x_j} = 0. \tag{9}$$

where T_s is the static temperature field

$$T_s = -\frac{(1-e^{\gamma z})}{(1-e^{\gamma})}.$$

Clearly, as γ approaches zero, Eqns (7)–(9) become identical to Eqns (1)–(3). For γ small, steady non-linear solutions are sought of the form

$$u_i = \sum_{\substack{n=1\\m=0}}^{\infty} \varepsilon^n \gamma^m u_i^{(n,m)},$$

$$\theta = \sum_{\substack{n=1\\m=0}}^{\infty} \varepsilon^n \gamma^m \theta^{(n,m)},$$

$$P = \sum_{\substack{n=1\\m=0}}^{\infty} \varepsilon^n \gamma^m P^{(n,m)},$$

$$R = \sum_{\substack{n=0\\m=0}}^{\infty} \varepsilon^n \gamma^m R^{(n,m)},$$

with the result that steady finite amplitude hexagonal (and only hexagonal) flows are possible at $R<R_c$ here is the critical Rayleigh number at which infinitesimal disturbances (satisfying the linearized version of Eqns (7)–(9) with $\gamma \neq 0$) are marginally stable.

A stability analysis of the steady non-linear solutions to Eqns (7)–(9) for R near R_c yields the regime diagram of Fig. 3(a). A laboratory experimental test of the theory yields the regime diagram of Fig. 3(b).

Although the modified perturbation described above appears to have been

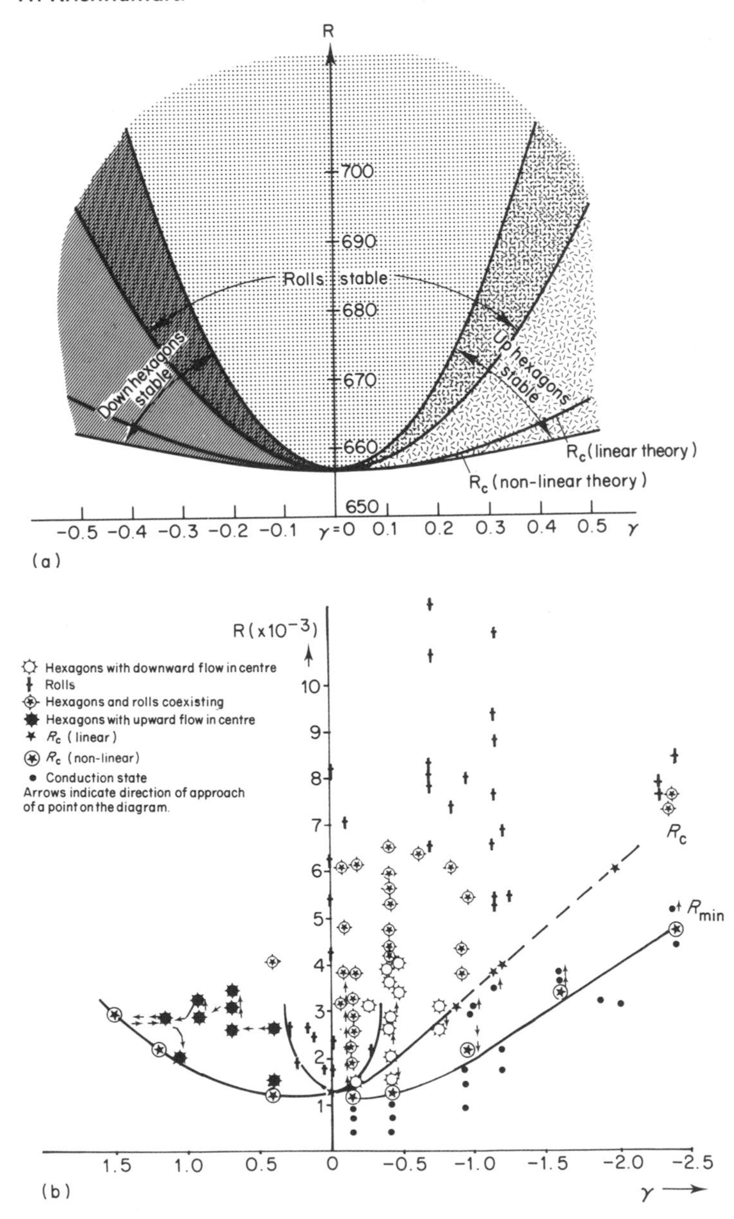

Figure 3. (a) Stability diagram, showing the regions of the R–γ plane where rolls, up- or down-hexagons are stable, for large Prandtl number. (b) Regime diagram of laboratory observations on the R–γ plane. Arrows indicate the sequence of experiments.

Figure 4. x–t representation of cellular convective flow at increasing Rayleigh numbers (from top to bottom), for Prandtl number of $0{\cdot}8 \times 10^3$.

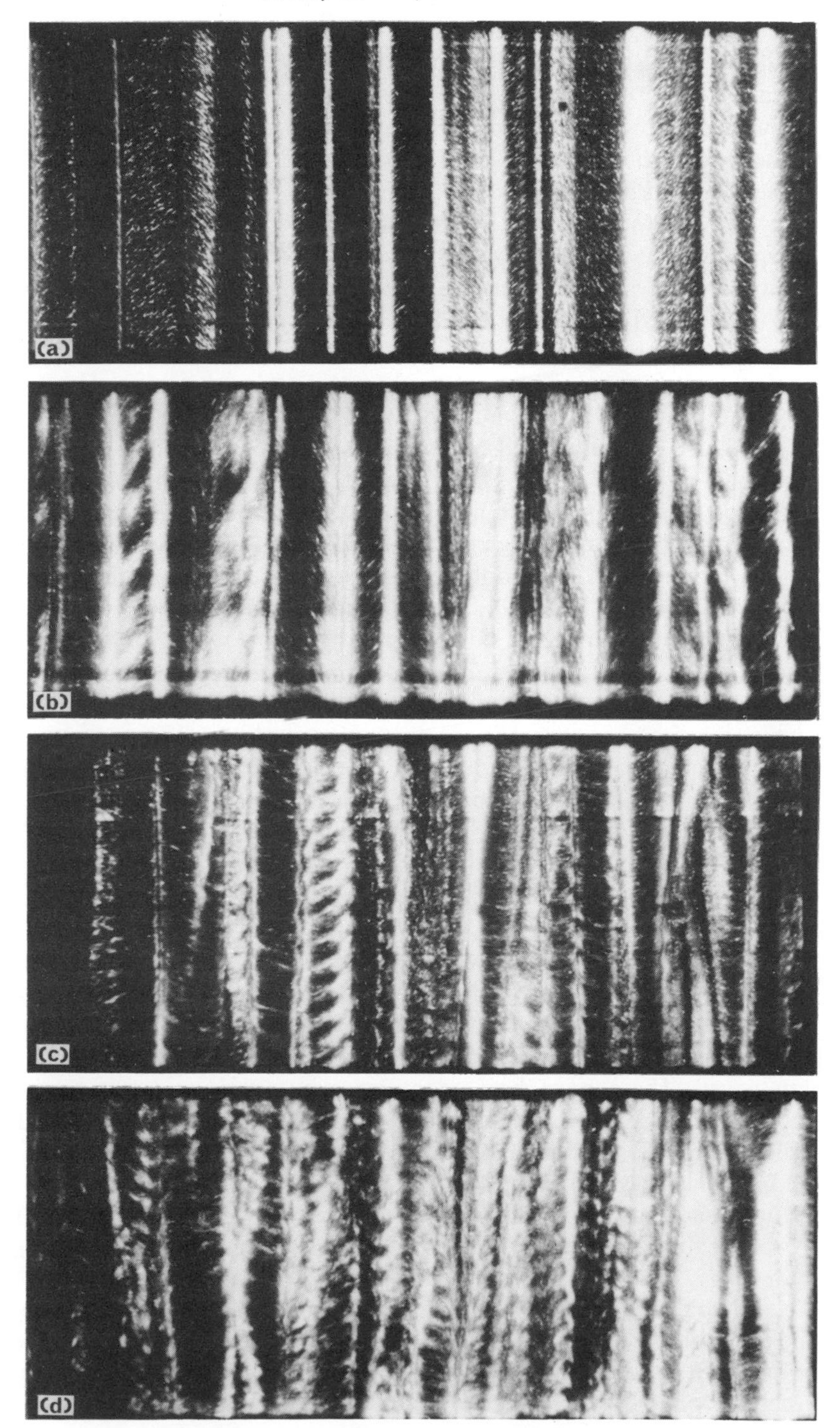
(a)
(b)
(c)
(d)

successful in studying the stability of small amplitude non-linear flows, it is difficult to imagine how it could be applied in general where amplitudes are no longer small.

Returning to the case $\gamma = 0$, the next two transitions that occur with increased Rayleigh number (for $Pr \gtrsim 7$) will be described very briefly. The first of these is a transition at $R = 13R_c$ in which the roll convection becomes unstable, leading to a three-dimensional steady flow. This transition is labelled II in Fig. 1, and has associated with it the second discrete change in slope of the heat flux curve (Krishnamurti, 1970a). The resulting flow is described as a roll having superposed upon it a smaller amplitude, smaller wavelength roll at right angles (Busse and Whitehead, 1971).

The next transition is from steady three-dimensional to time-dependent flow. The transition Rayleigh number is Prandtl number dependent and is labelled III in Fig. 1. It should be mentioned here that all these transitions except the one for $R = R_c$, $\gamma = 0$, are dependent on initial conditions; hysteresis is a common occurrence. Thus the transition point may depend upon previous history including previous Rayleigh number, and imperfections in flow pattern. Figure 1 refers only to transitions observed when the Rayleigh number was increased in small steps from lower values of R, and the flow pattern was allowed to evolve uncontrolled.

One of the most noticeable features of the time dependence is clearly brought out in an x–t representation of the flow. Here x refers to a coordinate line which is a fixed horizontal line in the fluid, and which is being observed as the time t increases. In Figs 4, 5 and 6, that line of fluid was defined by a narrow pencil of light (approximately 3 mm in diameter) illuminating the entire 50 cm width of the tank. (Only 45 cm of this line is seen in these photographs.) Except in the case of Fig. 6, the light beam was near the bottom of the fluid layer, and a similar photograph was obtained with the light beam near the top of the layer.

Figure 4(a) shows steady flow. The bright lines parallel to the t-axis represent cell boundaries which are unchanged with time. Figure 4(b) shows a time-periodic flow in which regions of strong shear (bright regions) repeatedly move from one cell boundary to the next, but always confined within a cell. These have been identified with regions of hot (or cold) temperature anomolies (Krishnamurti, 1970b) which are convected around by the cellular flow. As the Rayleigh number is increased these become more and more frequent (Figs 4(c) and (d)).

Figure 5 shows two x–t photographs, both at Rayleigh number 10^6. The upper one is for $Pr = 0{\cdot}8 \times 10^3$, the lower one is for water, $Pr = 7$. It is clear that for $Pr = 7$, $Ra = 10^6$, *there are no cell boundaries*! The flow consists of hot rising and cold sinking bubbles apparently not confined to any cell. These would appear more closely related to the bubbles of the Howard model (1966), than those hot or cold spots convected by a cellular flow. It should be pointed out that the "bubbles" in Fig. 5(b) probably should not be considered an initial

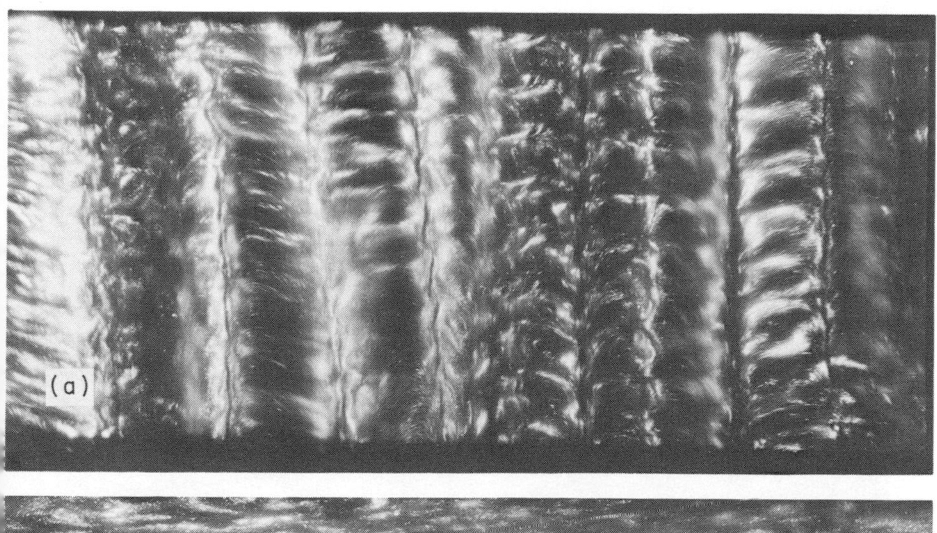

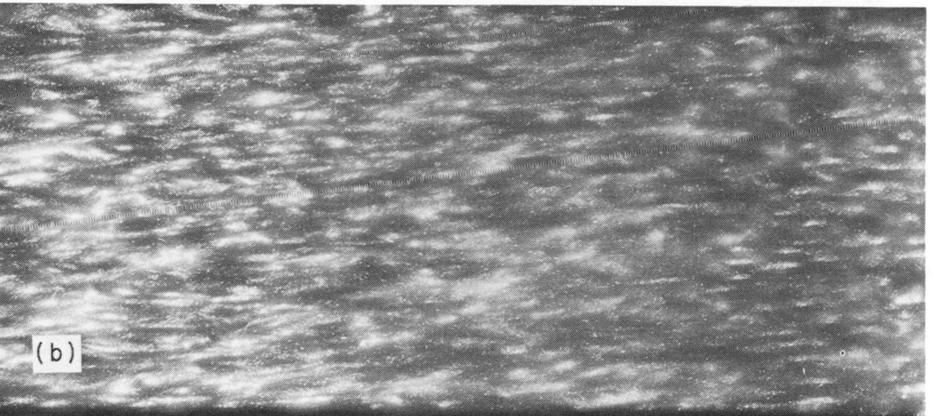

Figure 5. x–t representation at $R = 10^6$ for two different fluids. (a) $Pr = 0{\cdot}8 \times 10^3$; (b) $Pr = 7$.

transient state which would later become organized into cells; this flow pattern was seen to persist for many days.

Figure 6 shows an even more remarkable flow. These are x–t representations at $Pr = 7$, $Ra = 10^7$. Figure 6(a) is for a fluid line near the bottom of the layer, Fig. 6(b) for a fluid line near the top of the layer. Figure 6(a) shows all bubbles drifting from left to right near the bottom of the layer; 6(b) shows bubbles drifting from right to left near the top of the layer. Dye injected into the layer (Fig. 7) was followed, indicating a net Langrangian flow. There appears to be a very dramatic change of scale in the motion field. Further studies indicated that:

(1) Horizontal temperature gradients in the boundary were measured and found to be 2 to 3 orders of magnitude too small to explain the observed flow rate of some 45 cm in 3 minutes.

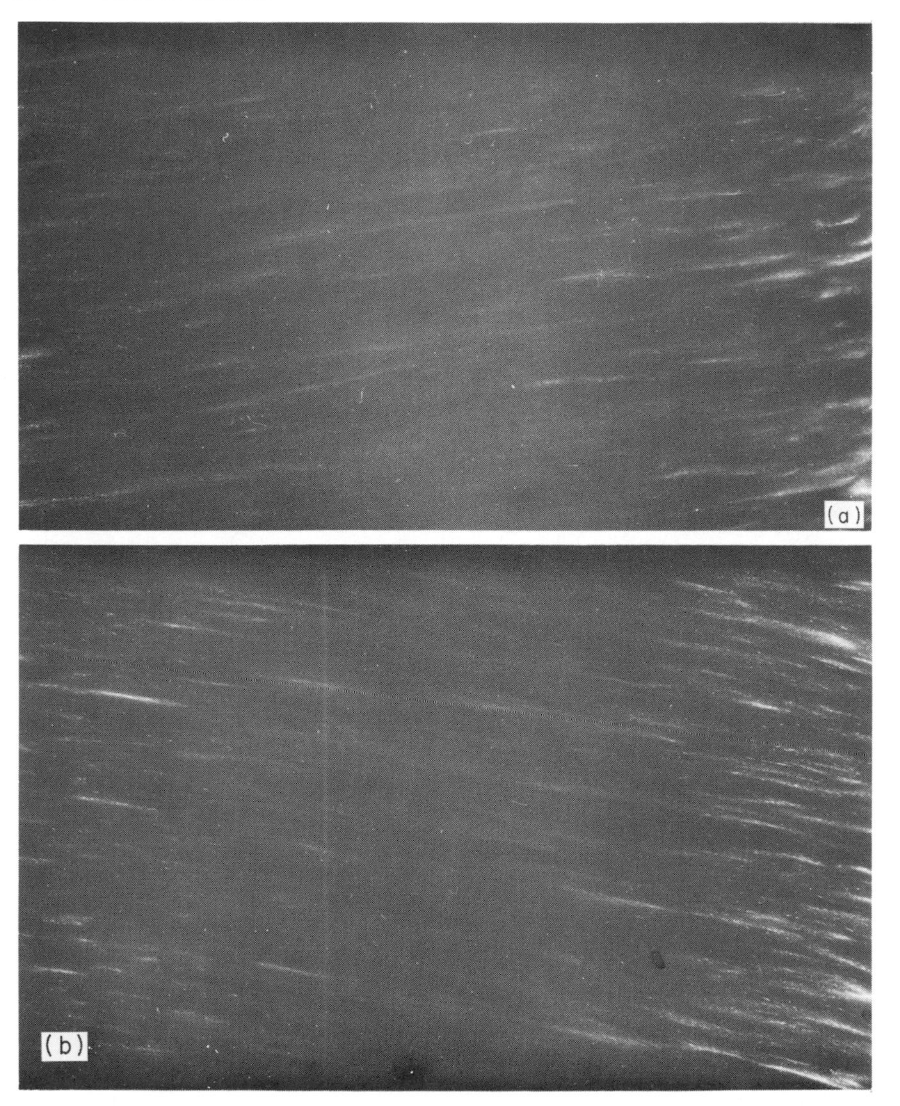

Figure 6. x–t representation at $R = 2{\cdot}4 \times 10^6$, $Pr = 7$, showing the large scale shearing flow. (a) For a line of fluid near the bottom of the layer; (b) for a line of fluid near the top of the layer. The layer is 50 cm × 50 cm × 5 cm deep.

(2) The tank was levelled to $\pm 0{\cdot}0005$ inches in 12 inches as with all the convection experiments.

Yet there was apparently some left to right asymmetry in the experimental apparatus which *selected* the *direction* of the flow, although it could not be considered to be driving the flow.

(3) The plumes (or elongated bubbles) were observed to tilt with height from lower right to upper left. The magnitude of the flow in these bubbles (u', w') was

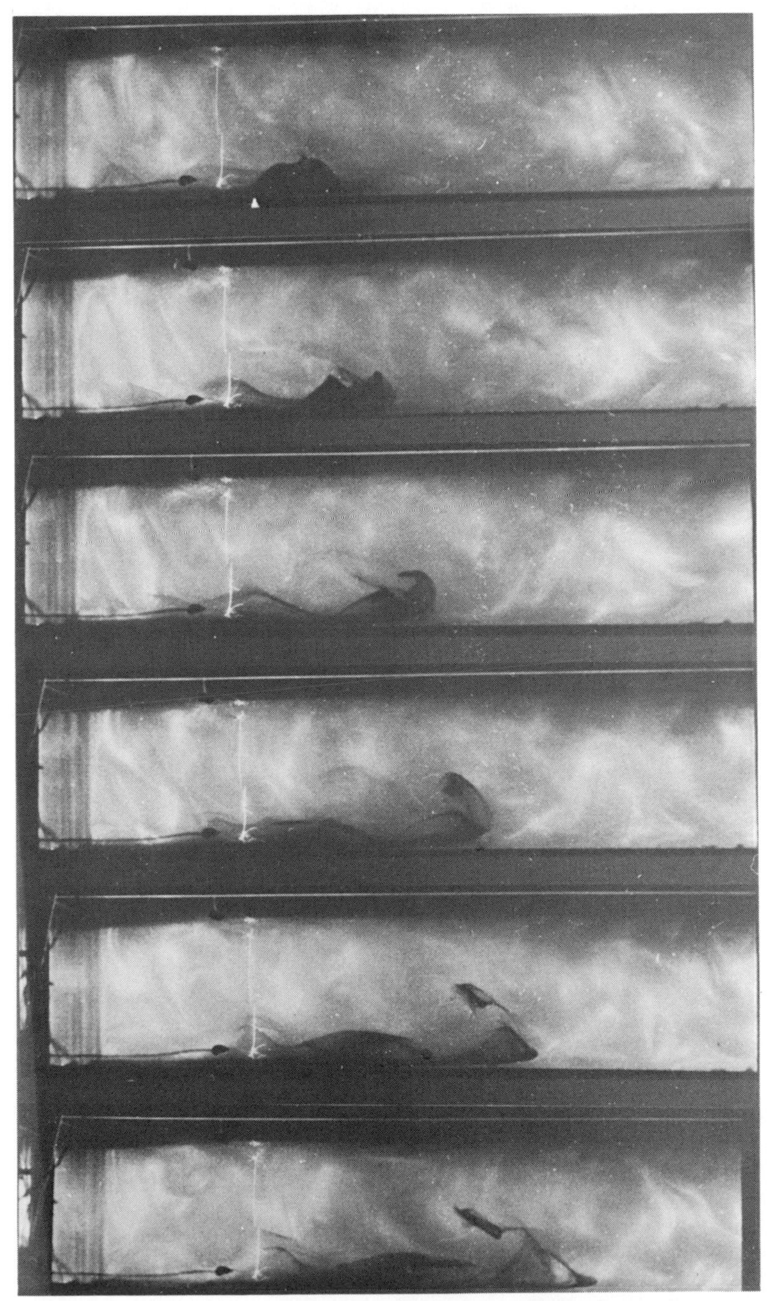

Figure 7. The net drift to the right of a patch of dark dyed fluid (thymol blue). Each photograph is of an x–z plane in the fluid, seen at successive times.

of such a magnitude that the Reynolds stress $(\overline{u'w'})$ (where the bar represents an x and time average) could balance the viscous dissipation of the large scale flow $\bar{u}$:

$$(\overline{u'w'}) - \nu \frac{d\bar{u}}{dz} = \text{constant}.$$

(4) For $Pr = 7$, the first disappearance of cells and appearance of bubbles seems to occur at $Ra \simeq 10^5$.

(5) At $R = 2 \times 10^6$, $Pr = 7$, bubbles would be observed for several days with no noticeable drift. Then the large-scale flow may be observed for several days. and then stop rather abruptly, leaving only bubbles once more.

(6) At $R = 2 \times 10^7$, $Pr = 7$, the large-scale flow was observed continuously for 7 or 10 days (the duration of the observations).

Discussion

The transition in flow from one like Fig. 5(a) to Fig. 5(b) reminds me of a phase change such as the melting of a solid. In Fig. 5(a), fluid parcels are always confined within their own cell boundaries. In Fig. 5(b), and especially in Fig. 6, fluid parcels are no longer bound or confined in space.

The observed shift in the scale of motion from small (horizontal scale comparable to layer depth) to a large horizontal scale has clearly some important implications. If such a shift in scale occurred in a shallow layer in the asthenosphere, for example, there would no longer be a need to explain large horizontal scales of motion observed at the Earth's surface by correspondingly deep convection.

Acknowledgement

This research has been sponsored by the Office of Naval Research, Fluid Dynamics Branch. This support is gratefully acknowledged.

References

Busse, F. H., 1967. The stability of finite amplitude cellular convection and its relation to an extremum principle. *J. Fluid Mech.* **30**, 625.

Busse, F. H. and Whitehead, J. A., 1971. Instabilities of convection rolls in a high Prandtl number fluid. *J. Fluid Mech.* **47**, 305–320.

Howard, L. N., 1966. *Proceedings of the Eleventh International Congress of Applied Mechanics*, 1109–1115.

Krishnamurti, R., 1968. Finite amplitude convection with changing mean temperature. 1, Theory. *J. Fluid Mech.* **33**, 445–455.

Krishnamurti, R., 1970. On the transition to turbulent convection. 1, The transition from two- to three-dimensional flow. *J. Fluid Mech.* **42**, 295–307.

Krishnamurti, R., 1970b. On the transition to turbulent convection. 2, The transition to time-dependent flow. *J. Fluid Mech.* **42**, 308–320.

Krishnamurti, R., 1975. On cellular cloud patterns. 1, Mathematical model. *J. Atmos. Sci.* **32**(7) 1353–1363.

Palm, E., Ellingsen, T. and Gjevik, B., 1967. On the occurrence of cellular motion in Benard convection. *J. Fluid Mech.* **30**, 651.

Schlüter, A., Lortz, D. and Busse, F., 1965. On the stability of steady finite amplitude convection. *J. Fluid Mech.* **23**, 129–144.

Segel, L. A. and Stuart, J. T., 1962. On the question of the preferred mode in cellular thermal convection. *J. Fluid Mech.* **13**, 289–306.

Convection in Multi-Layer Systems

A.-T. NGUYEN, F. A. KULACKI*

Department of Mechanical Engineering,
The Ohio State University, Columbus, Ohio, USA

Introduction

A goodly number of numerical and experimental studies now exist on natural convection in a horizontal layer containing a single-phase fluid with uniformly distributed heat sources (see the review given by Kulacki and Emara, 1977). These studies have established the relation between the Nusselt and Rayleigh numbers for essentially two-dimensional convection in layers with either two isothermal boundaries or an isothermal upper boundary and an adiabatic lower boundary. This relation has been verified by numerical solutions of the governing equations up to $Ra \approx 10^8$, and calculated temperature fields are in generally good agreement with available measurements. For the most part, numerical solutions are available for laminar convection and model Prandtl numbers ($0{\cdot}1 \lesssim Pr \lesssim 20$). In the geophysical context, several studies have appeared for fluid layers with non-uniform internal heating, large (infinite) Prandtl number, and temperature-dependent thermophysical properties. However, none of these, apparently, have been verified by suitable laboratory experiments. Thus, much remains to be done in the way of basic experimentation on convection in a fluid layer which more faithfully represents the situation expected in the Earth's mantle.

This chapter reports the preliminary results of an experimental study on convection in a system that is one step removed from the aforementioned single phase layer, namely a layer containing two immiscible, stably stratified sublayers. The upper sublayer contains no internal energy sources while the lower sublayer is heated only by volumetric energy generation. The lower boundary of the layer is adiabatic, while the upper boundary is isothermal. Both boundaries are rigid, i.e., zero slip. The internally heated sublayer is water, and the upper sublayer is silicone oil. Thus, the system under investigation here represents, in a limited sense, a well-differentiated upper–lower mantle. Measurements of overall Nusselt numbers at the upper surface as a function of sublayer depth ratio and Rayleigh numbers are

* Present address: Department of Mechanical and Aerospace Engineering, University of Delaware, Newark, Delaware, USA.

presented as the primary results of interest here. The heat transfer data are correlated in terms of the Rayleigh number of the heated sublayer, with sublayer depth ratio as a parameter. Data are also obtained on the time-averaged temperature profiles in the two sublayers.

The only reported study of convection in two-fluid layers like the one considered in the present work is that of Schramm and Reineke (1978). Their study considers a layer bounded from below and above by rigid, isothermal surfaces of equal temperatures. Numerical results for mean temperature fields compared well with interferometric measurements for a layer with $\rho_1/\rho_2 = k_1/k_2 = 0{\cdot}1$ and a modified Archimedes number, $gL^3/\nu_2\alpha_2$, of $1{\cdot}4 \times 10^{10}$. (The subscripts denote the sublayer numbers; see Fig. 1.) They also

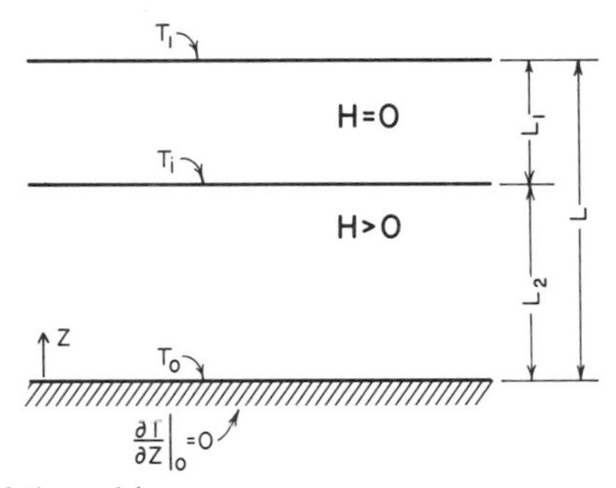

Figure 1. Geometry of the sublayers.

found that the ratio of the upward to downward heat transfer passes through a maximum at about $L_1/L_2 = 0{\cdot}33$. This is attributed to a limiting role played by Rayleigh–Bénard convection in the upper sublayer once fully developed convection is established there and the lower sublayer thickness is increased.

Apparatus and procedure

An apparatus similar to that used by Kulacki and Emara (1977) is used in the present study. The upper sublayer is General Electric Silicon Fluid (No. 96-5). Its thermophysical properties at 24·4 °C are $\rho_1 = 0{\cdot}918$ g/cm^3, $k_1 = 0{\cdot}00116$ W/cm °C, $\beta_1 = 1{\cdot}08 \times 10^{-3}$ °C, $\nu_1 = 0{\cdot}0506$ cm^2/s, $\alpha_1 = 0{\cdot}00068$ cm^2/s, and $Pr_1 = 74{\cdot}2$. Thus, for the present experiments, $\rho_1/\rho_2 = 0{\cdot}92$, $k_1/k_2 = 0{\cdot}19$ and $Pr_1/Pr_2 = 11{\cdot}8$, where the lower (heavier) sublayer is water containing some dissolved copper sulphate to make it electrically conductive. In the evaluation of the properties, those of the upper sublayer are calculated at the mean temperature, while those of the lower sublayer are determined at the temperature of the lower surface. Internal heat generation is provided in the lower sublayer by the passage of 60 Hz electric current horizontally between two vertical boundaries of the layer.

Time-averaged temperature profiles in the layer are obtained with a small diameter glass-encased thermocouple probe which can be positioned with a micrometer head to within 0·0025 cm of a desired depth in the layer. All of the experiments are started with the layer at thermal equilibrium, and the steady state is determined when the recorded output of the probe shows no variation with time.

Results

Nusselt numbers of steady convection for $0 \leqslant L_1/L_2 \leqslant 0{\cdot}433$ are presented in Fig. 2. The data are correlated in the form. $Nu = CRa_2^m$, where the Nusselt number is defined in terms of $T_0 - T_i$, the total layer thickness, and an effective conductivity defined as $k_e = L(L_1/k_1 + L_2/k_2)^{-1}$. Those correlation variables are convenient because they permit a comparison of the present data to those of Kulacki and Emara (1977) for the case $L_1/L_2 = 0$. The ranges of the Rayleigh and Prandtl numbers for each sublayer are given in Table I.

TABLE I *Non-dimensional parameters in the correlations of the steady-state heat transfer data*

L_1/L_2*	Ra_1	Ra_2	Pr_1	Pr_2
0·0	—	$1{\cdot}08 \times 10^4$–$2{\cdot}17 \times 10^{12}$	—	5·9–6·3
0·035–0·039	660–$2{\cdot}7 \times 10^6$	$4{\cdot}66 \times 10^7$–$2{\cdot}24 \times 10^{11}$	53·4–66·8	3·7–5·3
0·110–0·111	$4{\cdot}5 \times 10^3$–$1{\cdot}9 \times 10^8$	$1{\cdot}91 \times 10^6$–$1{\cdot}53 \times 10^{11}$	66·1–73·8	5·9–6·2
0·431–0·433	$9{\cdot}9 \times 10^3$–10^{10}	$1{\cdot}7 \times 10^4$–$4{\cdot}75 \times 10^{11}$	60·7–76·5	3·8–6·0

* The most frequently measured value is given in Fig. 2.
$Ra_2 = (g\beta_2/\alpha_2 \nu_2)(H_2 L_2^5/2k_2)$.
$Ra_1 = (g\beta_1/\alpha_1 \nu_1)(T_i - T_1) L_1^3$.

From Fig. 2, it can be seen that for small values of L_1/L_2, the overall Nusselt number exhibits a rapid reduction below that for $L_1/L_2 = 0$. This occurs in the present data for a range of values of Ra_1 that spans conduction to turbulent Rayleigh–Bénard convection. As the sublayer thickness ratio is increased, the reduction in the Nusselt number is less severe. For the present data, it appears that a relative minimum in the Nusselt number occurs at $L_1/L_2 \approx 0{\cdot}035$. This is illustrated in Fig. 3 for several values of Ra_2, where the dashed lines indicate the expected trends owing to the known limits at $L_1/L_2 = 0$. The present data are, of course, insufficient to determine whether a maximum in the heat transfer coefficient exists as was observed by Schramm and Reineke (1978) in a layer with downward and upward heat transfer. However, very recent measurements at $L_1/L_2 = 0{\cdot}75$ indicate that the overall Nusselt number asymptotically approaches the correlation for $L_1/L_2 = 0$. These measurements are not presented here owing to their tentative nature.

Steady-state temperature profiles are presented in Figs 4–7 for several values of L_1/L_2 and lower sublayer Rayleigh numbers, Ra_2, well within the turbulent regime of convection. It is evident that convection exists within the

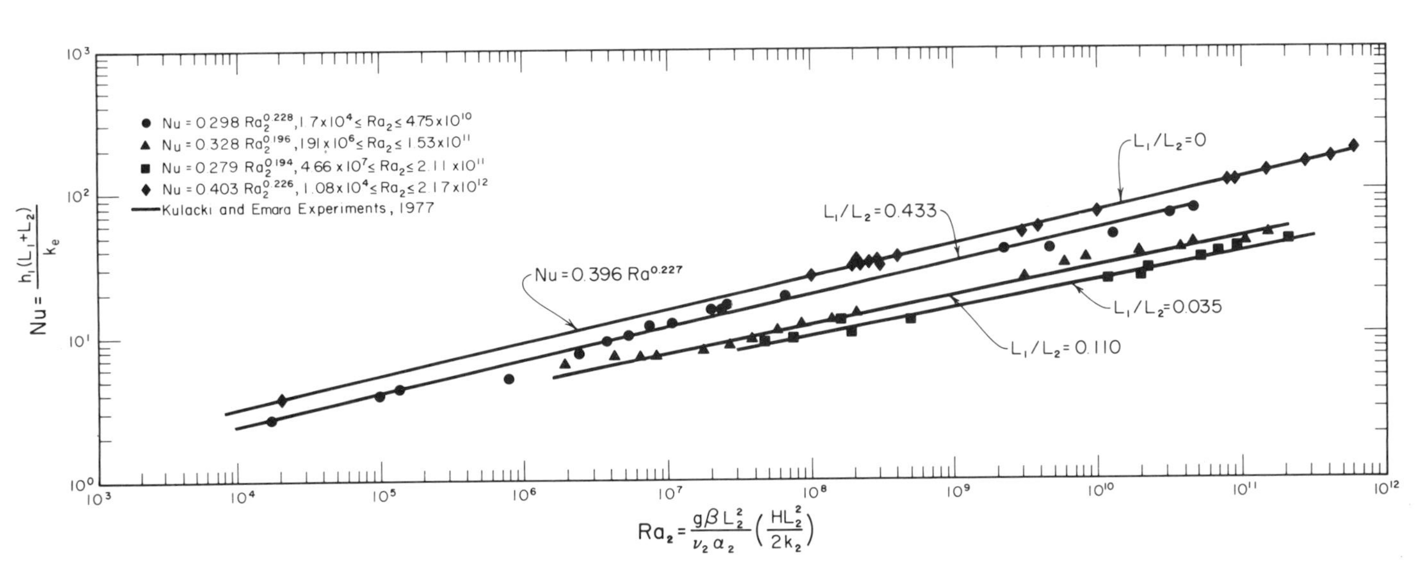

Figure 2. Steady-state heat transfer correlations.

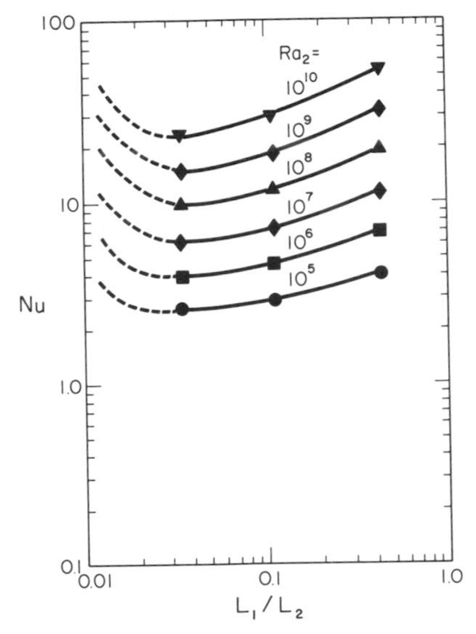

Figure 3. Variation of Nusselt number with sublayer thickness ratio at constant Rayleigh number.

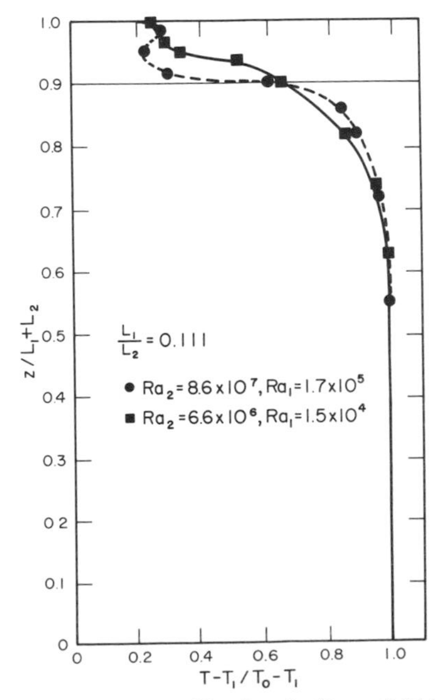

Figure 4. Steady-state temperature profiles for $L_1/L_2 = 0{\cdot}111$.

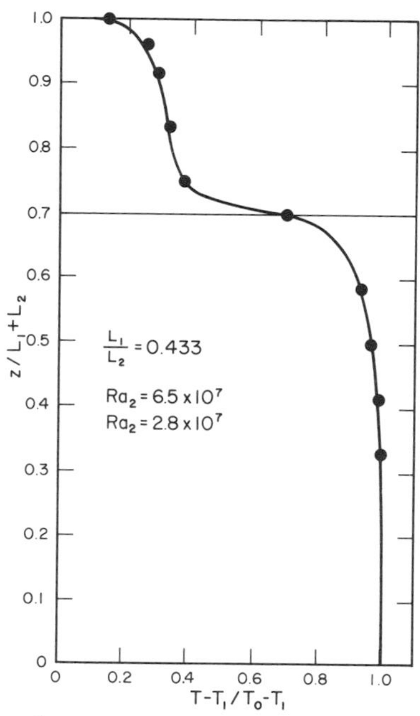

Figure 5. Steady-state temperature profile for $L_1/L_2 = 0{\cdot}433$.

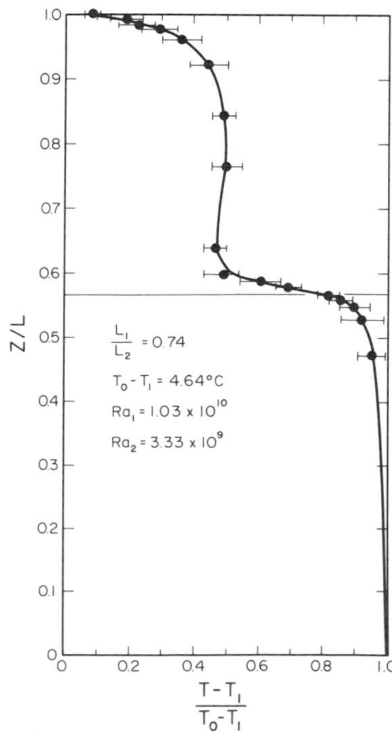

Figure 6. Steady-state temperature profile for $L_1/L_2 = 0{\cdot}74$.

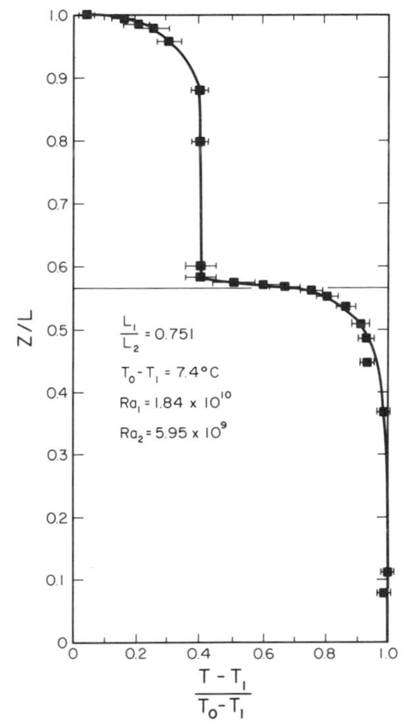

Figure 7. Steady-state temperature profile for $L_1/L_2 = 0{\cdot}751$.

upper sublayer in all of these experiments, and temperature profiles exhibit the characteristic "s" shape of Rayleigh–Bénard convection. However, in all of the present experiments, temperature profiles in the upper sublayer have been found to be asymmetric. This is a result of both probe interference and the different hydrodynamic boundary conditions at the sublayer interface and the upper surface of the layer.

Included in Figs 6 and 7 are bars through the data points showing the maximum variation about the mean. Near the upper surface and at the sublayer interface, turbulent fluctuations are seen to be much larger than in the core regions of either sublayer. Although probe interference may be a factor in these measurements, it is clear that thin boundary layers exist on either side of the sublayer interface, with intense eddy mixing being the dominant mode of upward heat transport. This transport process, however, does not produce observable deformations of the sublayer interface. This is probably a result of the stable stratification of the two sublayers and the minor role that upward bound eddies play in the heat transfer process in the internally heated sublayer.

For the most part, the present data correspond to the situation wherein turbulent convection exists in the lower sublayer, and laminar to weakly turbulent convection exists in the upper sublayer. (Chu and Goldstein (1973) present a discussion of interferometric measurements of the temperature fields of Rayleigh–Bénard convection for $7 \times 10^3 \leqslant Ra_1 \leqslant 7 \times 10^6$. In the present work, the sublayer interface lies hydrodynamically between a zero slip and a zero shear surface since $\mu_1/\mu_2 \approx 5$. This, however, is not expected to make a great deal of difference in the qualitative interpretation of the flow field in the upper sublayer, and the descriptions and interpretations of Chu and Goldstein can be applied here.) Thus, the data for the several values of L_1/L_2 can all be compared on a self-consistent basis. When this is done, it can be seen that, with the present correlation variables, no combination of parameters is possible that will easily combine all of the data. The turbulent transport process at the sublayer interface, therefore, is probably the controlling factor on the maximum temperature difference across the layer. In the present case, one can expect a complicated relation between the Prandtl numbers of the two sublayers, the sublayer thickness ratio, and Ra_2 to describe the eddy exchange process across the sublayer interfaces.

Acknowledgement

The support of the U.S. Nuclear Regulatory Commission under Contract NRC-04-74-149 and Department of Mechanical Engineering is gratefully appreciated.

References

Chu, T. Y. and Goldstein, R. J., 1973. Turbulent convection in a horizontal layer of water. *J. Fluid Mech.* **60**, 141–159.

Kulacki, F. A. and Emara, A. A., 1977. Steady and transient thermal convection in a fluid layer with uniform volumetric energy sources. *J. Fluid Mech.* **83**, 375–395.

Schramm, R. and Reineke, H. H., 1978. Natural convection in a horizontal layer of two different fluids with internal heat sources. *Proc. Sixth Int. Heat Transfer Conf.* **2**, 299–304, National Research Council of Canada.

Convection of a Very Viscous Fluid Heated from Below

D. J. TRITTON, D. M. RAYBURN, M. A. FORREST

School of Physics, The University, Newcastle upon Tyne, UK

1 Introduction

Theories of convection in the Earth's mantle and the interiors of other terrestrial planets are evidently dependent on an adequate knowledge and understanding of the basic fluid dynamics of thermal convection. Laboratory experiments play an important role in gaining this knowledge and understanding, and in this article we describe one such set of experiments. At the end we offer some comments on their geophysical implications.

The main reason why the very extensive existing work on convection does not provide all the information needed lies in significant differences in the parameter ranges from the applications (such as meteorology or engineering) that have motivated that work. One such difference is the high value of the Prandtl number (the ratio of the kinematic viscosity to the thermal diffusivity) of the mantle. It is this aspect on which the present work focuses. Another factor almost certainly of importance in the mantle is the variation of viscosity with temperature. However, this effect is deliberately minimized in the present experiments, as we take the view that it is best to gain an understanding of the basic phenomena with as few complications as possible.

In idealized form, the situation under consideration consists of an infinite upward-facing horizontal plate beneath a semi-infinite expanse of fluid; the plate is maintained at a higher temperature than the ambient fluid, so that convection occurs. There is no other cause of motion. This configuration has been extensively investigated for fluids of moderate Prandtl number (Thomas and Townsend, 1957; Croft, 1958; Townsend, 1959; Sparrow, Husar and Goldstein, 1970) and investigations of Bénard convection at high Rayleigh number also provide relevant information (Deardorff and Willis, 1967; Tritton, 1977, Section 4.3; Krishnamurti, 1973; Chu and Goldstein, 1973).

A predominant feature shown by these previous experiments is the generation of "thermals" by intermittent boundary layer instability. The fluid close to the boundary is heated by conduction and is thus lighter than the overlying ambient fluid. The consequent instability results in the ejection of the thermals – regions of hot fluid with a characteristic structure – upwards into the colder ambient fluid. (Other names used for these structures are "plumes" and "bubbles".) There are some differences between different observers about the degree of regularity of the motion – whether, in a given place, thermals occur periodically or randomly and whether they occur simultaneously or independently in different places (Tritton, 1977, Section 4.3). In general, however, the motion is usually regarded as a form of turbulence. In addition to the thermals, there are vigorous velocity and temperature fluctuations on smaller length-scales; presumably a typical turbulent "energy cascade" (Tritton, 1977, Sections 21.3, 21.4) occurs.

Howard (1964) proposed a heuristic model of the process of boundary layer instability and thermal formation. In addition to providing a useful understanding of the mechanics involved, this predicts some of the quantitative features.

The primary question which the present work attempts to answer concerns the extent to which the description of the flow developed for moderate Prandtl number still applies when the Prandtl number is very large. One cannot achieve in the laboratory a value of the Prandtl number even remotely comparable with that of the Earth's mantle ($\sim 10^{23}$). However, this is not so much of a disability as it might appear. The consequence of the Prandtl number being very large is that the Reynolds number of the convection becomes very small, and dynamical inertia is negligible. Once this has occurred, the equations of motion take an asymptotic form and the use of the appropriate scale factors allows extrapolation to even higher Prandtl number. One thus requires only a sufficiently high Prandtl number for the Reynolds number to be low, and we shall see that this can be achieved with a Prandtl number around 10^5.

It should be noted that the low Reynolds number does not imply that the equations of motion have become linear. Thermal inertia (appearing in the equations as a term of the form $\mathbf{u} . \nabla T$, where $\mathbf{u}$ is velocity and T temperature) remains significant and, since the temperature appears in the equation for the velocity through the buoyancy, this keeps the system fully non-linear.

Howard's model provided useful guidance in planning the experiments. The mechanism implied by it could in principle operate at any value of the Prandtl number. There might therefore be significant similarities between the flows at high and moderate values of the Prandtl number. On the other hand, there must also be significant differences; the energy cascade, in particular, is essentially a high Reynolds number process. One might therefore anticipate that, in the high Prandtl number flow, thermals can originate much as at lower Prandtl number but the details of the process and the structure of the thermals will be different. In particular, it was difficult to anticipate whether the occurrence of the thermals would be periodic in time and regular in position or whether it would be a stochastic process.

We have carried out two preliminary experiments on this topic, and have a more detailed and thorough study in progress. This article is primarily a survey of the results of the preliminary experiments. The background fluid dynamics is not developed in detail; in consequence, the article contains various assertions without their supporting evidence. A fuller treatment will be given elsewhere when we can also give more complete and rigorous results. The preliminary results are, however, of significant fluid dynamical interest as the first observation of a stochastic process (which one may or may not wish to call a form of turbulence, depending on how one chooses to define that word) occurring in conditions when the only or primary non-linearity is in the thermal equation.

2 Scaling

Although we are omitting the fluid dynamical details, it is of course essential that the results are presented in an appropriate non-dimensional form, both so that different experiments may be compared and so that the results may be scaled to the Earth's mantle. The ideal situation of a semi-infinite expanse of fluid above an infinite horizontal boundary maintained at a fixed temperature T_1, higher than the temperature T_0 of the ambient fluid, includes no length or time scale in its specification. Such scales are therefore provided by appropriate combinations of other specifying parameters, and it may be shown that the relevant length and time scales are respectively

$$L = [\kappa\nu/g\alpha(T_1 - T_0)]^{1/3},$$
$$\Psi = [\nu^2/g^2\,\alpha^2(T_1 - T_0)^2\,\kappa]^{1/3},$$

where g is the acceleration due to gravity and α, ν and κ are respectively the thermal expansion coefficient, the kinematic viscosity, and the thermal diffusivity of the fluid. The velocity scale is then given as

$$U = L/\Psi = [g\alpha(T_1 - T_0)\kappa^2/\nu]^{1/3}.$$

One may expect any characteristic temperature difference, length, time or velocity of the flow to be proportional to respectively $(T_1 - T_0)$, L, Ψ or U. From the equations of motion, it may be inferred that in general the constants of proportionality will depend on the Prandtl number,

$$Pr = \nu/\kappa,$$

but that they must take universal values if the dynamical inertia $(\partial\mathbf{u}/\partial t + \mathbf{u}\,.\,\nabla\mathbf{u})$ is negligible; i.e. they must have fixed asymptotic values as Pr becomes very large. (This assumes that the Boussinesq equations apply (Tritton, 1977, Section 13.2, Appendix to Ch. 14), which involves, among other approximations, the assumption that none of the fluid properties is temperature dependent.)

The numerical factors involved in relating typical flow lengths, periods, and velocities to the above scales are not necessarily around unity in order of magnitude, and may need to be considered in estimating corresponding orders of magnitude in a geophysical context. For example, Howard's model implies that the boundary layer thickness, δ, and the time, τ, between consecutive thermals in the same region are

$$\delta \simeq R_\delta^{1/3} L; \quad \tau \simeq R_\delta^{2/3} \Psi,$$

where R_δ is a local instability parameter (a form of Rayleigh number) with a value around 10^3.

(Howard's more quantitative formulation of his model also predicts that the active phase during which thermals are formed becomes very short compared with τ at large *Pr*. For reasons that we will discuss elsewhere, this prediction was open to question even before there was any experimental evidence – essentially, the time involved does not scale correctly as $Pr \to \infty$. However, it is consistent to discard this detail of the model whilst retaining its other features.)

3 The experiments

The convection apparatus is basically very simple. A tank with square horizontal cross-section, side 0·6 m, is filled to a depth of about 0·3 m with silicone oil. The base of the tank is made of an aluminium alloy and can be brought to a uniform and constant temperature by the pumping of thermostatted water through channels below it. The side-walls of the tank are of Perspex. The top surface of the fluid is free; we shall see below that rather poorly defined boundary conditions there do not matter.

Two preliminary experiments have been carried out using different grades of oil (Rayburn, 1973; French*, 1976). The parameters of the experiments are summarized in Table I. In principle, the Prandtl number is the only defining non-dimensional parameter for the ideal configuration formulated in Section 1. The Rayleigh number quoted in Table I is based on the depth of the fluid and is an indication of how well or poorly this ideal is approached (the Rayleigh number is infinite in the ideal configuration); this point will be discussed further in a moment. The Reynolds number in Table I is based on observed velocity and length-scales (see Section 5) – typical speed and size of a thermal. It is defined only in order of magnitude. We see that in experiment I, the dynamical inertia was still significant (although much smaller than in previous work) whilst experiment II achieved the aim of making this negligible.

There are two respects in which this system does not approximate well to the ideal configuration, so that the experiments have to be done in a special way. Firstly, the aspect ratio of 2 : 1 is not large enough for the side-walls to have no

* Now M. A. Forrest.

TABLE I *Experimental conditions*

	Experiment I	Experiment II
Nominal kinematic viscosity	350 cS	10 000 cS
Prandtl number	$2{\cdot}6 \times 10^3$	8×10^4
Rayleigh number	5×10^7	3×10^6
Reynolds number	1	10^{-2}
Method	Thermocouples	Schlieren

influence; if one waits for a (statistically) steady state to be reached before making observations, then a large-scale general circulation is established in the tank with a velocity large enough to distort the boundary layer processes. Secondly, information from Bénard convection (Chu and Goldstein, 1973; Long, 1976) indicates that the Rayleigh number was not high enough – certainly in experiment II and probably in experiment I – for the upper surface to have no influence on the flow in the vicinity of the lower boundary. Again in the steady state, cold thermals originating from the top boundary layer would penetrate the layer to interact with the lower boundary layer – or the convection might even have a cellular structure. Fortunately, both these difficulties can be largely overcome by making the observations during a transient, quasi-steady period. One starts with the whole system at room temperature and then raises the base temperature quickly to a value at which it is then maintained. One then makes the observations during the initial period in which heat is being transferred into the fluid rather than through it. The distinction is of little importance so far as the boundary layer dynamics are concerned. The duration of this initial period is long compared with the time-scale of these dynamics (e.g. τ as defined in Section 2) and so the unsteadiness in the imposed conditions has little influence on the processes under investigation. Also, the large-scale circulation has not yet established itself. The length-scale of the boundary layer processes is small compared with the horizontal dimensions of the tank and so the presence of the side-walls has little influence. Although the observations were thus made during a transient period, this lasted a few hours in experiment I and longer in experiment II.

The observational methods were different in the two experiments. In experiment I, apart from a few flow visualizations using dye, the observations consisted of temperature measurements with thermocouples. Several thermocouples were used simultaneously, but here we shall consider only the variation with time of the temperature at a single point. The range of positions of this point will be indicated in Section 5.

In experiment II, schlieren flow visualization (making use of the variation of refractive index with temperature – Merzkirch, 1974) was employed. Briefly, the system is a conventional Töpler-schlieren one with the exception that the knife-edge is replaced by a slit. Light deflected horizontally in either direction is thus lost. No part of the picture is brighter than in the absence of temperature gradients; a temperature gradient of either sign causes darkening. A symmetrical structure such as a thermal thus produces a symmetrical pattern.

We have not yet exchanged the two methods of observation between the two fluids or used the two simultaneously, although obviously this should be done.

Both methods of studying the flow provide information about the temperature field. Clearly the velocity field also has to be studied before the flow is fully understood, but this requires the development of new experimental techniques.

4 Observations

Figure 1 shows three sample lengths of very much longer oscillograms from experiment I. They have the form that one would expect as a result of the spasmodic occurrence of thermals at the position of observation: intermittent peaks are superimposed on an only very weakly fluctuating background temperature. (This background temperature rises gradually during the course of an experiment as the ambient fluid is warmed up; this trend is most apparent in the top trace of Fig. 1, which corresponds to a higher $(T_1 - T_0)$ than the other two examples.) There is thus both an important similarity to and an important

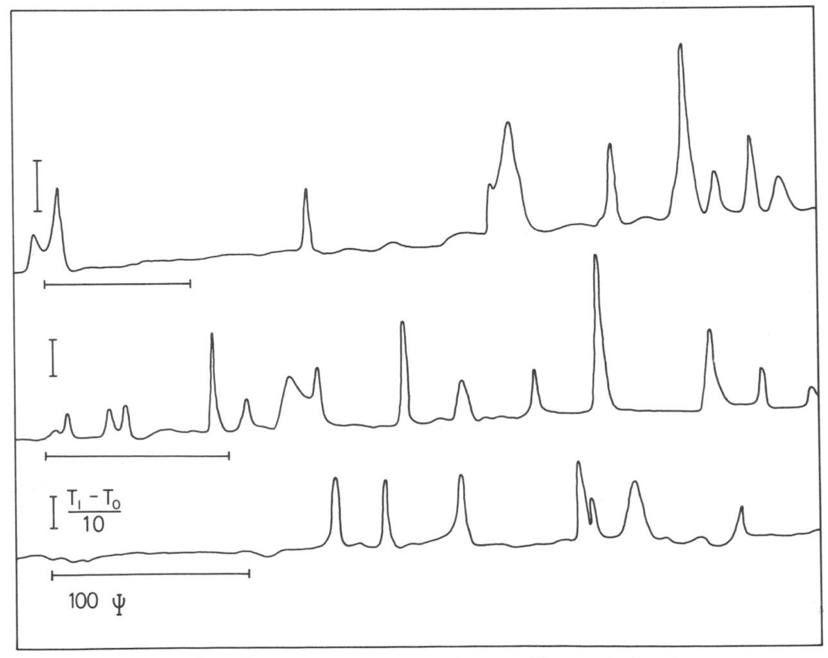

Figure 1. Short samples of temperature versus time oscillograms for three separate runs in experiment I. Temperature variation and time-scales are indicated, but absolute vertical positions of traces are of no significance.

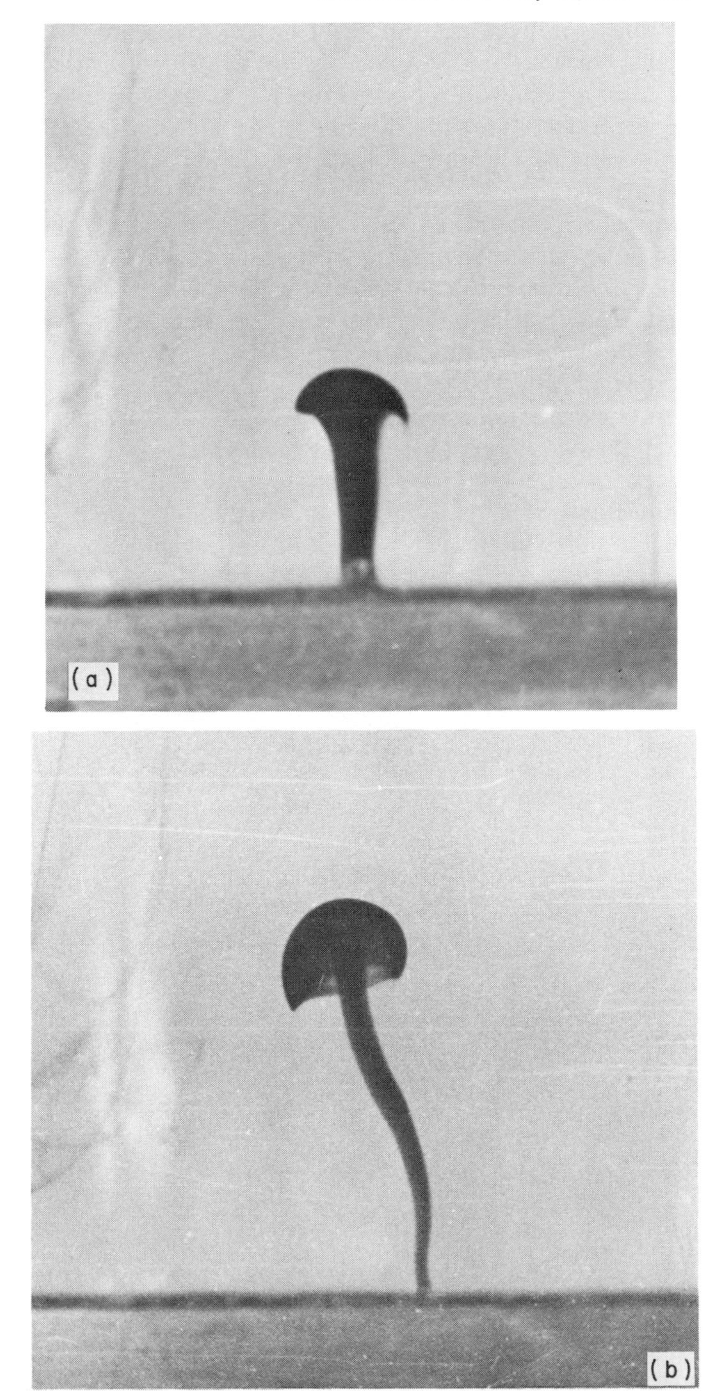

Figure 2. Dye visualization in experiment I.

difference from corresponding time/temperature traces in lower Prandtl number fluids, such as the observations in air with Townsend's (1959) apparatus. These also showed periods of quiescence interspersed with shorter bursts of higher temperature activity. But during these bursts there were rapid turbulent fluctuations in contrast with the smooth rise and fall of the temperature for most peaks in Fig. 1.

The interpretation of the peaks as indicating the occurrence of thermals was confirmed by injecting a small quantity of dyed silicone oil close to the heated bottom boundary. Each time dye was injected, it remained for a while as an effectively stagnant patch; it then erupted in a localized region with the formation of a thermal. Figure 2 shows photographs of the same thermal at two stages of its growth. The "mushroom head" is believed to be an inertial effect, illustrating the fact that the Reynolds number in experiment I was not small enough for such effects to be wholly negligible.

It is helpful in interpreting the observations of experiment II to have comparison pictures of thermals of known structure. Figures 3 and 4 show schlieren pictures of convection above a localized heat source in two different

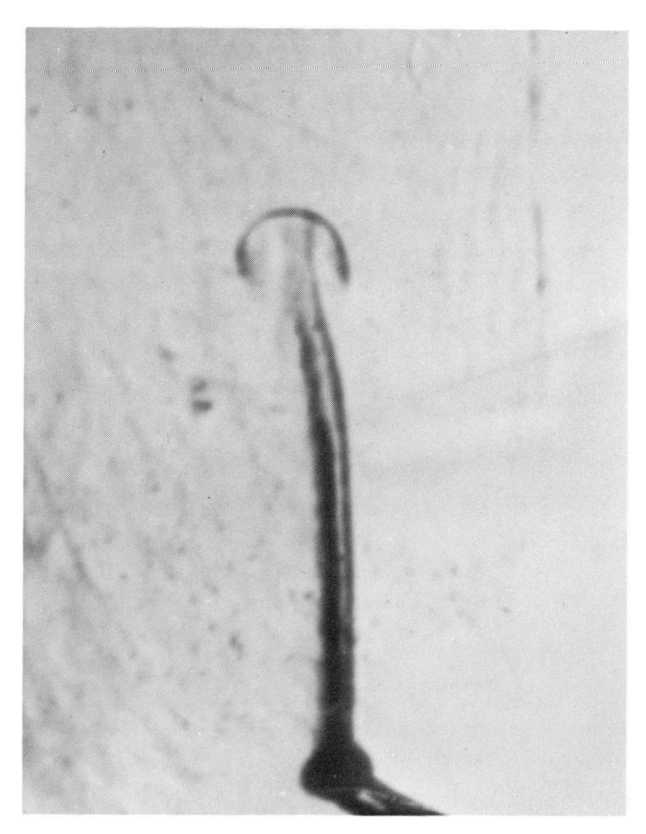

Figure 3. Convection from local heat source in water.

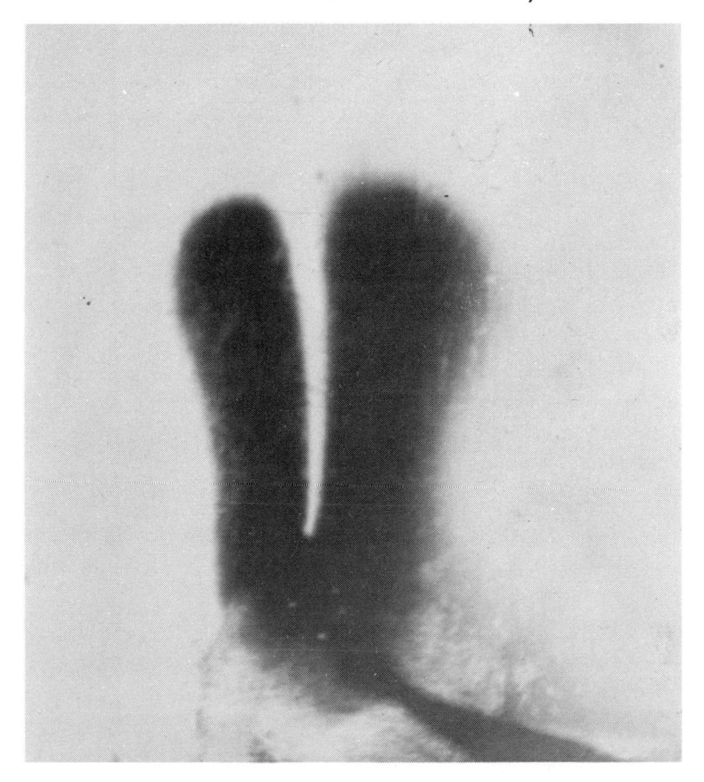

Figure 4. Convection from local heat source in fluid of experiment II.

fluids; in each case, the heat source was switched on a short time before the photograph was taken so that one sees the head of the thermal. In Fig. 3 the fluid was water (Prandtl number $\simeq 6$); the thermal shows the mushroom head characteristic of high Reynolds number starting flows. Figure 4 was taken with the silicone oil used in experiment II and there is negligible "mushrooming". The characteristic features are two dark bands on either side of a central vertical undarkened strip (as one expects from the mode of operation of the schlieren system). This is what one looks for in identifying thermals in the main experiment.

Figure 5 shows examples of schlieren visualizations of the convection over the heated horizontal surface. In Fig. 5(a), there is just one thermal in the field of view. In Fig. 5(b), there are two, apparently side-by-side although in fact they are almost certainly separated by a larger distance in the line of sight. In Fig. 5(c), there are again two thermals, an older one with its head extending above the field of view and a younger one growing in front of or behind it. Figure 5(d) shows a rather complicated pattern produced when there are several thermals in the field of view. In isolation a figure such as this is difficult to analyse, but when it is seen as one of a time sequence of pictures in which the

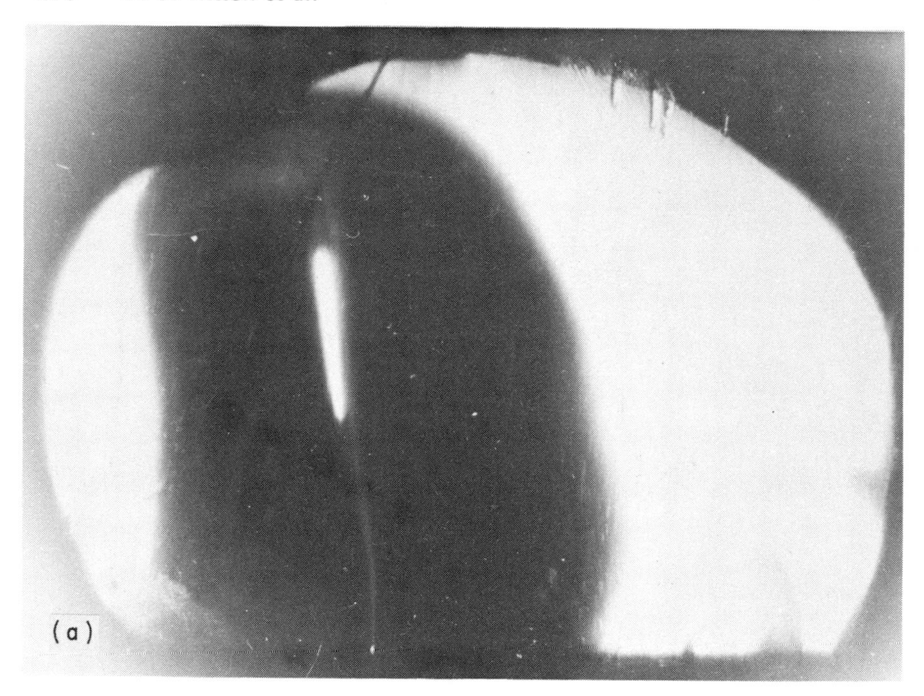

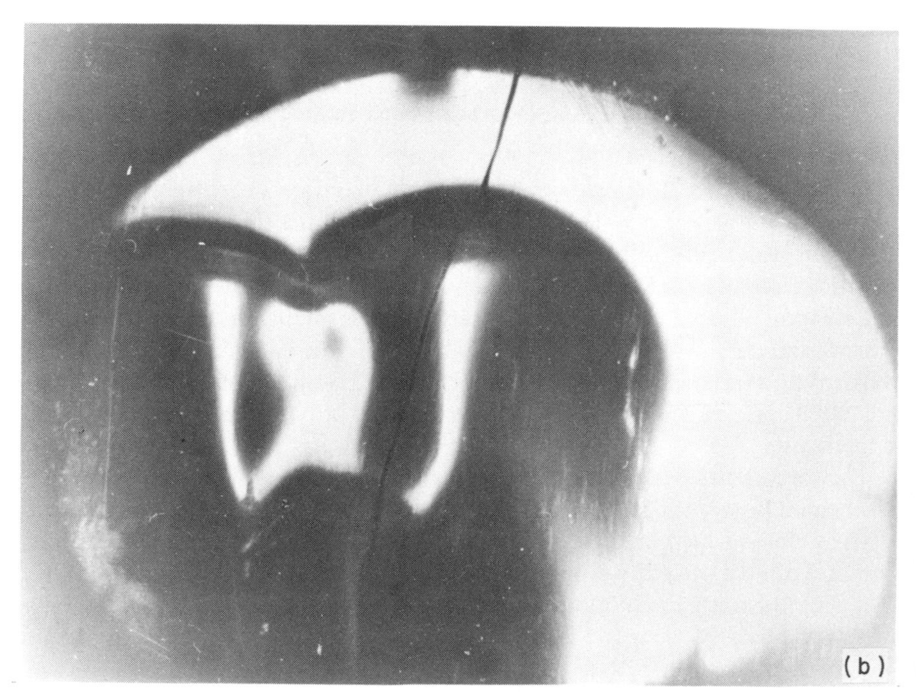

Figure 5. Examples of schlieren pictures from experiment II.

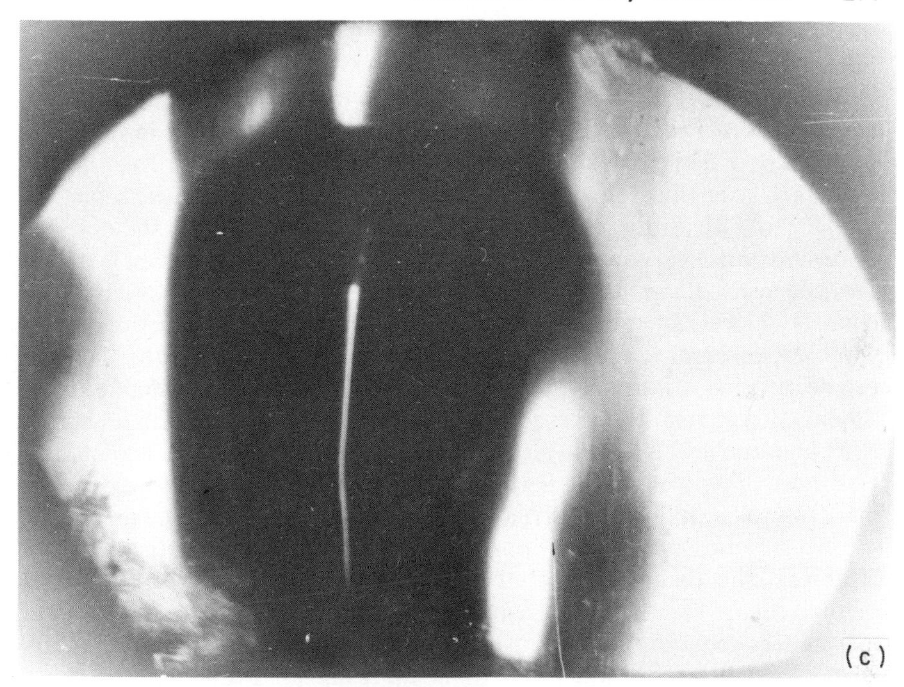
(c)

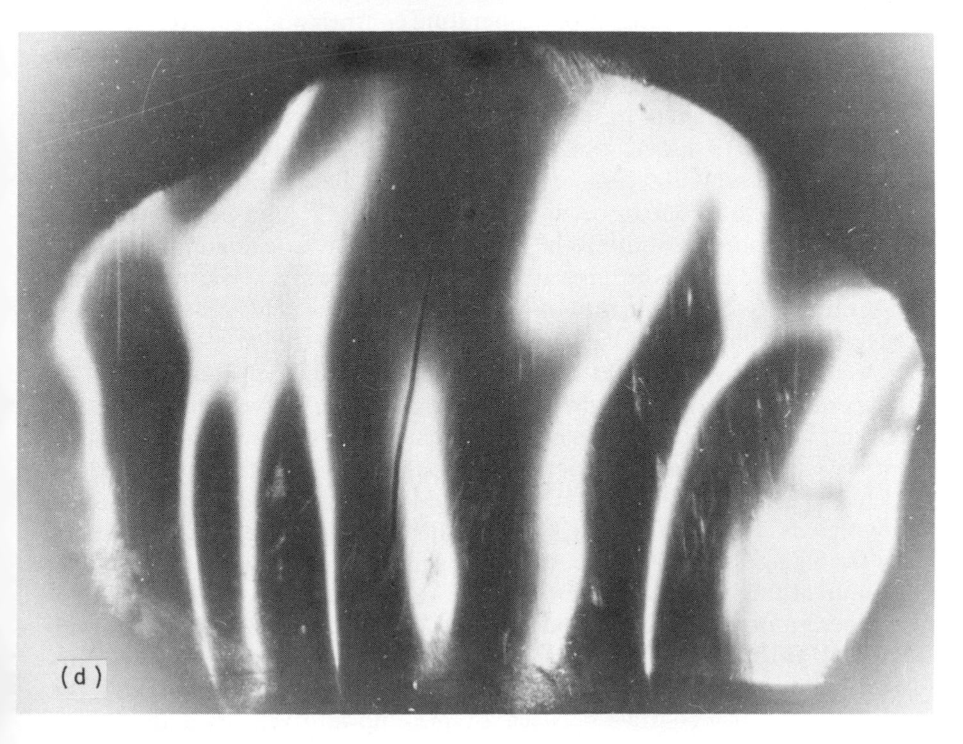
(d)

evolution of each thermal can be traced, there is usually little ambiguity in interpretation.

The schlieren system responds to horizontal temperature gradients. It also acts more weakly as a shadowgraph, thus darkening the picture in regions of large second derivative of the temperature in the vertical. The thermal boundary layer thus shows as a dark region at the bottom of each picture.

The qualitative structure of the convection is apparent from the observations summarized above. As at lower Prandtl number, the dominant feature is the occurrence of thermals. The temperature field outside the boundary layer consists of local hot columns within considerably larger expanses of fluid at effectively ambient temperature. In the schlieren observations, the thermals can be seen to grow out of the boundary layer; each has a well-defined front (slightly but not markedly broader than the main stem below it) that appears as a bulge on the boundary layer and then rises steadily into the ambient fluid. For most of its lifetime the height of the thermal is much larger than the boundary layer thickness; its dynamics should clearly be described in terms of its own structure rather than just in terms of the evolution of the boundary layer instability (in contrast with the Howard model). The process by which a thermal ends is much less apparent than that by which it starts (and is one of the matters requiring further investigation). It appears to fade out simultaneously at all heights rather than to have a rising tail. Presumably there is a gradual weakening – decline of temperature contrast and velocity – throughout the thermal. However, in terms of the schlieren observations its disappearance is relatively abrupt compared with its total lifetime; there is usually little ambiguity in saying whether a thermal is present or not.

In contrast to lower values of the Prandtl number, the thermals are themselves essentially laminar structures (for both grades of viscosity). However, they appear to occur in a random way – an aspect that will be considered more quantitatively in Section 5. The oscillograms show no obvious periodicity in the times at which a thermal occurs at a given place, and the variability of the number of thermals within the schlieren field of view suggests that there is no highly regular spatial structure at a given time.

As noted in Section 3, the observations made to date all concern only the temperature field. We have no direct information on the velocity field. Some indirect information may be inferred from changes in the temperature field (we estimate below a characteristic velocity magnitude), but this is very incomplete. In particular, it should not be supposed that rising fluid is confined to the hot regions identified in the experiments. It seems almost certain, particularly for the higher Prandtl number, that the rising hot fluid carries with it a larger volume of fluid at ambient temperature – just as a body in low Reynolds number motion produces an extended velocity field (Tritton, 1977, Sections 8.2, 9.4). Thus it may be that, although the temperature field has a very skew distribution between hot and cold regions, the velocity field is not so skew – i.e. there is not much contrast between a typical upward velocity and a typical downward one (which would again be a difference from behaviour at lower Prandtl number).

5 Quantitative results

Whilst the emphasis so far has been on the qualitative features of the flow, some quantitative results may be inferred.

The oscillograms of experiment I have been examined to give information about the frequency of occurrence and duration of the thermals. As can be seen from Fig. 1, there is little ambiguity in the identification of the peaks. There are a few cases where subjective decisions are involved – very small bumps that one might or might not include and double peaks that might be counted as one or two thermals. Arbitrary but consistent procedures were adopted. However, the number of such cases was sufficiently small that different arbitrary decisions would not significantly change any of the following conclusions. The principal results are included in Table II. Some explanation of the quantities quoted is given below.

TABLE II *Main quantitative results*

		S.D.
Experiment I		
Average time between thermals	60Ψ	(60Ψ)
Average thermal duration	$12{\cdot}5\Psi$	$7{\cdot}5\Psi$
Experiment II		
Average thermal duration	17Ψ	$8{\cdot}5\Psi$
Average speed of thermal head	$8{\cdot}2\,U$	$1{\cdot}7\,U$
Typical thermal width	$25\,L$	

The distribution of thermals in time is of particular interest. Figure 6 shows a histogram of the number of thermals occurring in a sequence of intervals each of 200Ψ duration. The average number of thermals is 3·33, and the smooth curve on Fig. 6 is the Poisson distribution for this mean. It is a good representation of the observations. It is clearly an oversimplification to say that the production of thermals is a Poissonian process; amongst other complications are the finite duration of thermals and the fact that we have ignored all interactions with thermals in neighbouring positions. Nevertheless, Fig. 6 provides strong support for supposing that the convection is stochastic. (The choice of a period of 200Ψ for this representation is evidently arbitrary. It is convenient as giving a sufficiently large mean combined with a sufficiently large number of samples.)

The average interval between thermals quoted in Table II comes from the above analysis. The standard deviation has not been evaluated directly from the data (as it is much quicker to count events than to measure intervals), but the figure quoted in parentheses is the one that would be given by a true Poisson distribution.

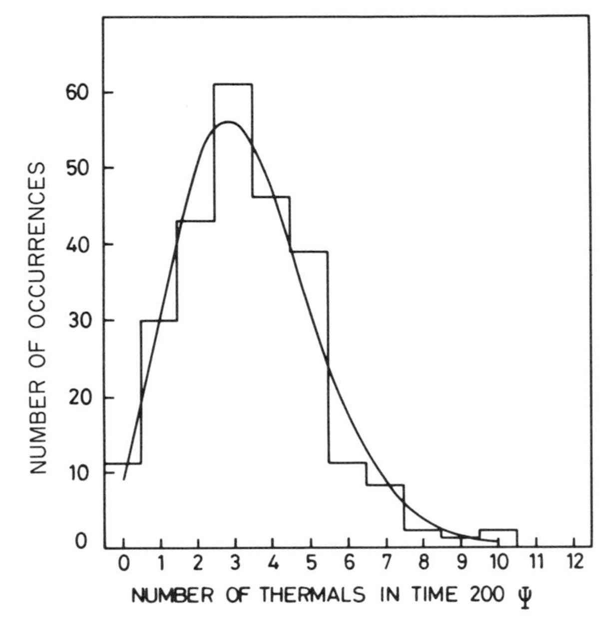

Figure 6. Histogram of number of thermals occurring in time 200Ψ (experiment I).

The average duration of the thermals and its standard deviation are derived from accumulated measurements of individual peaks. It is difficult to assign a precise start and end to a peak. The thermal duration was therefore defined as $2\Delta t$, where Δt is the width of the peak at half its maximum height above the background.

The standard deviation quoted in Table II represents, of course, genuine variability, not experimental error. It is thus given as the square root of the variance (not the standard deviation of the mean). This remark applies also to all the subsequent entries in Table II.

The above conclusions are based entirely on analysis of the temperature fluctuations at one point. Data from different values of z/L, where z is the height above the base, have been combined in the averages. All observations used were from the range $15 < z/L < 90$. Although we do not yet have enough data for a full analysis of trends with z/L, a check was made that there were no obvious ones that would have made it inappropriate to average in this way. No trend in the average number of thermals with z/L could be detected. There were signs that the average thermal duration was a bit shorter in the range $25 < z/L < 55$ than outside it, but this is not conclusive.

The results for experiment II quoted in Table II are based on identifying and observing the development of individual thermals through a sequence of schlieren photographs. A thermal duration thus corresponds to the time it is so identifiable, from its emergence at the bottom of the picture to its fading out.

Generally, the top of the thermal has extended beyond the top of the field of view (at $z/L \simeq 55$), so it is the fading of the lower portion that is detected.

The correspondence between the mean thermal durations for the two experiments is very satisfactory. The small difference can, of course, be accounted for by the quite different procedural definitions of a thermal duration in the two experiments, and provides no evidence for a trend with Prandtl number. The large fractional standard deviations in the thermal durations are again indicative of the stochastic nature of the convection. This is particularly true in experiment II where the thermal is observed as a whole, and the spread cannot be accounted for (as it might have been in experiment I) by differences in probe position relative to the centre of the thermal. Figure 7 shows a histogram of thermal durations in experiment II.

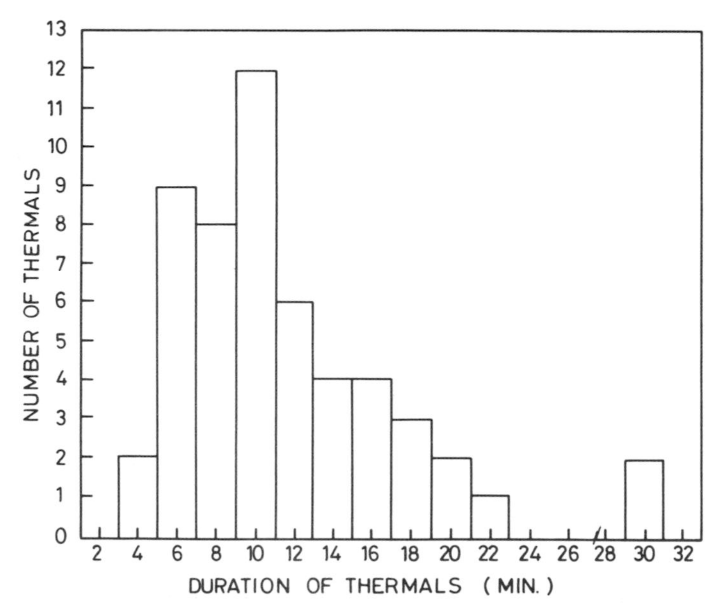

Figure 7. Histogram of thermal durations inferred from schlieren photographs taken at 2 minute intervals (1 min = $1{\cdot}45\Psi$).

The rate of rise of the top of a thermal is easily measured from a schlieren sequence and thus provides the one readily determinable velocity. To give the result quoted in Table II, an average velocity was determined for each thermal during the time its top was in the field of view (omitting some thermals where complexity of the picture made the determination difficult); the mean was then obtained by averaging over thermals. The standard deviation corresponds to the spread during the second averaging. Whilst it will have some contribution from the inaccuracy of the measuring process, it is primarily an indication of the variability from one thermal to another. Although the velocity determined is a Lagrangian average of a special point in the thermal, it is reasonable to suppose that it provides a good order of magnitude indication of a typical flow velocity involved in the convection.

Compared with the preceding results, the figure given in Table II for the typical thermal width is based much more on sample measurements than on any systematic averaging procedure. The quantity sampled was the full distance between the outer edges of the dark column produced by a thermal. It is thus an approximate measure of the diameter of the column of hot fluid. (As noted earlier, the diameter of the column of rising fluid may be quite different.) Just where the observed edge occurs in relation to the temperature field depends, however, on the details of the schlieren optics*.

It is worth noting that the typical thermal width so estimated is of the same order of magnitude as the thickness ($R_\delta^{1/3} L$) of the thermal boundary layer from which it arises.

6 Application to mantle convection

How do the foregoing observations aid an understanding of convection in the Earth's mantle? In the authors' opinion, the most important point is a rather general one. Evidence that the convection pattern in the Earth has changed does not necessarily imply a change in the parameters governing the convection; instabilities can produce changes whilst the imposed conditions are remaining constant†. This statement is a commonplace of fluid dynamics. However, it has not always been appreciated during the development of ideas about mantle convection (although there have been papers in which the point was well recognized; e.g. Knopoff, 1964; Tozer, 1965; Jones, 1977). Alternatively, it is sometimes argued that the generalization is not applicable to the mantle because of the very low Reynolds number. The fact that a system in which the only non-linearity is thermal inertia can give rise to a form of turbulence is thus of significance for mantle convection theories.

We would wish to be cautious about any more specific analogy between our observations and mantle convection. It is our general opinion that the number of acceptable suggestions for the structure of mantle convection has outrun the available evidence for discriminating between them. However, it seemed of interest to scale the laboratory results to possible mantle conditions. The outcome of such a scaling is given in Table III (explained in more detail below). The appearance of a time-scale of 10^8 years and a velocity of a few centimetres per year encourage further speculation. It is at least possible that the pattern

* If, as in these experiments, the slit width is set so that the central bright band of a thermal image is much narrower than the full thermal width, then darkening is produced by a temperature gradient significantly smaller than the maximum gradient present. The measured width will thus correspond to the flanks of the temperature profile. (But it must be remembered that this profile is an average over the light path through the thermal.)

† Undoubtedly the history of the Earth has included changes in the mantle parameters, but one should not necessarily expect specific developments in the convection pattern to be correlated with these changes.

and temporal development of mantle convection have a dynamical resemblance to those of our laboratory situation. By this it is meant that the same basic dynamical process is occurring, although the detailed consequences in the mantle will be affected by complications not present in the experiment. We therefore proceed to further explanation of Table III and some discussion of its possible implications.

TABLE III *Application to Earth's mantle*

Surface heat flux	$H \sim 10^{-6}$ cal/cm^2 s	
Expansion coefficient	$\alpha \sim 10^{-5}$/K	
Kinematic viscosity	$\nu \sim 10^{21}$ cm^2/s	
Thermal diffusivity	$\kappa \sim 10^{-2}$ cm^2/s	
Thermal conductivity	$k \sim 10^{-2}$ cal/cm s K	
	Estimate from above values	Scales as:
Time between thermals	$\tau \sim 5 \times 10^{15}$ s $\sim 10^8$ yr	$(k\nu/Hg\alpha\kappa)^{\frac{1}{2}}$
Thermal duration	$\sim 10^{15}$ s $\sim 3 \times 10^7$ yr	$(k\nu/Hg\alpha\kappa)^{\frac{1}{2}}$
Boundary layer thickness	$\sim 10^7$ cm $\sim 10^2$ km	$(k\nu\kappa/Hg\alpha)^{\frac{1}{4}}$
Temperature difference	$(T_1 - T_0) \sim 10^3$ K	$(H^3\nu\kappa/k^3 g\alpha)^{\frac{1}{4}}$
Typical velocity	$\sim 10^{-7}$ cm/s ~ 3 cm/yr	$(Hg\alpha\kappa^3/k\nu)^{\frac{1}{4}}$
Average area per thermal	$\sim 10^{16}$ cm^2 $\sim 10^6$ km^2	$(k\nu\kappa/Hg\alpha)^{\frac{1}{2}}$

In the experiments, a certain temperature difference between the boundary and the ambient fluid is imposed; the heat transfer between the boundary and the fluid is then determined by this. In applying the results to the Earth, it seems better to impose on the calculations a value of the heat transfer and allow the corresponding temperature difference to be inferred; there are much stronger observational constraints on the former. In the forthcoming experiments, we plan to measure the heat transfer. In the meantime, we have used indirect evidence on the relationship between the heat transfer per unit area, H, and $(T_1 - T_0)$: extrapolation from other experiments and interpretation of our results in terms of the Howard model. This may be checked from our observations of the boundary layer thickness, δ, since

$$H \sim k(T_1 - T_0)/\delta,$$

(k being the thermal conductivity).

The top part of Table III lists (in the units that are probably most familiar to most readers) the values of H and the material properties from which the convection parameters have been evaluated. Since we shall be indicating the effect of varying these quantities, we shall not discuss the reasons for our choice of values.

The lower part of Table III lists the values of various quantities associated with mantle convection as given by scaling from our laboratory system. The algebraic expressions are given to show how different choices of the original quantities would affect the results. For example, some readers may consider that the value of H is too close to the surface heat flux, in view of the heat generated in the crust, and that a lower value should be used; the consequent modifications are readily seen.

The numerical orders of magnitude quoted are based primarily on scaling the quantitative laboratory measurements. Thus the fact that these differ from the orders of magnitude of the algebraic expressions comes mainly from the fact that the numbers in Table II are quite large (although the re-writing of the scales in terms of H instead of $(T_1 - T_0)$ via the relationship

$$H \sim k\left[\frac{g\alpha(T_1 - T_0)^4}{R_\delta \nu\kappa}\right]^{1/3} \sim \frac{k}{10}\left[\frac{g\alpha(T_1 - T_0)^4}{\nu\kappa}\right]^{1/3}$$

also contributes).

In comparing the time-scales with the evidence for changes in the pattern of mantle convection, it seems to us that $\tau \sim 10^8$ years is the most relevant time-scale. This would govern the interval between two episodes of similar behaviour. (The figure for τ derives mainly from experiment I, in which the Reynolds number was not as low as desired. It is thus worth noting that experiment II also showed evidence of a time-scale of around 50Ψ, but further experiments are needed before a precise definition and value can be formulated.) Before doing the experiments, we had thought it possible that the thermals would occur with a regular period and simultaneously over an extended region, and that this might be related to the suggestion that changes in mantle convection have been correlated on a global scale. The observation that the process is stochastic makes it more relevant to evidence for local changes than to that for global changes.

The boundary layer thickness δ is of interest not only as indicating the depth over which most of the variation of the horizontally averaged temperature occurs, but also as indicating whether this type of convection is relevant. One requires that the boundary layer instability should be able to occur without strong influence from other boundaries. Essentially this means that the Rayleigh number of the whole system should be high enough. A convenient form of this criterion is that the boundary layer thickness must be small compared with the overall depth of the convecting region.

The vertical temperature change across the boundary layer will be the major part of $(T_1 - T_0)$. Outside the boundary layer, the thermals will give rise to instantaneous horizontal temperature variations. Typically the temperature difference between the middle of a thermal and the ambient fluid may be $0.1(T_1 - T_0)$ to $0{\cdot}3(T_1 - T_0)$ (cf. Fig. 1), but further experiments are needed to make this statement more precise.

The value quoted in Table III for the average area per thermal is intended as an indication of how many thermals are likely to be found in a given area. It is to be compared with the total surface area of the Earth of 5×10^8 km^2 and of the core–mantle boundary of $1{\cdot}5 \times 10^8$ km^2. It is based on the average number of thermals within the field of view of the schlieren system of experiment II. This is again a point on which further experiments are needed to provide more precise information.

So far we have not specified the region of the mantle in which these processes might be occurring. It should perhaps be pointed out that, although in the laboratory it is convenient to use a hot upward-facing boundary, the situation at a cold downward-facing one is exactly analogous; the instability will generate cold downgoing thermals. Thus the processes may occur at the top or bottom of the convecting region, or at both. They do require the boundary of the region to be sharp, at least compared with the boundary layer thickness. One would thus not expect to find upgoing thermals if mantle convection is restricted to an upper region by a gradual increase in viscosity with depth.

In suggesting that this type of convection might relate to the episodicity of tectonic activity, we were envisaging that it would be the boundary layer at the top of the mantle that exhibited the instability. It seems to us quite likely that a boundary layer of this kind exists there; indeed, it becomes almost certain if a substantial fraction of the heat coming out of the Earth is also coming out the mantle, if the convecting region extends through a substantial fraction of the total depth of the mantle, and if the viscosity is towards the lower end of the range of suggested values. The heat transferred from the mantle can be generated within the mantle without affecting this conclusion; only a small fraction of the heat would be generated within the boundary layer, and so the boundary layer dynamics would be only slightly modified.

On the hypothesis that convection extends throughout the mantle, there is also the possibility that an unstable boundary layer is present at the bottom of the mantle. Jones (1977) has suggested this as a mechanism that might cause variations in the heat transfer from the core and thus changes in the frequency of geomagnetic reversals. This particular proposal would require the thermals to occur synchronously over the whole boundary, so in that respect it is not supported by our experiments. The more general question of whether a boundary layer of this type occurs there depends on the core–mantle heat transfer; the proposal requires a minimum of perhaps 5% of the Earth's heat to be generated within the core, comparable with values discussed in the context of a thermally driven dynamo (Gubbins, 1976).

This suggestion that hot upgoing thermals may be generated may prompt the reader to ask what is the connection of these ideas with the "plume" model of mantle dynamics (Morgan, 1972; Deffeyes, 1972). We have deliberately referred to the structures observed in our experiment as "thermals" rather than "plumes" to avoid giving readers with a geophysical background a preconceived view of their nature. The laboratory thermals, at least in a fluid of constant viscosity, differ from the proposed mantle plumes, at least as formulated by their proponents, in that the latter have narrow columns of rising material as well as of hot material. Probably the main inference to be drawn from the experiments is again a rather general one: narrow columns of hot material can arise at an extended (in two dimensions) heat source; if crustal "hot spots" are to be accounted for by such columns in the mantle, this does not necessarily imply, as is sometimes argued, hot spots at the core–mantle boundary – requiring in turn another stage of explanation. However, we conclude by emphasizing that the connection with "plumes" is peripheral; evidence that hot-spots should be differently interpreted would not, in our view, affect the above comments on the relationship between the laboratory experiments and mantle convection.

Acknowledgements

We are grateful for most useful discussions with Dr P. A. Davies, Dr C. W. Titman and Miss L. J. Rickards; the last named is carrying out the more comprehensive experiments mentioned.

References

Chu, T. Y. and Goldstein, R. J., 1973. Turbulent convection in a horizontal layer of water. *J. Fluid Mech.* **60**, 141–159.

Croft, J. F., 1958. The convective regime and temperature distribution above a horizontal heated surface. *Q. J. R. Met. Soc.* **84**, 418–427.

Deardorff, J. W. and Willis, G. E., 1967. Investigation of turbulent thermal convection between horizontal plates. *J. Fluid Mech.* **28**, 675–704.

Deffeyes, K. S., 1972. Plume convection with an upper-mantle temperature inversion. *Nature* **240**, 539–544.

French, M. A., 1976. Non-inertial convection in a fluid heated from beneath. M.Sc. dissertation, Univ. Newcastle upon Tyne.

Gubbins, D., 1976. Observational constraints on the generation of the Earth's magnetic field. *Geophys. J. R. Astr. Soc.* **47**, 19–39.

Howard, L. N., 1964. Convection at high Rayleigh number. Proceedings of the 11th International Congress of Applied Mechanics, Munich, 1109–1115, Springer-Verlag, Berlin.

Jones, G. M., 1977. Thermal interaction of the core and the mantle and long-term behaviour of the geomagnetic field. *J. Geophys. Res.* **82**, 1703–1709.

Knopoff, L., 1964. The convection current hypothesis. *Rev. Geophys.* **2**, 89–122.

Krishnamurti, R., 1973. Some further studies on the transition to turbulent convection. *J. Fluid Mech.* **60**, 285–303.

Long, R. R., 1976. Relation between Nusselt number and Rayleigh number in turbulent thermal convection. *J. Fluid Mech.* **73**, 445–451.

Merzkirch, W., 1974. *Flow Visualization*, Academic Press, London and New York.

Morgan, W. J., 1972. Deep mantle convection plumes and plate motions. *Bull. Am. Ass. Petrol. Geol.* **56**, 203–213.

Rayburn, D. M., 1973. High Rayleigh number convection in a high Prandtl number fluid heated from below. M.Sc. dissertation, Univ. Newcastle upon Tyne.

Sparrow, E. M., Husar, R. B. and Goldstein, R. J., 1970. Observations and other characteristics of thermals. *J. Fluid Mech.* **41**, 793–800.

Thomas, D. B. and Townsend, A. A., 1957. Turbulent convection over a heated horizontal surface. *J. Fluid Mech.* **2**, 473–492.

Townsend, A. A., 1959. Temperature fluctuations over a heated horizontal surface. *J. Fluid Mech.* **5**, 209–241.
Tozer, D. C., 1965. Heat transfer and convection currents. *Phil. Trans. R. Soc.* **A258**, 252–271.
Tritton, D. J., 1977. *Physical Fluid Dynamics*, Van Nostrand Reinhold, Wokingham.

Numerical Simulation of Mantle Convection

U. KOPITZKE*

Institut für Geophysik und Meteorologie, Technische Universität, Braunschweig, West Germany

---

Introduction and method

The main evidence we have today about mantle convection comes from surface data, i.e. plate motion and some related data like heat flux, ocean-floor topography, etc. Only very few direct observational indications exist about the flow beneath the plates. However, mantle temperatures must be influenced very much by convection, since convectional heat transfer is usually more efficient than conductive transport. Some good estimates of upper mantle temperatures are available, derived from pyroxene geothermometry, electrical conductivity, and from olivine–spinel transformation. By comparing these "observational" data with temperatures of different convection models we can hope to find out more about the form of mantle convection.

One important point of controversy is the depth of mantle convection. Several circumstances could confine the flow to the upper mantle (very high viscosity in the lower mantle, chemical heterogeneity). While the idea of upper mantle convection (or convection in two stages in the upper and lower mantle) was favoured in the past, there is now increasing evidence for whole mantle convection (e.g. Elsasser, Olson and Marsh, 1979).

In order to examine further this problem a dynamical convection model – using the finite element technique – was set up, and models of upper mantle and deep mantle convection were compared. The general design is the same as in previous models: a rectangular enclosure containing a Newtonian fluid is heated from within by a bottom heat flux; an initial temperature estimate is made; and, by iteration, a steady state is achieved. Some peculiarities which may be of special importance for the mantle temperatures are included in the model: frictional heating, adiabatic gradient, and depth-dependence of thermal conductivity. Until now more sophisticated *dynamical* models of mantle convection (e.g. DeBremaecker, 1977) failed to incorporate the lithosphere in a reasonable manner. The surface velocity in such models varied

* Now U. Christensen.

steadily instead of being uniform; thus, the upper boundary layer did not behave like a rigid plate. This seems to be due to improper rheological modelling of the lithosphere and to a coarse numerical resolution at the active margins of the plate. In the present model the numerical grid is finest in the upper and marginal boundary layers. The viscosity is simply assumed to be depth-dependent, the lithosphere is simulated by a layer of high viscosity (up to 10^{25} poise). However, it seems adequate to introduce "zones of weakness" with reduced viscosity at the plate margins. They are justified by the upwelling of hot, partially molten material at the spreading centre and, perhaps, by the enlarged stress level in the bending area of lithospheric subduction, which could reduce the effective viscosity via non-linear creep, plastic deformation, transient creep, fracture, etc. This concept of weak zones proved to be successful: the surface velocity is uniform over nearly the whole length of the plate (variation only a few per cent), and the heat flux profile agrees very closely with theoretical and observational data.

The Rayleigh number in the models was always fitted in order to make the overturn time of the plate equal to the mean value of the Earth's plate system. Thus, a realistic simulation of heat balance and heat release from the mantle is ensured.

Results

In the upper mantle model the depth of convection is confined to 643 km. In Fig. 1 different temperature–depth profiles of the model are shown. The temperature at the bottom of the lithosphere (1100 °C) is in accordance with

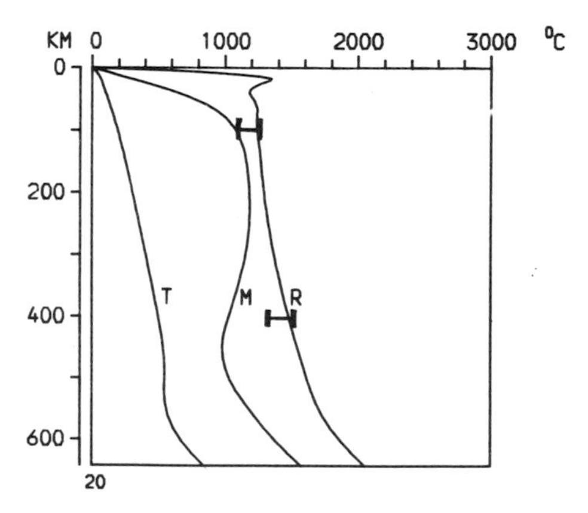

Figure 1. Temperature–depth profiles, upper mantle convection model. T = trench region, M = middle, R = ridge. The middle geotherm is representative for most of the length of the cell except narrow marginal regions. The following "observational" values are included:
100 km 1100–1250 °C (Mercier and Carter, 1975), pyroxene–geotherm.
400 km 1300–1500 °C (Gebrande, 1975), olivine–spinel transformation.

oceanic pyroxene data. However, below 200 km the thermal gradient becomes negative and at a depth of 400 km the temperature is only 1040 °C. This is too low compared with estimates from olivine–spinel transformation, which predict values of about 1400 °C (Fusijawa, 1968; Graham, 1970; Gebrande, 1975).

The depth of the deep mantle model is 1750 km, which is about 60% of the whole mantle depth. However, from a thermal point of view it may be considered to be similar to a whole mantle model. The heating of the cell is mainly a volume process and the cooling a surface process, and a 1750 km deep plane model has nearly the same volume content as a 2900 km deep spherical sector of the same surface area.

Figure 2 shows the streamline pattern and temperature distribution of the deep convection cell. Although the viscosity increases from 3×10^{21}

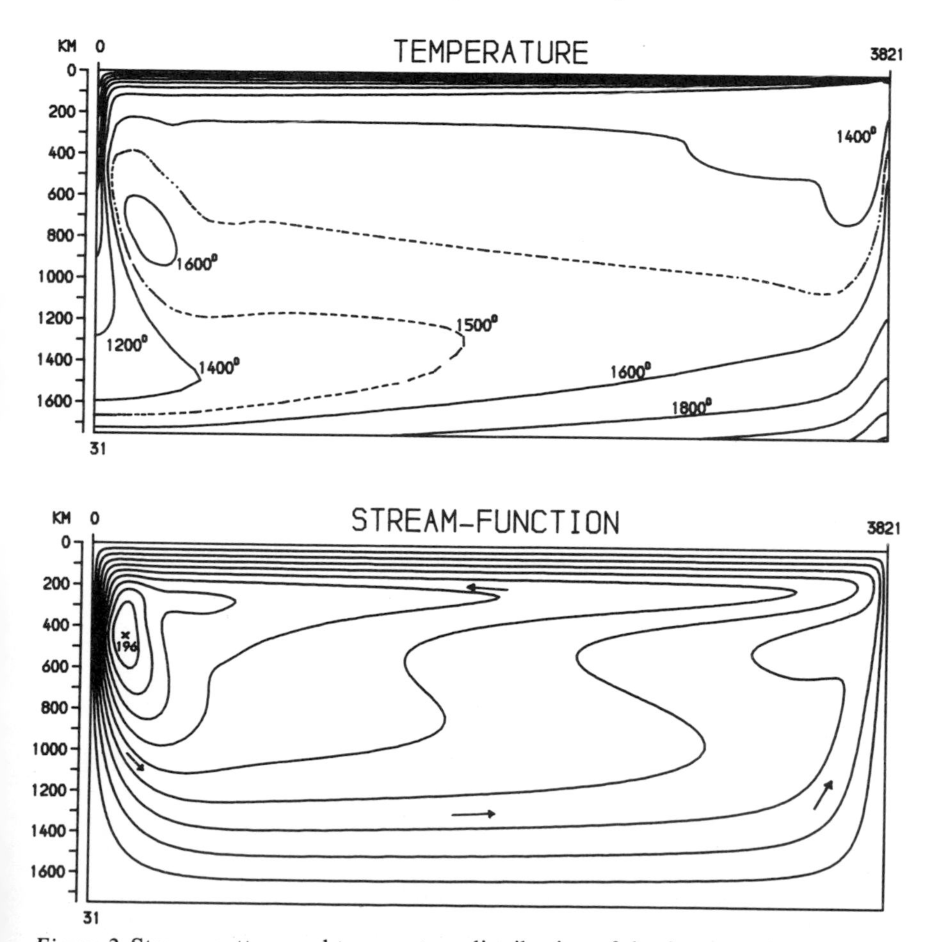

Figure 2. Stream pattern and temperature distribution of the deep mantle convection model.

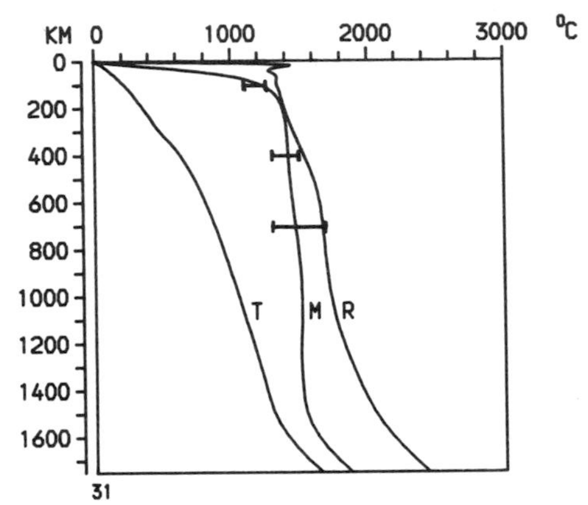

Figure 3. Temperature–depth profiles, deep mantle convection model. T = trench, M = middle, R = ridge. "Observational" values included:
100 km 1100–1250 °C (Mercier and Carter, 1975), pyroxene–geotherm.
400 km 1300–1500 °C (Gebrande, 1975), olivine–spinel transformation.
700 km 1300–1700 °C (Watt and O'Connell, 1978), petrological model.

poise (asthenospheric minimum) to more than 10^{24} poise in the lower mantle, the return flow is neither inhibited from penetrating the lower mantle, nor concentrated at intermediate depths. Temperatures (Figs 2 and 3) are now more satisfactory. The mean temperature at 100 km is 1240 °C and at 400 km depth 1420 °C – in good agreement with the "observational" values. The temperature gradient is slightly higher than zero, but subadiabatic in most parts of the cell. In the lower mantle temperatures remain rather low, except in the lower thermal boundary layer.

Conclusions

The following conclusions may be drawn.

(1) A comparison of temperatures in upper and deep mantle models favours whole mantle convection.

(2) A moderate increase of viscosity in the lower mantle (10^{24} poise) does not hinder whole mantle convection.

(3) The temperature gradient below the lithosphere might be subadiabatic and temperatures in the lower mantle could be rather low (1500–2000 °C). A thermal boundary layer at the core–mantle boundary (more substantial than in the present model) must then ensure that the melting temperature is exceeded in the outer core. However, the result given by this model may possibly change if a more realistic model of whole mantle convection (e.g. 2900 km deep and spherical) can be applied.

References

DeBremaecker, J.-Cl., 1977. Convection in the Earth's mantle. *Tectonophysics* **41**, 195–208.

Elsasser, W. M., Olson, P. and Marsh, B. D., 1979. The depth of mantle convection. *J. Geophys. Res.* **84**, 147–155.

Fusijawa, H., 1968. Temperature and discontinuities in the transition layer within the Earth's mantle: geophysical application of the olivine–spinel transition in the Mg_2SiO_4–Fe_2SiO_4 system. *J. Geophys. Res.* **73**, 3281–3294.

Gebrande, H., 1975. Ein Beitrag zur Theorie thermischer Konvektion im Erdmantel mit besonderer Berücksichtigung der Möglichkeit eines Nachweises mit Methoden der Seismologie. Diss., Inst. für Allgemeine und Angewandte Geophysik, Univers. München, West Germany.

Graham, E. K., 1970. Elasticity and composition of the upper mantle. *Geophys. J. R. Astr. Soc.* **20**, 285–302.

Mercier, J. C. and Carter, N. L., 1975. Pyroxene geotherms. *J. Geophys. Res.* **80**, 3349–3362.

Watt, J. P., and O'Connell, R. J., 1978. Mixed oxide and perovskite structure model mantles from 700–1200 km. *Geophys. J. R. Astr. Soc.* **54**, 601–630.

Viscosity Estimates of the Lithosphere and Asthenosphere

U. R. VETTER, J. F. STREHLAU, R. O. MEISSNER

Institut für Geophysik, Neue Universität, D 2300 Kiel, West Germany

Introduction

The concept of plate tectonics is based on the postulate of a rigid lithosphere overlying a fluid-like asthenosphere. The hypothesis that parts of the Earth's interior behave as a viscous fluid was derived from the early principle of isostatic equilibrium. More than 30 years ago, seismological observations by Gutenberg led to the inference of a zone of distinctly lower seismic velocities and higher absorption of seismic wave energy at depths of roughly 100–200 km. This low-velocity channel has been attributed to a decrease in rigidity and a temperature approaching the melting point. It has since been identified by the term "asthenosphere".

More recently, heat flow studies and theoretical models of convection have been applied to the thermal boundary layer concept: the lithosphere is defined as a conductively cooling layer above a certain isotherm, e.g. 1400 K (Parmentier, Turcotte and Torrance, 1975). (See also Pollack and Chapman, 1977.) Since seismic velocities and absorption are temperature-dependent, the seismically and thermally defined lithosphere and asthenosphere appear to be physically equivalent.

A further approach to the definition of lithosphere and asthenosphere employs the rheological properties of the crust and upper mantle. These properties can be estimated from modelling observations of long-term surface loads, such as ice coverage and deglaciation (Cathles, 1975), volcanic seamounts (McNutt and Menard, 1978) and topographic rises at ocean trenches (Melosh, 1978). Results of these studies indicate that the "effective elastic plate thickness" decreases with increasing age of the load, reaching only 30–50% of the "seismic" lithosphere for old loads (Bodine, Steckler and Watts, 1979).

Thus, different definitions of lithosphere and asthenosphere seem to apply for different time constants of loading, varying from short-term seismic stresses (periods of < 1–100 s) to long-term tectonic stresses (periods of 10^5 yr or longer).

The "effective rheologic" thickness of the lithosphere does not correspond to the "thermal" or "seismic" thickness. It appears that there is no unique definition of the lithosphere–asthenosphere system which would apply at all time-scales and to all tectonic situations present in the Earth's crust and upper mantle. However, since mechanisms of plastic creep presumably govern all tectonic processes, one may suggest the following terminology: the asthenosphere is characterized by the ability of steady-state creep to occur on a global scale, whereas the lithosphere is that depth region which behaves rigidly to a first approximation and where creep is definitely confined to localized regions, for instance to the lower crust in warm areas. The boundary between lithosphere and asthenosphere most likely is not a sharp jump, but a zone of gradual transition as a function of temperature, stress, and other physical and mineralogical parameters.

In this paper, we have used available data for steady-state creep in olivine to estimate viscosity variations in the upper mantle. We have attempted to interpret these viscosity–depth models for an application of the lithosphere–asthenosphere terminology. However, we will not assign a specific thickness to this rheologically defined lithosphere because its thickness depends on local tectonic situations and the particular parameters chosen for modelling them.

Creep laws

Laboratory measurements provide valuable indications of the behaviour of rocks under different conditions of stress, temperature, pressure, strain rate and water content. An important result of such experiments is that it is found that different creep laws describe the creep response to stress in different temperature and pressure regimes. Various authors have compiled deformation mechanism maps for certain minerals, as for instance Stocker and Ashby (1973) for olivine, and Rutter (1976) for calcite and quartz. These deformation maps show in general four different types of creep processes in the temperature range investigated. Two grain-size dependent creep processes with strain rate $\dot{\varepsilon}$ proportional to stress σ (Coble creep and Nabarro–Herring creep) dominate at low stresses and low strain rates, Nabarro–Herring creep prevailing in the high-temperature field; two dislocation creep/glide processes with $\dot{\varepsilon}$ proportional to σ^n operate at high stresses. For tectonic stress levels a creep law with $\dot{\varepsilon} \sim \sigma^3$ is considered by many to be valid over a wide temperature range. We should emphasize that laboratory strain rates are some orders of magnitude higher than tectonic strain rates; also, stresses in the experiments are in most cases higher than tectonic stresses usually regarded as ambient in the Earth. Therefore, the creep data used here for viscosity calculations are extrapolated to lower strain rates and stresses. The general creep law may be written (Kohlstedt, Goetze and Durham, 1976) as

$$\dot{\varepsilon} = C\sigma^n \exp(-Q+pV/RT) \quad (1)$$

with $\dot{\varepsilon}$ = creep rate, C = constant, σ = shear stress, Q = activation energy, V = activation volume, p = lithostatic pressure, R = gas constant and T = temperature in K. With Weertman's (1970) transformation this creep law becomes a function of the ratio of melting temperature T_m to ambient temperature T:

$$\dot{\varepsilon} = C\sigma^n \exp(-g\,T_m/T). \tag{2}$$

T/T_m is called the homologous temperature; g is a material-dependent number proportional to the activation energy of the creep process. Kohlstedt and Goetze (1974) in their laboratory experiments found an average value of $g = 29$ for dislocation creep in olivine single crystals corresponding to an average activation energy of 125 kcal/mol (see also Weertman and Weertman, 1975).

It is assumed in the following that only two creep laws are important in the temperature regime $T > 0{\cdot}5\,T_m$ and for tectonic stresses (0–2 kbar): (a) the Nabarro–Herring creep law (NH)

$$\dot{\varepsilon} = C_1\,\sigma \exp(-gT_m/T) \tag{3}$$

with $C_1 = 2 \times 10^{-4}$ cm s/g for melting temperature and a grain size of 0·5 cm, and $C_1 = 4{\cdot}5 \times 10^{-3}$ cm s/g for melting temperature and a grain size of 1 mm (see also Vetter and Meissner, 1977; Vetter, 1978); and (b) the dislocation creep law

$$\dot{\varepsilon} = C_3\,\sigma^3 \exp(-gT_m/T) \tag{4}$$

with $C_3 = 4{\cdot}2 \times 10^{11}/\mathrm{kbar}^3\,\mathrm{s}$ (Kohlstedt and Goetz, 1974). Our assumption has been motivated by the contention of some authors for a linear, Newtonian upper mantle (e.g. Cathles, 1975), and by the argument of others for a non-Newtonian mantle (e.g. Post and Griggs, 1973). Nabarro–Herring and power law creep are independent deformation mechanisms. The faster creep rate under a given stress–temperature condition will always dominate.

Effective viscosities of the upper mantle

Using the relation

$$\eta \approx \sigma/\dot{\varepsilon} \tag{5}$$

and creep data of olivine, effective viscosities η for the two creep processes mentioned above have been calculated. For dislocation creep, creep rates of $10^{-14}, 10^{-15}, 10^{-16}$/s were applied in accordance with conditions in the 50–400 km depth range. Given these values of creep rate, the stress varies between about 0·1 bar for $T > 0{\cdot}8\,T_m$ and 100 bars for $T \approx 0{\cdot}5$–$0{\cdot}6\,T_m$. (For more information see Vetter, 1978, Figs 3 and 4.) The calculations according to Eqn (3) (NH-creep) were performd using three different grain sizes between 1 mm and 1 cm, consistent with estimates for upper mantle rocks. Figure 1 shows a

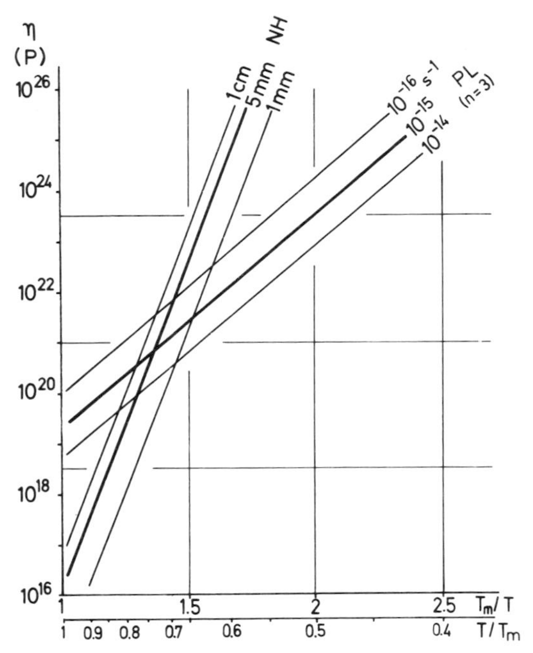

Figure 1. Effective viscosity (in poise) versus temperature on the basis of olivine creep data (Kohlstedt and Goetze, 1974). NH: Nabarro–Herring creep, grain sizes 1 mm to 1 cm. PL: Power law creep, strain rates 10^{-14}–10^{-16}/s.

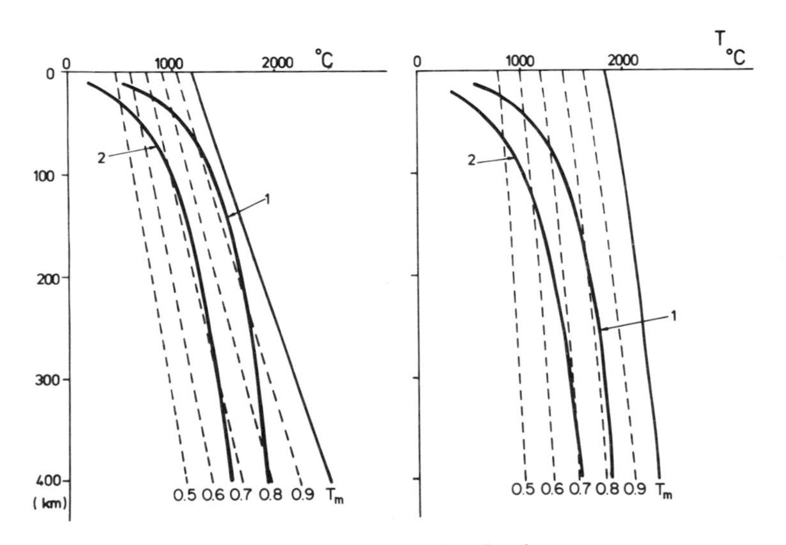

Figure 2. Temperature–depth and melting point–depth curves.
Left: T_m = dry pyrolite solidus (Stocker and Ashby, 1973).
Right: T_m = forsterite$_{90}$ melting (Carter, 1976). (1) Geotherm of a young ocean (Griggs, 1972); (2) geotherm of a continental shield (Clark and Ringwood, 1964).

plot of viscosity versus homologous temperature for the two creep laws. The transition from power law creep to linear NH-creep seems to take place at a stress of about 1 bar and $T = 0{\cdot}7$–$0{\cdot}8\ T_m$, depending on grain size and creep rate. A value of ε around 10^{-14}/s, a grain size between 5 mm and 1 cm and thus a transition temperature of approximately $T = 0{\cdot}8\ T_m$ may be reasonable values, at least below oceans where stresses may be less than 1 bar.

In Figure 2, published continental and oceanic geotherms as well as fractions of two mantle melting curves are shown. We have chosen the dry pyrolite solidus and the forsterite$_{90}$ melting curve as probable lower and upper bounds for the mantle melting curve. Using constant creep rates for oceanic and continental plates, the temperature data were converted into viscosity–depth profiles compiled in Figs 3 and 4. The dry pyrolite solidus yields a rather "hot" mantle with a pronounced low-viscosity channel, especially below oceans. The forsterite$_{90}$ melting curve results in a moderate-temperature mantle model with no low-viscosity channel in the upper 400 km.

These models can be improved by variable creep rates. A value of $\dot{\varepsilon} = 10^{14}$/s may be adequate below "fast"-moving oceanic plates, but may become smaller with decreasing depth because of decreasing temperatures. For this reason, additional calculations of viscosity–depth profiles for a young ocean and a continental shield with creep rates of 10^{-14}/s at depth decreasing upwards to 10^{-17}/s have been carried out. For $T \geqslant 0{\cdot}8\ T_m$ the Nabarro–Herring creep law was applied, in all other temperature regimes the power creep law with an exponent $n = 3$. The results are compiled in Figs 5 and 6.

Discussion

Our viscosity estimates indicate for the upper mantle values of 10^{19}–10^{22} poise below oceans and 10^{21}–10^{23} poise below continents. These are effective viscosities for a "dry", purely kinematic plate model far from plate boundaries under stresses originating from the mechanical plate movement. The generally higher viscosity of the continental lithosphere and asthenosphere compared to the oceanic ones were based on thermal models only. The viscosity below the continents may be further increased if a basalt depletion by means of large volcanic activity has taken place, as is observed in many shield areas (Jordan, 1979). Such a basalt depletion would lead to a decrease of the iron content and hence the density, but to an increase of seismic velocity and viscosity. Consequently, it increases the tendency for a viscosity difference between oceanic and continental asthenosphere to exist and for a growing stability of continental shield areas.

Deviations from these estimates can occur for several reasons. A local water saturation, for example, would decrease the melting point strongly and may result in an increase of the homologous temperature. This effect may even allow steady-state creep in parts of the lower continental crust, particularly in

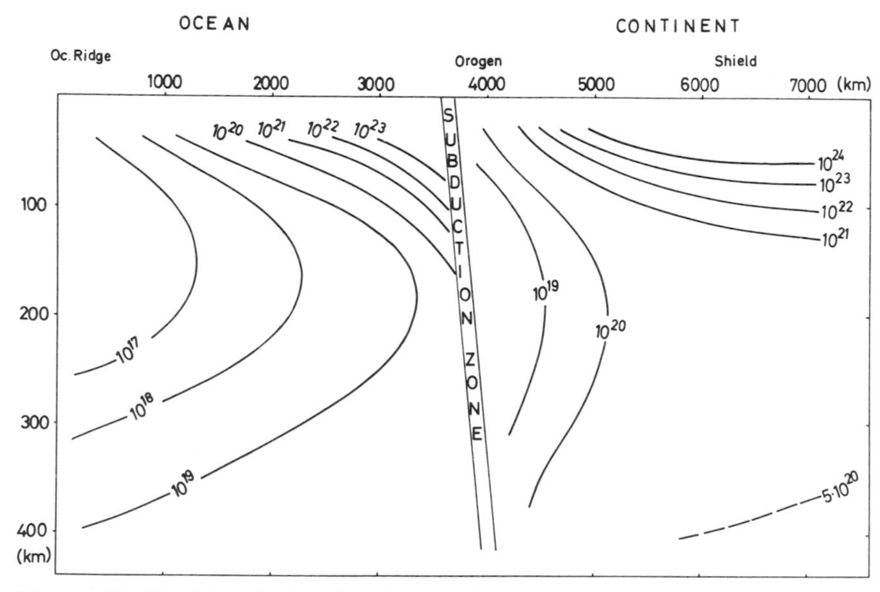

Figure 3. Profile of the effective viscosity calculated for a dry pyrolite mantle; NH creep for $T \geqslant 0{\cdot}8\ T_m$ and grain size 5 mm; PL creep for lower temperatures and constant strain rates (2×10^{-14}/s for the ocean, 4×10^{-15}/s for the continent).

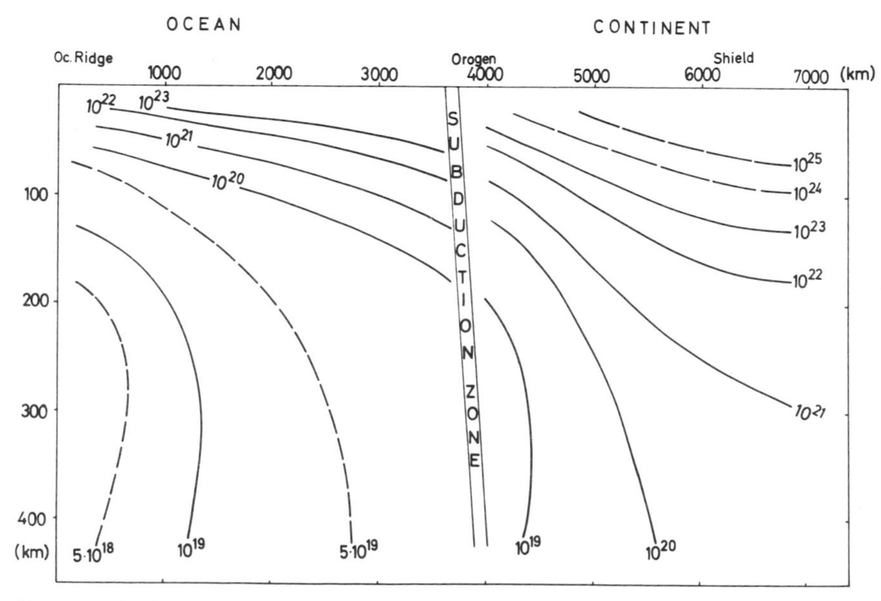

Figure 4. As for Fig. 3 for a forsterite$_{90}$ mantle.

Young Ocean

T_m : dry pyrolite solidus $\qquad$ T_m : forsterite$_{90}$ melting

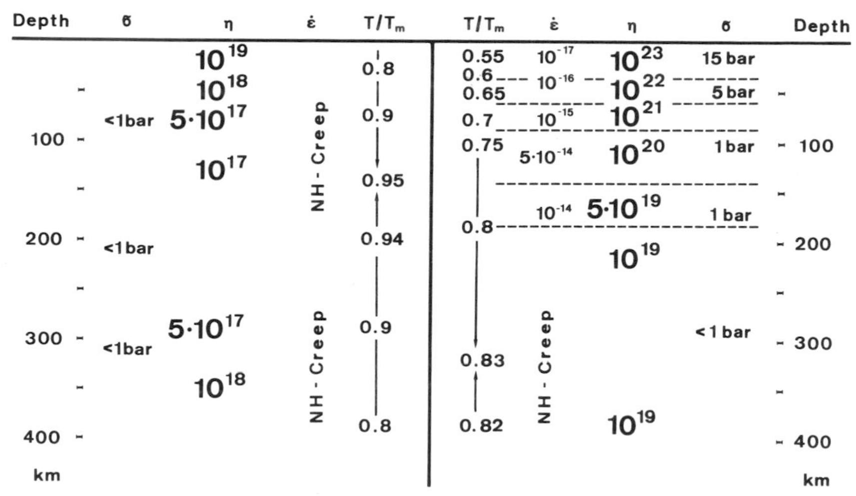

Figure 5. Effective viscosity versus depth for a young ocean (two mantle melting curves, NH creep (grain size 5 mm) for $T \geqslant 0{\cdot}8\ T_m$, PL creep with variable strain rate for lower temperatures).

Continental Shield

T_m : dry pyrolite solidus $\qquad$ T_m : forsterite$_{90}$ melting

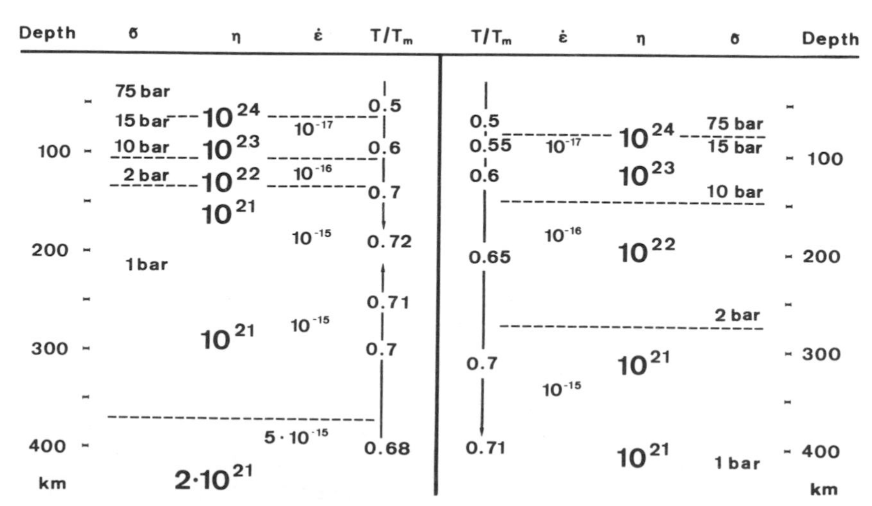

Figure 6. As for Fig. 5 for a continental shield.

thick and warm crusts, as is suggested by the occurrence of roots below young mountain ranges and of antiroots below grabens and depressions (Meissner, 1974). Important viscosity anomalies may exist at plate margins, i.e. along spreading centres, subduction zones and transform faults. Stresses may concentrate along parts of these tectonic zones and thus cause deformation in the form of transient creep as observed by episodic aseismic crustal movements preceding and following large earthquakes (Savage and Prescott, 1978). Physically, a fault zone is characterized by a reduced grain size, reduced shear modulus, higher water content, and thus by a reduced effective viscosity. The formal application of a steady-state creep law (dislocation creep) to small time steps of an aseismic creep event along a fault zone results in effective viscosities varying with time. At the beginning of the creep process the "effective fault zone viscosity" may be as low as 10^{14-15} poise. These effects cause deformations to be localized along such "zones of weakness". Stress release by earthquakes occurs in the upper, often termed "brittle", section of faults, mostly between 5 and 20 km, but may extend to depths of 60 km, depending on the temperature and water content. Thermally, the seismic depth region may be defined by temperatures below 0·5 to 0·6 T_m.

Static stress concentrations within plates, such as those below seamounts or below sediment loads at passive continental margins, may change the creep conditions to the extent that deformation becomes dominated by a different creep mechanism, e.g. Coble creep (Goetze, 1978). More data will be necessary to include this mechanism in our modelling efforts.

Conclusions

We have derived viscosity–depth profiles from an extrapolation of laboratory creep data for olivine to low strain rates using a linear Nabarro–Herring creep law and a non-linear dislocation creep law together with two models of mantle melting. We have obtained viscosity estimates varying from about $10^{18\pm1}$ poise for young oceans to about $10^{22\pm1}$ poise below old continental shields. In our definition, the lithosphere corresponds to the region which moves in the form of "rigid" plates over the Earth's surface; localized creep deformation may occur at plate boundaries or within plates only under certain conditions of temperature, stress, or mineralogy. Our estimates indicate that the rheology of crust and upper mantle can be very inhomogeneous. Accordingly, the rheological lithosphere–asthenosphere boundary varies considerably in depth and may even be located in the lower crust in warm areas. The asthenosphere, defined as the region with viscosities certainly less than about 10^{22}–10^{23} poise, corresponding to large-scale continuous convective movements, seems to have different viscosities below continents and oceans, at least down to 300 km. Below that zone the mantle viscosity may take on a uniform value around 10^{22} poise as assumed by Peltier and Andrews (1976) and Cathles (1975) for the whole mantle.

References

Bodine, J. H., Steckler, M. S., and Watts, A. B., 1979. Observations of flexure and the rheology of the oceanic lithosphere. (Abstract) *Trans. Am. Geophys. Union* **60**, 393. mantle. *Rev. Geophys. Space Phys.* **2**, 35–68.

Carter, N. L., 1976. Steady-state flow of rocks. *Rev. Geophys. Space Phys.* **14**, 301–360.

Cathles III, L. M., 1975. *The Viscosity of the Earth's Mantle.* Princeton University Press, Princeton, NJ.

Clark, S. P. Jr. and Ringwood, A. E., 1964. Density distribution and constitution of the mantle. *Rev. Geophys. Space Phys.* **2**, 35–68.

Goetze, C., 1978. The mechanism of creep in olivine. *Phil. Trans. R. Soc. Lond.* **A288**, 99–199.

Griggs, D. T., 1972. The sinking lithosphere and the focal mechanism of deep earthquakes. In *Nature of the Solid Earth* (Ed. Robertson, E. C.) 361–384, McGraw-Hill, New York.

Jordan, T. H., 1979. The deep structure of the continents. *Scientific American* 1/79, 70–82.

Kohlstedt, D. L. and Goetze, C., 1974. Low-stress high-temperature creep in olivine single crystals. *J. Geophys. Res.* **79**, 2045–2051.

Kohlstedt, D. L., Goetze, C. and Durham, W. B., 1976. Experimental deformation in single crystal olivine with application to flow in the mantle. In *Physics and Chemistry of Minerals and Rocks* (Ed. Strens, R. G.) 35–50, Wiley, London.

McNutt, M. and Menard, H. W., 1978. Lithospheric flexure and uplifted atolls. *J. Geophys. Res.* **83**, 1206–1212.

Meissner, R. O., 1974. Viscosity–depth structure of different tectonic units and possible consequences for the upper part of converging plates. *J. Geophys. (Z. Geoph.)* **40**, 57–73.

Melosh, H. J., 1978. Dynamic support of the outer rise. *Geophys. Res. Lett.* **5**(5), 321–324.

Parmentier, E. M., Turcotte, D. L. and Torrance, K. E., 1975. Numerical experiments on the structure of mantle plumes. *J. Geophys. Res.* **80**, 4417–4424.

Peltier, W. R. and Andrews, J. T., 1976. Glacial–isostatic adjustment. I, the forward problem. *Geophys. J. R. Astr. Soc.* **46**, 605–646.

Pollack, H. N. and Chapman, D. S., 1977. On the regional variation of heat flow, geotherms, and lithosphere thickness. *Tectonophysics* **38**, 279–296.

Post, R. L. Jr and Griggs, D. T., 1973. The Earth's mantle: evidence of non-Newtonian flow. *Science* **181**, 1242–1244.

Rutter, E. H., 1976. The kinetics of rock deformation by pressure solution. *Phil. Trans. R. Soc.* **A283**, 203–219.

Savage, J. C. and Prescott, W. H., 1978. Asthenosphere readjustment and the earthquake cycle. *J. Geophys. Res.* **83**, 3369–3376.

Stocker, R. L. and Ashby, M. F., 1973. On the rheology of the upper mantle. *Rev. Geophys. Space Phys.* **11**, 391–426.

Vetter, U. R., 1978. Stress and viscosity in the asthenosphere, *J. Geophys.* **44**, 231–244.

Vetter, U. R. and Meissner, R. O., 1977. Creep in geodynamic processes. *Tectonophysics* **42**, 37–54.

Weertman, J., 1970. The creep strength of the Earth's mantle. *Rev. Geophys. Space Phys.* **8**, 145–168.

Weertman, J. and Weertman, J. R., 1975. High temperature creep of rock and mantle viscosity. *Ann. Rev. Earth Planet. Sci.* **3**, 293–315.

Plasticity and Anelasticity of Olivine and Forsterite from Laboratory Experiments

M. DAROT, Y. GUEGUEN, C. RELANDEAU

Laboratoire de Tectonophysique, 2 Chemin de la Houssinière, 44072 Nantes Cedex, France

J. WOIRGARD

E.N.S.M.A., rue Guillaume VII, 86034 Poitiers Cedex, France

Introduction

Although the elastic properties of olivine have been investigated for a long time and are of fundamental importance in improving our knowledge of the Earth's mantle, the importance of plastic and anelastic properties of this mineral has been realized only in the last decade. This occurred at the end of the sixties when plate tectonics emerged as a new framework for the Earth sciences. Ten years later, a large amount of data is available on olivine single crystals and polycrystals, covering a broad range of stress–temperature conditions. The first data on olivine plasticity were obtained at high stresses with the Griggs apparatus (Griggs, 1967; Raleigh, 1968). Data in a geophysically more interesting range of stresses (10–100 MPa) were obtained later (Kohlstedt and Goetze, 1974; Durham and Goetze, 1972a; Jaoul *et al.*, 1979; Darot, 1980). However, very low stresses data (0·1–10 MPa) are not yet available. Investigation of anelastic properties is less developed (Gueguen, Woirgard and Darot, 1980).

Published data and new data are reviewed and discussed in this chapter. Results on plasticity of single crystals, polycrystals and on anelasticity are successively considered.

Plasticity of olivine and forsterite

The final goal of the study of olivine plasticity is the determination of flow laws and their extrapolation to upper mantle deformation. Ashby and Verrall

(1978) reviewed the possible micromechanisms of deformation in olivine. Two major groups of mechanisms can be distinguished; diffusion mechanisms (diffusional creep) and dislocation mechanisms (recovery creep). In general the strain rate can be described by a function $\dot{\varepsilon} = \dot{\varepsilon}(\sigma, T, P, P_{(O_2)}, X)$ in which σ is the differential stress, T the temperature, P the confining pressure, $P_{(O_2)}$ the oxygen partial pressure and X stands for other possible relevant parameters such as stoichiometry, FeO activity, etc.

Single crystals

Published flow laws

Most of the experiments were carried in a steady-state regime under atmospheric pressure in a dead load creep machine. The general form used for the strain rate was

$$\dot{\varepsilon} = A\sigma^n \exp(-Q/RT),$$

in which σ is the differential stress and Q the activation energy. Table I summarizes the flow law data for single crystals. No marked effect of $P_{(O_2)}$ was found on forsterite whereas the olivine strain rate varied as $P_{(O_2)}{}^{1/6}$ (Kohlstedt and Hornack, 1980; Poumelec *et al.*, 1980). The influence of impurities or non-stoichiometry is still unknown, but the data available on olivine and pure forsterite do not show significant differences.

New data. Creep experiments have been carried out on synthetic forsterite single crystals cut in three orientations $[1\,1\,0]_c$, $[1\,0\,1]_c$, $[0\,1\,1]_c$*. Two types of experiments were performed. Increasing stress step runs ($10 < \sigma < 100$ MPa) and temperature step runs (1400 °C < T < 1650 °C) enabled parameters n and Q to be determined. A dead load creep machine with two distinct atmospheres (i.e. working atmosphere : adjustable H_2/CO_2 mixture, resistor protecting atmosphere : argon) was used for these experiments (Darot, 1980). The results are plotted in Figs 1 and 2. As shown in Fig. 2 the activation energy for the $[1\,1\,0]_c$ orientation is found to be stress dependent. For the $[1\,0\,1]_c$ orientation the situation is more complex because of interaction between different families of dislocations. The three investigated orientations lead to two different flow laws. From a mechanical point of view the $[0\,1\,1]_c$ orientation is equivalent to the $[1\,1\,0]_c$ orientation, the only difference between the two being that the resolved shear stress on the active glide plane is much smaller in the $[0\,1\,1]_c$ orientation than in the $[1\,1\,0]_c$ orientation.

Creep on the $[1\,1\,0]_c$ orientation follows the expression

$$\dot{\varepsilon} \propto f(\sigma) \exp\left[-\frac{Q(\sigma)}{RT}\right],$$

* The subscript "c" refers to an imaginary cubic lattice.

TABLE I

Material	σ (MPa)	T (°C)	$P_{(O_2)}$	Orientation*	n	Q (kcal/mol)	Reference
San Carlos olivine	5–150	1400–1650	Not kept constant			126 ± 2 (Not corrected for $P_{(O_2)}$ effect)	K–G, 1974
San Carlos olivine	10–180	1150–1600	Not kept constant	$[1\,1\,0]_c$ $+[1\,0\,1]_c$ $+[0\,1\,1]_c$	$3{\cdot}6 \pm 0{\cdot}3$ $3{\cdot}7 \pm 0{\cdot}2$ $3{\cdot}5 \pm 0{\cdot}3$	Not measured	D–G, 1977a
San Carlos olivine	100		Constant	$[1\,0\,1]_c$		107 ± 13	K–H, 1980
Synthetic forsterite	8–60	1450–1650		$[1\,0\,1]_c$	$3{\cdot}6 \pm 0{\cdot}3$	135 ± 15	D–G, 1977b
Synthetic forsterite	20–100	1400–1650		$[1\,1\,0]_c$	$2{\cdot}5 \pm 0{\cdot}1$	110 ± 14	This study

* The subscript "c" refers to an imaginary cubic lattice. The given orientations correspond to the direction of the differential stress.

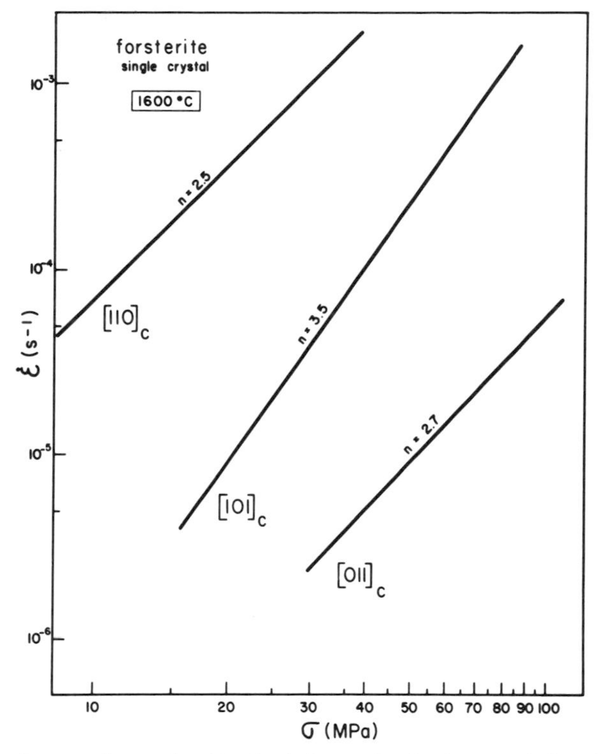

Figure 1. Log $\dot{\varepsilon}$ versus log σ plot. Synthetic forsterite single crystals $[1\,1\,0]_c$, $[1\,0\,1]_c$ and $[0\,1\,1]_c$. It is important to notice that n values for $[1\,1\,0]_c$ and $[0\,1\,1]_c$ are similar.

If $\sigma > 20$ MPa, the stress dependence of $Q(\sigma)$ is low and $f(\sigma) = \sigma^{2\cdot5\pm0\cdot1}$ then the flow law is

$$\dot{\varepsilon} = 1{\cdot}06 \times 10^6\, \sigma^{2\cdot5} \exp\left[\frac{-110 \pm 14}{RT}\right].$$

If $\sigma < 20$ MPa, the stress dependence is important for both $f(\sigma)$ and $Q(\sigma)$, and no simple analytic form can be attributed to $f(\sigma)$.

Under the same experimental conditions ($20 < \sigma < 100$ MPa, $T \geqslant 1400$ °C) $\dot{\varepsilon}_{[1\,1\,0]_c}$ is larger than $\dot{\varepsilon}_{[1\,0\,1]_c}$. The flow law corresponding to the $[1\,0\,1]_c$ orientation has the form $\dot{\varepsilon} \propto \sigma^{3\cdot5} \exp(-Q/RT)$. The stress exponent is higher for $[1\,0\,1]_c$ than for $[1\,1\,0]_c$. Determining the activation energy requires more measurements than those available now, especially if one attempts to analyse the stress dependence.

Diffusion data

A complete set of diffusion data on iron, magnesium, oxygen and silicon is now available (Buening and Busek, 1973; Reddy and Cooper, 1978; Poumelec *et al.*, 1980). Table II gives the self-diffusion data for the olivine components. Silicon has been found to be the slowest diffusing species in forsterite.

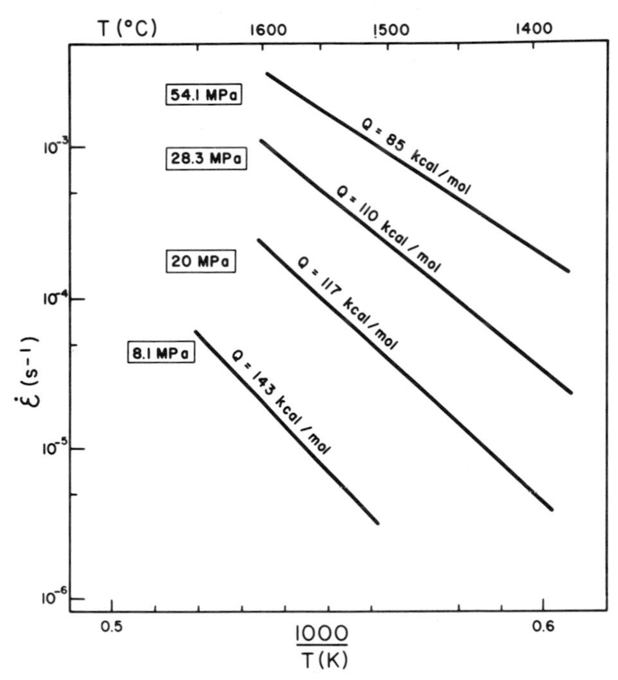

Figure 2. Log $\dot{\varepsilon}$ versus $1/T$ plot. Synthetic forsterite single crystals $[1\,1\,0]_c$. The fan arrangement of straight lines indicates a stress dependence of the activation energy.

TABLE II *Inter-diffusion and self-diffusion data for olivine elements*

$ID_{Fe-Mg} = 1{\cdot}7 \times 10^{-12} \exp\left(\frac{-58}{RT}\right)$	(cm^2/s) Buening and Buseck (1973)
$SD_{Si} = 1{\cdot}5 \times 10^{-6} \exp\left(\frac{-90 \pm 15}{RT}\right)$	(cm^2/s) Poumellec *et al.* (1980)
$SD_o = 10^{-4} \exp\left(\frac{-77 \pm 15}{RT}\right)$	(cm^2/s) Jaoul *et al.* (1980)

Moreover, annealing experiments (Ricoult, 1978) show that even at high temperatures the kinetics of dislocation annealing is very slow. Comparison between activation energies for creep and diffusion shows that diffusion cannot be the unique controlling mechanism since for low stresses Q reaches values much larger than for silicon self-diffusion.

Microstructures

Dislocation microstructures have been extensively studied by Gueguen (1979a) in peridotite xenoliths from basalts and kimberlites. Durham (1975), Jaoul *et al.* (1979) and Darot and Gueguen (1980) described typical dislocation microstructures in deformed olivine and forsterite single crystals. Figure 3 shows specific dislocation microstructures associated with the main glide

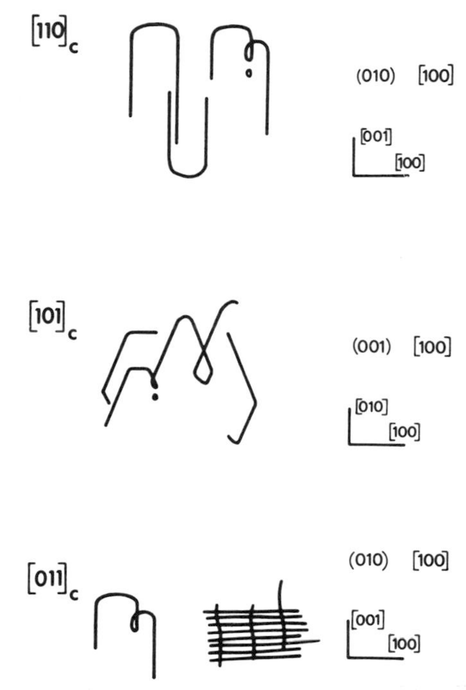

Figure 3. Characteristic dislocation features in glide planes for deformations with various stress directions.

$[1\,1\,0]_c$; long straight edges are more numerous than short bent screws. $[1\,0\,1]_c$; long "zig-zagging" mixed dislocations in $\{1\,1\,0\}$ directions, short bent screws are less developed. $[0\,1\,1]_c$; dislocation loops composed of long straight edges and short bent screws $\mathbf{b} = [1\,0\,0]$ and $\mathbf{b} = [0\,0\,1]$, $\mathbf{b} = [1\,0\,0]$ screws are more numerous.

systems. Such configurations are also found in naturally deformed olivine crystals. Observations of microstructures show that edge dislocations of the (0 1 0) [1 0 0] glide system are dominant. This emphasizes the importance of this system in high temperature deformation. Linearity of these dislocations, even at high temperatures implies the existence of "deep energy valleys". Gueguen (1979b) proposed a model of climb dissociation to explain such energy valleys. The stress dependence of the activation energy found on the $[1\,1\,0]_c$ deformations is consistent with this model, since motion of split dislocation implies recombination of partials, a process depending on the stress.

Aggregates

Extrapolation of single crystals data to polycrystals is not straightforward. Thus direct data obtained from olivine aggregates are of great interest. In the case of polycrystals, plastic flow may result from dislocations or from diffusion mechanisms. This last case includes Nabarro–Herring creep where vacancies diffuse through the grain, and Coble creep where diffusion takes place along the grain boundary (Ashby and Verral, 1978).

Published data. Most published data (Carter and Avé Lallemant, 1970; Post, 1973; Zeuch and Green, 1977) were obtained in the solid medium Griggs apparatus. Stresses achieved with this apparatus are higher than 100 MPa which is large compared with stresses estimated for the upper mantle (Nicolas, 1978). For this reason, several authors have attempted to obtain creep data on polycrystals at lower stresses (Schwenn and Goetze, 1977; Berckemer, Auer and Drisler, 1979). Schwenn's data are not direct creep data but hot-pressing data. Berckemer *et al.*'s data were obtained with natural peridotite specimens at temperatures close to or above the solidus (Gueguen and Darot, 1980) so that partial melting may have modified the mechanical behaviour of their rock samples.

New data. To avoid this kind of problem, sintering of pure forsterite has been performed at 1700 °C and low porosity (3–5%) aggregates have been obtained.

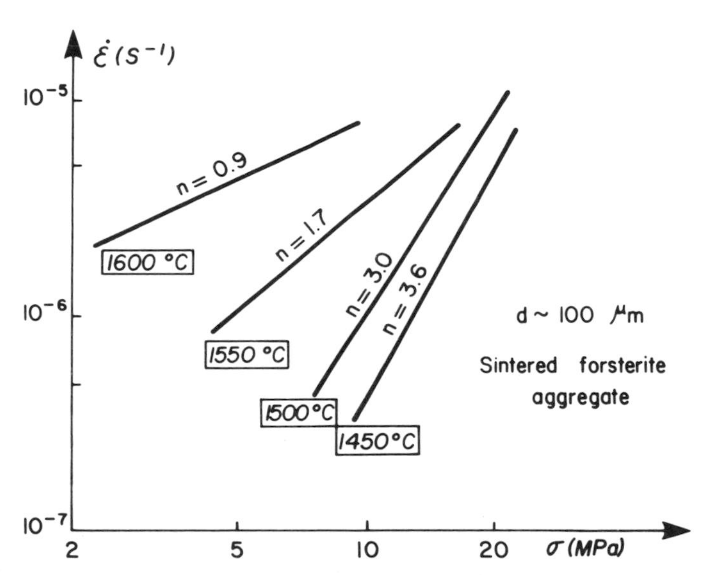

Figure 4. Log $\dot{\varepsilon}$ versus log σ plot. Sintered forsterite aggregate. The mean grain size is 100 μm. The stress exponent is decreasing for increasing temperatures hence decreasing stresses.

Deformation experiments have been conducted on specimens of grain size 20–200 μm between 1450 and 1600 °C (Relandeau, 1980). Maximum strains between 2 and 5% were achieved depending on the grain size, for stresses ranging between 3 and 20 MPa. Beyond these strain values, the strain rate tends to accelerate and the specimen yields by porosity-weakening (Goetze, 1978). Figure 4 gives the results for 100 μm grain size specimens. The variations of the stress exponent with temperature suggest a progressive change of plastic flow mechanisms between 1500 and 1550 °C for this grain size. It is important to note that in the same σ–T conditions below 1500 °C, the polycrystal and the single crystal strain rates are very similar. This comparison is illustrated in Fig. 5. At least it indicates that polycrystals deformation may occur in the

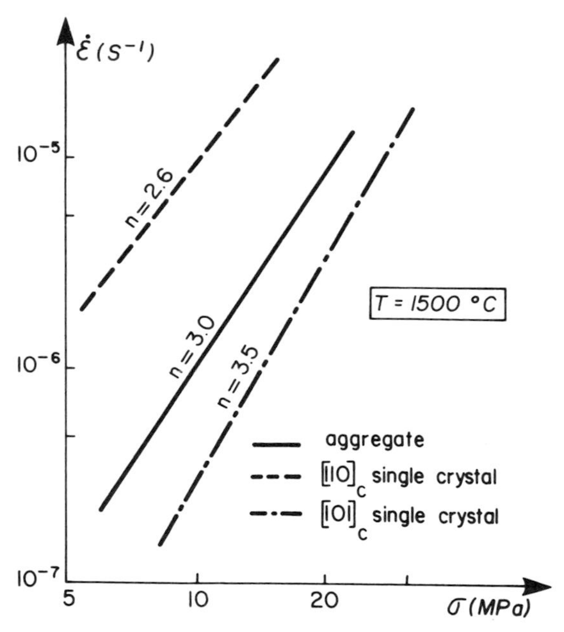

Figure 5. Log $\dot{\varepsilon}$ versus log σ plot. Comparison of strain rates for $[1\,1\,0]_c$ and $[1\,0\,1]_c$ forsterite single crystals and sintered forsterite aggregate at constant temperature $T = 1500$ °C.

dislocation creep domain (recovery creep). At 1600 °C, the stress exponent is close to 1, which possibly reflects a diffusion mechanism. There is not yet sufficient data, however, to propose a numerical creep law for this regime.

Anelasticity of olivine and forsterite

Despite the fact that attenuation in the Earth is an important problem (Liu, Anderson and Kanamori, 1976), and that Q profiles derived from seismology (Anderson and Hart, 1978) indicate a low-Q zone in the upper mantle, very few data are available on attenuation of acoustic waves at low frequencies and high temperatures in Earth materials. This accounts for a great part of the difficulty of such experiments.

Experimental data

Woirgard and Gueguen (1978) have reported data obtained between 0 °C and 1100 °C at low frequencies (1–10 Hz) on (a) synthetic, undeformed polycrystalline forsterite and (b) naturally deformed peridotite (Fig. 6). They have observed a peak of attenuation in the second case at 930 °C. Annealing of the

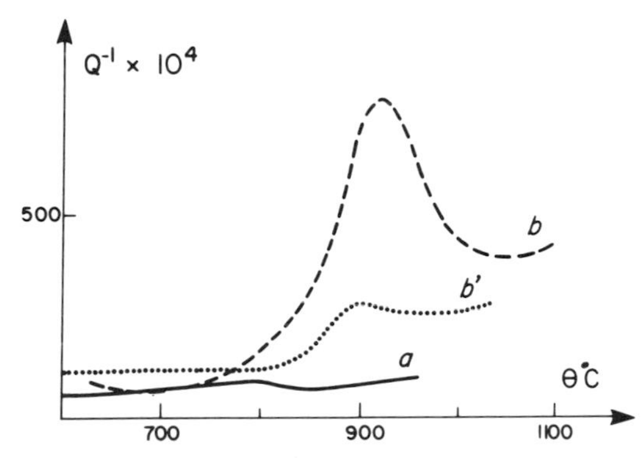

Figure 6. Q^{-1} results for peridotite and forsterite aggregate at constant frequencies (flexion pendulum). (a) Synthetic forsterite (1·8 Hz); (b) naturally deformed peridotite (5 Hz); (b′) naturally deformed peridotite after 5 h annealing at 1100 °C.

sample results in a considerable reduction of the peak. This behaviour suggests a dislocation effect but confirmation must wait for additional data on single crystals.

Berckemer *et al.* (1979) have used transient creep data to calculate Q for natural forsterite samples. The use of such an indirect method is not satisfactory for two reasons: (a) very similar transient creep behaviour may correspond to very different Q spectra; (b) the assumption of linear behaviour is required for such calculations, but this assumption may not be valid. Non-linear effects have been observed in metals at high temperatures, even at strain amplitudes as low as 10^{-6}–10^{-7} (Woirgard *et al.*, 1980).

Possible microscopic mechanisms

It seems now that dislocation effects can explain the totality of the peaks observed at high temperatures on most materials (Woirgard, 1976). Our knowledge of the dislocations microstructures in olivine and forsterite at high temperatures (see previous main section) suggests two possibly important mechanisms of attenuation by dislocation in that mineral.

Oscillation of a straight dislocation segment between two energy "valleys" can result in high attenuation at a frequency dependent on the temperature: $\nu = \nu_0 \exp(W/RT)$, where W is the double kink energy formation. This is the "Bordoni peak". The fact that long, straight edge dislocations are dominant in olivine crystals deformed at high temperatures (Fig. 4) indicates the potential importance of this mechanism.

Dislocation climb could also result in high attenuation. However bulk diffusion is so low (Poumelec *et al.*, 1980) that an absorption effect at low frequencies by this mechanism would require some short-cuts like pipe diffusion.

In any case, dislocation relaxation implies a frequency dependence for Q^{-1} since such processes are thermally activated. The relaxation time varies with temperature: $\tau = \tau_0 \exp(E/RT)$. Due to the rapid increase of T with depth in the upper mantle, a peak of attenuation (and therefore a drop of velocity) is expected between 100 and 400 km. Data on single crystals are of fundamental importance to give us the values of parameters such as E, the activation energy. They are expected over the next few years.

Conclusions

The progress made in understanding olivine and forsterite behaviours is illustrated by experimental results obtained during the last five years. Several important conclusions can be drawn from these data for single crystals:

(1) The activation energy for creep is stress dependent at high temperatures: it increases with decreasing stress.

(2) In olivine, the creep rate is $P_{(O_2)}$ dependent ($P_{(O_2)}{}^{1/6}$) but in forsterite it is not.

(3) Silicon is the slowest diffusing species in forsterite; it is about two orders of magnitude slower than oxygen.

(4) The microstructures indicate that creep is controlled by edge dislocations.

These results demonstrate that olivine plasticity is more complex than previously thought. They suggest two possible creep controlling mechanisms. Creep could be glide controlled as suggested by the microstructures, but the edge dislocations would be dissociated out of their glide planes. In this case diffusion would be required for kink formation. A second possibility is that creep is climb controlled but the activation energy for climb is the sum of two terms which are the diffusion term and the jog formation term.

Plastic and anelastic properties of olivine could be related through dislocation processes. Dislocation relaxation is expected to produce attenuation peaks at high temperatures and low frequencies either as a result of dislocation glide (Bordoni peak) or dislocation climb (pipe diffusion). Internal friction data on single crystals are required and expected in the near future.

Acknowledgements

We are grateful to A. Nicolas for criticisms and review of the manuscript. This work was supported by the C.N.R.S. (ERA 547).

References

Anderson, D. L. and Hart, R. S., 1978. Attenuation models of the Earth. *Phys. Earth Planet. Int.* **16**, 289–306.

Ashby, M. F. and Verrall, R. A., 1978. Flow and fracture in the upper mantle. *Phil. Trans. R. Soc.* **A288**, 58–95.

Berckhemer, H., Auer, F. and Drisler, J., 1979. High temperature anelasticity and elasticity of mantle peridotite. *Phys. Earth Planet. Int.* **20**(1), 48–59.

Buening, D. K. and Buseck, P. R., 1973. Fe–Mg lattice diffusion in olivine. *J. Geophys. Res.* **78**, 6852–6862.

Carter, N. L. and Avé Lallemant, H. G., 1970. High temperature flow of dunite and peridotite. *Bull. Geol. Soc. Am.* **81**, 2181–2202.

Darot, M., 1980. Déformation expérimentale de l'olivine et de la forsterite. Thèse d'Etat, Nantes.

Darot, M. and Gueguen, 1980. High temperature creep of forsterite single crystals. II, Microstructural data and controlling mechanisms. *J. Geophys. Res.* (in press).

Durham, W. B., 1975. Plastic flow of single-crystal olivine. PhD. thesis, M.I.T., Cambridge, MA.

Durham, W. B. and Goetze, C., 1977a. Plastic flow of oriented single crystals of olivine. I, Mechanical data. *J. Geophys. Res.* **82**, 5737–5753.

Durham, W. B. and Goetze, C., 1977b. A comparison of creep properties of pure forsterite and iron-bearing olivine. *Tectonophysics* **40**, 15–18.

Goetze, C., 1978. The mechanism of creep in olivine. *Phil. Trans. R. Soc. Lond.* **A288**, 99–119.

Griggs, D. T., 1967. Hydraulic weakening of quartz and other silicates. *Geophys. J. R. Astr. Soc.* **14**, 19–31.

Gueguen, Y., 1979a. Les dislocations dans l'olivine des péridotites. Thèse d'Etat, Nantes.

Gueguen, Y., 1979b. High temperature olivine creep: evidence for control by edge dislocations. *Geophys. Res. Lett.* **6**(5), 357–360.

Gueguen, Y. and Darot, M., 1980. Comments on high temperature anelasticity and elasticity of mantle peridotite. *Phys. Earth. Planet. Int.* (in press).

Gueguen, Y., Woirgard, J. and Darot, M., 1980. Attenuation mechanisms and anelasticity in the upper mantle. *Proceedings of the XVII General Assembly of the IUGG*, Canberra, 1979.

Jaoul, O., Gueguen, Y., Michaut, M. and Ricoult, D., 1979. A technique for decorating dislocations in forsterite. *Phys. Chem. Minerals* **5**, 15–19.

Jaoul, O., Froidevaux, C., Durham, W. B. and Michaut, M., 1980. Oxygen self-diffusion in forsterite: implications for high temperature creep mechanism (in press).

Kohlstedt, D. L. and Goetze, C., 1974. Low stress high temperature creep in olivine single crystals. *J. Geophys. Res.* **79**, 2045–2051.

Kohlstedt, D. L. and Hornack, F., 1980. Effect of oxygen partial pressure on the creep of olivine. *Proceedings of the XVII General Assembly of the IUGG*, Canberra, 1979.

Liu, H. P., Anderson, D. L. and Kanamori, H., 1979. Velocity dispersion due to anelasticity; implications for seismology and mantle compositions. *Geophys. J. R. Astr. Soc.* **47**, 41–58.

Nicolas, A., 1978. Stress estimates from structural studies in some mantle peridotites. *Phil. Trans. R. Soc.* **A288**, 49–57.

Post, R. L., 1973. The flow laws of Mt Burnett dunite. *Tectonophysics* **42**, 75–110.

Poumelec, M., Jaoul, O., Froidevaux, C. and Havette, A., 1980. Silicon diffusion in forsterite: a new constraint for understanding mantle deformation. *Proceedings of the XVII General Assembly of the IUGG*, Canberra, 1979.

Raleigh, C. B., 1968. Mechanisms of plastic deformation in olivine. *J. Geophys. Res.* **73**, 5311–5406.
Reddy, K. P. R. and Cooper, A. R., 1978. Oxygen self-diffusion in forsterite and $MgAl_2O_4$. *Am. Ceram. Soc. Bull.* **57**, 310.
Relandeau, C., 1980. Fluage de la forsterite polycristalline. Thèse de Docteur Ingénieur, Paris.
Ricoult, D., 1978. Recuit expérimental de l'olivine. Thèse de 3ème Cycle, Nantes.
Schwenn, M. B. and Goetze, C., 1977. Creep of olivine during hot pressing. *Tectonophysics* **48**, 41–60.
Woirgard, J., 1976. Modèle pour les pics de frottement interne observés à haute tempèrature sur les monocristaux. *Phil Mag.* **33**, 623–637.
Woirgard, J. and Gueguen, Y., 1978. Elastic modulus and internal friction in enstatite, forsterite and peridotite at seismic frequencies and high temperatures. *Phys. Earth Planet. Int.* **17**, 140–146.
Woirgard, J., Gerland, M. and Rivière, A., 1980. High temperature and low frequencies damping of aluminium single crystals (in press).
Zeuch, D. H. and Green, H. W., 1977. Naturally decorated dislocations in olivine from peridotite xenoliths. *Contrib. Mineral. Petrol.* **62**, 141–51.

The Electrical Conductivity of the Lower Mantle

V. COURTILLOT, J. DUCRUIX, J. L. LE MOUËL,
J. ACHACHE

Départment des Sciences de la Terre, Université Paris VII, and Laboratoire de Géomagnétisme Interne (LEGSP-LA 195), Institut de Physique du Globe, Université Paris VI, France

Introduction

Analysis of annual mean values of geomagnetic field components from over 80 world observatories for the period 1947–1977, using very simple but efficient methods for the isolation of "true" (internal) secular variation, leads to two major results:

(1) The 11-year variation is clearly isolated and a good estimate of the Earth's reponse for such long periods is obtained.

(2) A secular variation acceleration is found to have occurred in the late 1960s over much of the northern hemisphere. Other possible accelerations ("impulses") around 1840 and 1910 are also reported.

Both observations constrain the mantle conductivity distribution, one "from the top", the other "from the bottom". The conclusion is that it is unlikely that conductivity exceed 100/Ω m for any appreciable thickness in the lower mantle. This is consistent with early conductivity estimates (e.g. Runcorn, 1955; Banks, 1969) but lower, by two orders of magnitude or more, than recently proposed values (Alldredge, 1977; Stacey, McQueen, Smylie, Rochester and Conley, 1978).

A survey of possible conduction mechanisms and estimated values of associated parameters in the deep mantle leads to temperatures around 3500 K, in very good agreement with recent temperature models (Stacey, 1977; Elsasser, Olson and Marsh, 1979), with obvious implications regarding whole mantle convection. However, physical parameters for the electronic semi-conduction process are so uncertain that we may in reverse use conductivity and temperature estimates to constrain the corresponding activation energy of the lower mantle material.

Review

Electrical conductivity is one of the key physical parameters of the Earth's interior which geophysicist would like to determine as accurately as possible. Indeed, conductivity can be expected to provide information regarding the thermal state of the mantle, or possibly its mineral assemblages. The study of the Earth's conductivity is undertaken by analysing time variations of the Earth's magnetic field recorded at the surface; in parallel, electrical conductivity of mineral assemblages at high temperatures and high pressures is studied in the laboratory in the hope of being able to account for the magnetic observations.

In the analysis of time variations of the geomagnetic field, the classical theory of electromagnetic induction is used. There are two basically different kinds of sources for these variations: variations with shorter periods (say smaller than a few years) arise from the circulation of electrical currents in the ionosphere and above. These currents in turn induce currents in the conducting Earth. With sufficient spatial coverage of magnetic field components for a given frequency the inducing (external) and induced (internal) parts can be separated and their relative importance measured. The important parameter is the penetration depth of the currents (and fields), which is related to pulsation ω and conductivity σ through the simple relation:

$$d = (2/\omega\mu_0\sigma)^{1/2}, \qquad (1)$$

where μ_0 is the magnetic permeability of free space. For a variation with a period of 1 day (cf. the solar daily variation) and an average conductivity of $0{\cdot}1/\Omega$ m, the penetration depth is on the order of 500 km at most. If one could isolate variations with the solar 11-year period depths of 1500 km could be reached (using an average conductivity of $40/\Omega$ m). This explains the effort of researchers towards isolating longer and longer periods in observatory records, which implies finding data with the best possible quality; in particular, continuity and base-line control for all components.

The second source of geomagnetic variation, which is responsible for longer period variations, termed "secular variation", lies in the fluid core. If variations with a given "period" are known to have an internal origin and yet can be observed at the surface of the Earth, this implies that the penetration depth must be larger than the whole thickness of the mantle, thus setting upper bounds on average mantle conductivity.

This short paper gives a brief summary of earlier work on this subject and an account of our own efforts, which are given in much more detail in Ducruix, Courtillot and Le Mouël (1980).

Probing the mantle "from the top", using external variations, Lahiri and Price (1939) were able to propose a first rough model of mantle conductivity in 1938. In this classic paper, they use the different response of the Earth for periods of 1 day (and its harmonics) and 27 days (and harmonics) to suggest an increase in conductivity at a depth of a few hundred kilometres. Banks (1969), using a (by then) much larger data set, with periods as long as one year, was able to propose a refined model going all the way down to the core–mantle interface. Parker (1970) however resumed the analysis, using Banks' data and an inverse problem approach. He was able to show that the resolution provided by the data set worsens with depth, becoming almost unacceptably bad at a depth of 1200 km. In the next years, the first successful attempts to isolate the 11-year solar cycle related variation were made by Courtillot and Le Mouël (1976a, b), Alldredge (1976) and Harwood and Malin (1977). More recent and detailed analyses are found in Ducruix *et al.* (1980), Yukutake and Cain (1979) and Alldredge, Stearns and Sugiura (1979). Although the solar cycle effect was clearly present in Z component data of Courtillot and Le Mouël (1976a, b), these authors pointed out the apparently very large amplitude of the effect in the Z component with respect to the horizontal component and showed that it could not be reconciled with simple induction models (see also Courtillot and Le Mouël, 1979). It was then discovered that this effect was due to a peculiar behaviour of the internal secular variation, to be discussed shortly. When a proper model of this internal secular variation is built, the 11-year effect is perfectly well isolated and the ratio of internal to external contributions can be estimated fairly precisely (Fig. 1). In a first discussion of this important datum, Ducruix *et al.* (1980) show that it is particularly sensitive to the conductivity values in the middle zone of the mantle. The upper mantle conductivity is determined from shorter period external variations. It is found that the middle mantle conductivity lies close to the values proposed by Banks (1969) and MacDonald (1957), more precisely in the order of 10–30/Ω m. It is unlikely that longer external periods can be isolated and thus unlikely that the mantle can be probed in this way much deeper than the first 2000 km.

Fortunately, some information on the remaining part of the mantle can be obtained through a careful study of internal secular variation (i.e. probing the mantle "from the bottom"). It had been suggested by Walker and O'Dea (1952) that a case could be made for the occurrence of sudden changes in the value of secular variation (i.e. in the slope of the component data). These changes were called secular variation "impulses" and were used by Runcorn (1955) in order to impose upper bounds on the average mantle conductivity. However, Alldredge (1975) and Courtillot and Le Mouël (1976a, b) independently came to the conclusion that most, if not all, "impulses" arose from an improper representation of magnetic secular variation by a sequence of short straight-line segments. It was in fact the 11-year variation which was being modelled by the sequence of "impulses"! But, as more data accumulated, Courtillot, Ducruix and Le Mouël (1978a) were able to recognize the first unmistakable occurrence of a secular variation "impulse" (one should rather say a secular acceleration step) in most of the northern hemisphere in the late 1960s. The "impulse" is particularly clear in the European observatory data (Fig. 2), but is also found elsewhere (see Courtillot *et al.*, 1978a). Evidence for earlier "impulses" in 1910 and possibly 1840 is also suggested, and the correlation

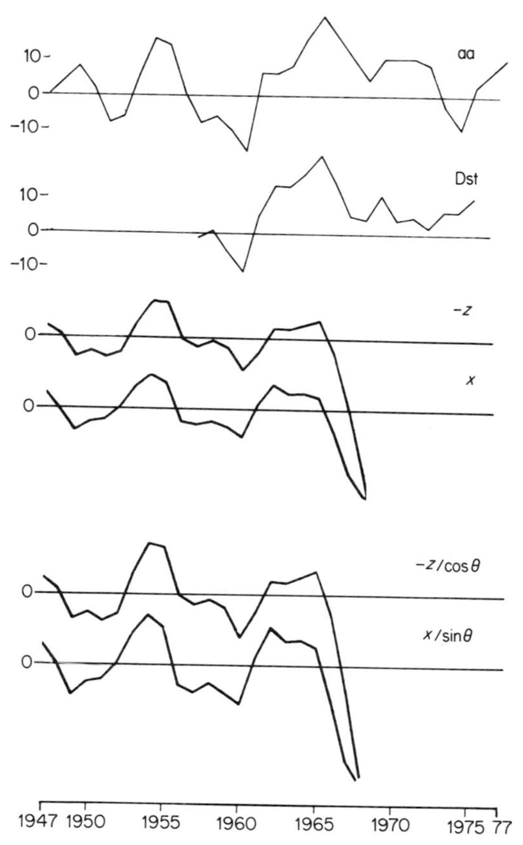

Figure 1. Isolation of the 11-year (solar cycle related) variation. From top to bottom are shown: the aa and Dst indices, the mean residual X (horizontal north component) and $-Z$ (vertical component, reversed in sign) curves for the ten European observatories listed in the caption of Fig. 2, and the mean values of ($X/\sin\theta$) and ($-Z/\cos\theta$) which take into account the geomagnetic colatitude of each observatory.

between these three geomagnetic events and the occurrence of sharp minima in the Earth's rotation rate is also striking. The occurrence of the impulse is particularly clear in the east component Y (or declination D). Practically, it implies that most field models and field maps of these elements are severely in error (see also Courtillot, Le Mouël and Leprêtre, 1978b; Galdeano, Courtillot and Le Mouël, 1980). The data of Fig. 2 imply that a very good representation of internal secular variation is given by two adjoining parabolic segments. The intensity of the "impulse" can tentatively be measured by the distance between the second parabolic segment and the extrapolation of the first one at a given time (see also Courtillot, Ducruix and LeMouël, 1979). This is shown in Fig. 3. One can then, as a first approximation, use Runcorn's (1955) plane case computation, yielding the diffusion of a signal through a plane layer of constant thickness L and conductivity σ. We have used a slightly modified time signature of the signal at the core–mantle interface (Achache *et al.*, 1980).

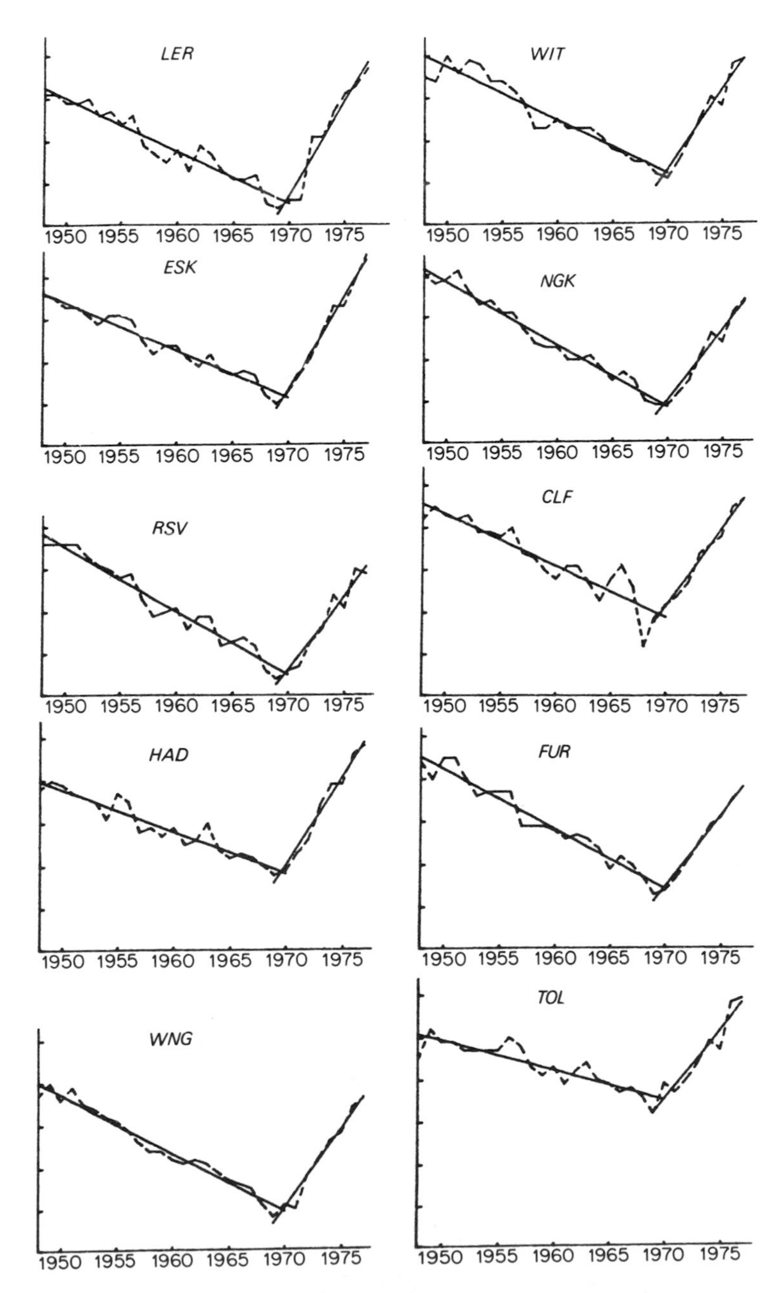

Figure 2. The secular variation (first differences of the raw data) in the following ten European observatories, is shown as a dashed line: Lerwick, Eskdalemuir, Rude Skov, Hartland, Wingst, Witteveen, Niemegk, Chambon-la-Forêt, Fürstenfeldbruck, Toledo. Tick spacing on the vertical axes is 10 nT/y. Solid straight lines are least-squares fits to the data, respectively for the periods 1948–1969 and 1969–1977.

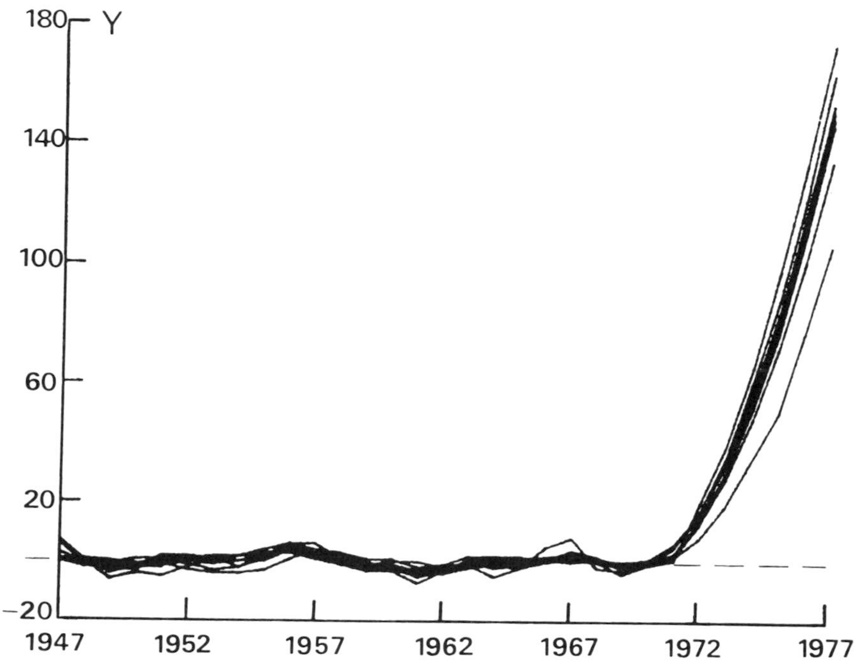

Figure 3. Differences between annual mean values of the *Y* (east) components and secular variation parabolae fitted to the data between 1947 and 1969. The ten European observatories of Fig. 2 are used.

It is seen from Fig. 4 that the important parameter is the delay between the occurrence of the "impulse" and the intersection of the new (asymptotic) variation with the extrapolated older regime observed at the Earth's surface: this parameter is $\mu_0 \sigma L^2/2$, related to the Cowling constant. The delay which we find is certainly less than 4 years and the corresponding upper bound for average conductivity (with $L = 2000$ km) is $60/\Omega$ m. Indeed, the fit is quite good as seen in Fig. 4. When due account is made for sphericity and finite extent of the sources (Smylie, 1965), this upper bound still is no higher than $150/\Omega$ m.

We are in the process of using this whole (and in part new) data set in order to construct properly a conductivity model (or rather investigate the scope of acceptable conductivity models). For the external variations, one can use the method first proposed by Bailey (1970) (Achache, 1979). However, a first-order model can already be proposed (Fig. 5) and is not likely to be modified drastically. It is seen in this model that, in fair agreement with the models of Banks and McDonald, conductivity probably does not exceed $100/\Omega$ m (or $200/\Omega$ m) for any appreciable depth range in the lower mantle. This is in contradiction (by orders of magnitude) with recent results of Alldredge (1977) and Stacey *et al.* (1978). We have explained elsewhere in detail why we thought that Alldredge's interpretation was not correct (Courtillot and Le Mouël,

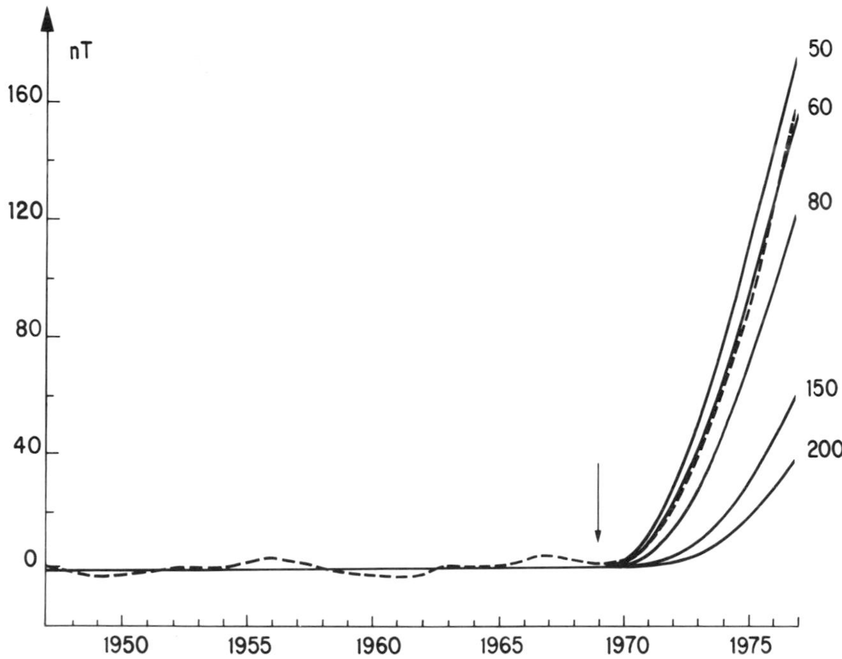

Figure 4. Solid lines represent the response $Y(L, t)$ for various values of the conductivity σ (/Ω m) of a plate of thickness $L = 2000$ km modelling the middle and lower mantle. The date of the secular variation "impulse" is taken to be 1969 as indicated by the vertical arrow. The dashed line is the average of the residual curves of Fig. 3.

1979). Stacey *et al.* (1978) base their analysis on the assumption that the spatial geomagnetic spectrum at the core is white; if one accepts our data and analysis, one can conclude that this spectrum is not white, but rather pink.

We are now in a position to see whether available laboratory experiments on electrical conductivity of likely mantle minerals can give us information on the thermodynamical properties of the Earth's deep mantle. Until very recently, most geophysicists agreed that, under the conditions of the middle and lower mantle, electronic semiconduction should be dominant (e.g. Tozer, 1959; Thomsen, 1976). This result has been challenged by Mao (1973). In any case, electrical conductivity is a thermally activated process, with a given activation energy E and pre-exponential parameter σ_0:

$$\sigma = \sigma_0 \exp(-E/kT). \tag{2}$$

Not much is known on values of E and σ_0 below the olivine–spinel transition and even less below the spinel–oxides transition. Values for σ_0 and E are extrapolations from lower pressure experiments on olivine (or germanium or silicon) and range from 5×10^4 to $7 \times 10^5/\Omega$ m and from 2 to 8 eV respectively (Stacey, 1969; Duba, 1972; Duba, Heard and Schock, 1974; Volarovich and Parkhomenko, 1976). Using "preferred" values of $5 \times 10^5/\Omega$ m and 2·5 eV, close to those of Stacey and Duba, and our estimate for conductivity, yields lower mantle temperatures on the order of 3500 K, not very different from those of Wang (1972), Stacey (1977) and Elsasser *et al.* (1979). The implications of these temperature estimates regarding whole mantle convection are discussed by Elsasser *et al.* (1979). However, the study of magnesiowüstites at

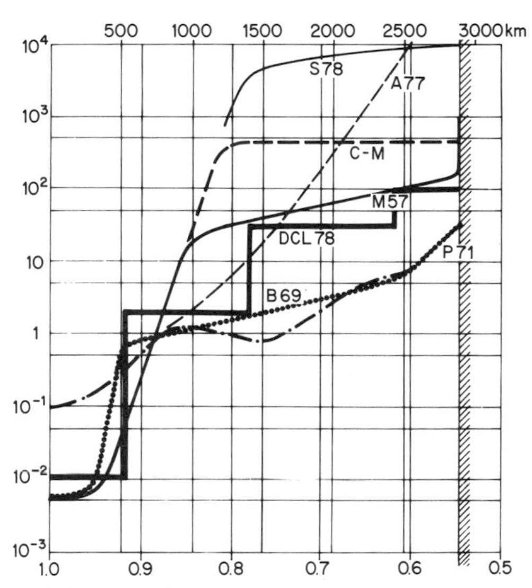

Figure 5. The conductivity models of MacDonald (M57), Banks (B69), Parker (P70), Cantwell–MacDonald (C–M), Alldredge (A77) and Stacey *et al.* (S78), together with our proposed model (DCL78). The horizontal axis displays the ratio r/a of the radial distance to the Earth's radius (the upper axis shows the corresponding depths in km). The vertical axis displays the conductivity ($/\Omega$ m), on a logarithmic scale.

high pressures by Mao (1973) yields a very low activation of 0·37 eV and $\sigma_0 = 10^{3 \cdot 34}/\Omega\,\mathrm{m}$. This would preclude any significant determination of temperature and also be in contradiction with intrinsic semiconduction. As more experiments are performed and the race towards still higher pressures goes on, it is hoped that this apparent contradiction may be solved.

On the other hand, following Ducruix *et al.* (1980), we can differentiate Eqn (2) in order to evaluate the relative errors in the various terms σ, σ_0, E and T. It is found that, at present, σ and T are the parameters which seem to be the least in error (of the order of 10%). We can then use determinations of σ and T to constrain values of σ_0 and E. With our figures, one obtains:

$$E(\mathrm{eV}) = 0{\cdot}26 \ln \sigma_0 - 1{\cdot}2. \qquad (3)$$

It is hoped that this (however indirect) observational constraint can help the interpretation of laboratory experiments and refine the knowledge of the mineralogical composition of the lower mantle.

Acknowledgements

This paper was also presented at the Seminar on Geophysics and Solid State Physics, organized in Paris by J. P. Poirier. IPGP Contribution NS 392.

References

Achache, J., 1979. L'inversion des données géomagnétiques de très longue période. Thèse de 3ème cycle, Université Paris VI.

Achache, J., Courtillot, V., Ducruix, J. and Le Mouël, J. L., 1980. The late 1960s secular variation impulse: further constraints on deep mantle conductivity. *Phys. Earth Planet. Int.* **23**, 72–75.

Alldredge, L. R., 1975. A hypothesis for the source of impulses in geomagnetic secular variation. *J. Geophys. Res.* **80**, 1571–1578.

Alldredge, L. R., 1976. Effects of solar activity on annual means of geomagnetic components. *J. Geophys. Res.* **81**, 2990–2996.

Alldredge, L. R., 1977. Deep mantle conductivity. *J. Geophys. Res.* **82**, 5427–5431.

Alldredge, L. R., Stearns, C. O. and Sugiura, M., 1979. Solar cycle variations in geomagnetic external spherical harmonic coefficients. *J. Geomag. Geoelectr.* **31**, 495–508.

Bailey, R. C., 1970. Inversion of the geomagnetic induction problem. *Proc. R. Soc.* **A135**, 185–194.

Banks, R. J., 1969. Geomagnetic variations and the electrical conductivity of the upper mantle. *Geophys. J. R. Astr. Soc.* **17**, 457–487.

Courtillot, V., Ducruix, J. and Le Mouël, J. L., 1978a. Sur une accélération récente de la variation séculaire du champ magnétique terrestre. *C. r. Acad. Sci. Paris* **D287**, 1095–1098.

Courtillot, V., Ducruix, J. and Le Mouël, J. L., 1979. Réponse aux commentaires de L. R. Alldredge "Sur une accélération récente de la variation séculaire du champ magnétique terrestre". *C. r. Acad. Sci. Paris* **B289**, 173–175.

Courtillot, V. and Le Mouël, J. L., 1976a. On the long period variations of the Earth's magnetic field: from 2 months to 20 years. *J. Geophys. Res.* **81**, 2941–2950.

Courtillot, V. and Le Mouël, J. L., 1976b. Time variations of the Earth's magnetic field with a period longer than two months. *Phys. Earth Planet. Int.* **12**, 237–240.

Courtillot, V. and Le Mouël, J. L., 1979. Comments on: "Deep mantle conductivity". *J. Geophys. Res.* **84**, 4785–4790.

Courtillot, V., Le Mouël, J. L. and Leprêtre, B., 1978b. Réseau magnétique de répétition de la France: campagne 1977. *IPGP Observations Magnétiques*, No. 35, Paris.

Duba, A., 1972. Electrical conductivity of olivine. *J. Geophys. Res.* **77**, 2483–2495.

Duba, A., Heard, H. C. and Schock, R. N., 1974. Electrical conductivity of olivine at high pressure and under controlled oxygen fugacity. *J. Geophys. Res.* **79**, 1667–1673.

Ducruix, J., Courtillot, V. and Le Mouël, J. L., 1980. The late 1960s secular variation impulse, the eleven year magnetic variation and the electrical conductivity of the deep mantle. *Geophys. J. R. Astr. Soc.* **61**, 73–94

Elsasser, W. M., Olson, P. and Marsh, B. D., 1979. The depth of mantle convection. *J. Geophys. Res.* **84**, 147–155.

Galdeano, A., Courtillot, V. and Le Mouël, J. L., 1980. La cartographie magnétique de la France au 1.7.1978. *Ann. Géophys.* **36**, 85–106.

Harwood, J. W. and Malin, S. C. R., 1977. Sunspot cycle influence on the geomagnetic field. *Geophys. J. R. Astr. Soc.* **50**, 605–619.

Lahiri, B. N. and Price, A. T., 1939. Electromagnetic induction in non-uniform conductors, and the determination of the conductivity of the Earth from terrestrial magnetic variations. *Phil. Trans. R. Soc.* **237**, 509–540.

MacDonald, K. L., 1957. Penetration of the geomagnetic secular field through a mantle with variable conductivity. *J. Geophys. Res.* **62**, 117–141.

Mao, H. K., 1973. Observations of optical absorption and electrical conductivity in magnesiowüstite at high pressures. *Carnegie Institute Ann. Rep. Geophys. Lab.* 554–557, Washington.

Parker, R. L., 1970. The inverse problem of electrical conductivity in the mantle. *Geophys. J. R. Astr. Soc.* **22**, 121–138.

Runcorn, S. K., 1955. The electrical conductivity of the Earth's mantle. *Trans. Am. Geophys. Union* **36**, 191–198.

Smylie, D. E., 1965. Magnetic diffusion in a spherically-symmetic conducting mantle. *Geophys. J. R. astr. Soc.* **9**, 169–184.

Stacey, F. D., 1969. *Physics of the Earth*, Wiley, New York.

Stacey, F. D., 1977. A thermal model of the Earth. *Phys. Earth Planet. Int.* **15**, 341–348.

Stacey, F. D., MacQueen, H. W., Smylie, D. E., Rochester, M. G. and Conley, D., 1978. Geomagnetic secular variation and lower mantle electrical conductivity. *Trans. Am. Geophys. Union* **59**, 1027.

Thomsen, L., 1976. Structure et constitution du manteau et du noyau. In *Traité de Géophysique Interne, Vol. 2* (Eds. Coulomb, J. and Jobert, G.), 355–401, Masson, Paris.

Tozer, D. C., 1959. The electrical properties of the Earth's interior. In *Physics and Chemistry of the Earth, Vol. 3* (Eds. Ahrens *et al.*), 414–436, Pergamon, Oxford.

Volarovich, M. P. and Parkhomenko, E. I., 1976. Electrical properties of rocks at high temperatures and pressures. In *Geoelectric and Geothermal Studies.* Ch. 3 (Ed. Adam, A.), KAPG Geophysical Monograph, Budapest.

Walker, J. B. and O'Dea, P. L., 1952. Geomagnetic secular-change impulses. *Trans. Am. Geophys. Union* **33**, 797–800.

Wang, C. Y., 1972. Temperature in the lower mantle. *Geophys. J. R. Astr. Soc.* **27**, 29–36.

Yukutake, T. and Cain, J. C., 1979. Solar cycle variations of the first degree spherical harmonic components of the geomagnetic field. *J. Geomag. Geoelectr.* **31**, 509–544.

Seismic Anomalies in the Lower Mantle and at the Core–Mantle Boundary

D. J. DOORNBOS*

Vening Meinesz Laboratory, University of Utrecht, the Netherlands

Introduction

Several studies on the thermodynamics of the mantle have suggested that the observed surface features (i.e. plate tectonics, and the preceding concept of continental drift) may be explained both by a model of shallow (upper mantle) and a model of deep (wholly mantle) convection, or a combination of the two (e.g. Richter and McKenzie, 1978; Oxburgh and Turcotte, 1978; Elsasser, Olson and Marsh, 1979). Of course, the conditions favouring any of the different models, and the implications arising from them, are not the same, but the critical parameters are not directly observable. Today, estimates of parameters like viscosity and chemical homogeneity are such that the possibility of whole mantle convection must be admitted (Davies, 1977; O'Connell, 1977; Sammis, Smith, Schubert and Yuen, 1977). Independent considerations concerning planetary evolution (Tozer, 1972) have led us to think that this is indeed a plausible possibility. This would support the early explanations of continental drift (see, for example, Heiskanen and Vening Meinesz, 1958), although it has become apparent that at the high Rayleigh numbers expected for the mantle, the convection pattern may not be simple (Tritton, this volume p. 267), but is to be regarded as having a stochastic character.

In these circumstances, which general non surface features of the convection process can be tested against geophysical data? In this regard, one may identify two implications: (1) Lateral variations in the mantle; given the uncertainties with regard to the convection pattern, it is at present difficult to predict a pattern of variations, or the magnitude of the anomalies. (2) If the mantle is heated from below, a thermal boundary layer is expected at the bottom. This

* Present address: NTNF/NORSAR, Postbox 51, 2007 Kjeller, Norway.

feature has been discussed by Jones (1977) and Elsasser, Olson and Marsh (1979). These authors give similar estimates of the average thickness (~ 70 km) and temperature gradient ($\sim 13°$/km) in the layer. In addition, the core–mantle boundary itself may be affected; the details of the effect are not known, but some dynamically produced topography would not be unexpected (Jacobs, 1975; Hide, 1977).

The geophysical data which have been and are being used to study the above mentioned features comprise gravity anomalies and seismic anomalies. The use of gravity data has been discussed by Runcorn (1972) and others, and correlations of the low degree harmonics with other geophysical data have been reported (Hide and Malin, 1970; Dziewonski, Hager and O'Connell, 1977). The problems of uniqueness and resolution, and consequently the important role of *a priori* assumptions in the interpretation of the gravity data, are well known (cf. Kahn, 1971). In this chapter we will discuss seismic anomalies. In the next section the relevant data and methods will be briefly surveyed. Then we will discuss how seismic wave phenomena due to interaction with the core–mantle boundary can be analysed to infer properties of the boundary region, and the results obtained so far will be summarized.

Seismic data and methods

A survey of seismic wave phenomena, their ray representation, and the resulting data, is given in Fig. 1. The most accessible data base is formed by the travel times, primarily of P waves (Fig. 1(a)). The classical velocity models of the Earth have been derived from them (Jeffreys, 1970) and more recently, these data have been used to study lateral variations. In principle, the inverse problem is both non-linear and non-unique (a travel time anomaly can be caused by a velocity anomaly anywhere along the ray). In practice, the problem is linearized and non-uniqueness has been reduced by combining data from many sources and receivers. Thus, so called three-dimensional inversion of travel time data has been developed and applied to the upper mantle following in particular Aki, Christofferson and Husebye (1977), and applied to the lower mantle by Dziewonski *et al.* (1977) and Sengupta (1975). Results from the latter authors suggest that there are large-scale lateral variations in the lower mantle, in particular in the lowest few hundred km. As the authors themselves indicate, the results should be regarded as tentative. Apart from the linearization procedure and the discretization of the model, practical limitations in these studies are set by non-uniform sampling of the Earth, errors in source parameters (location and origin time) and the concept of "station corrections", the latter being assumed to represent the effect of near-station anomalies; usually, they are estimated from the data themselves.

The eigenfrequencies of free oscillations have also been used in the inverse problem (e.g. Gilbert and Dziewonski, 1975) and the ray-mode duality predicts the same sampling of the mantle as for their corresponding (multiply reflected)

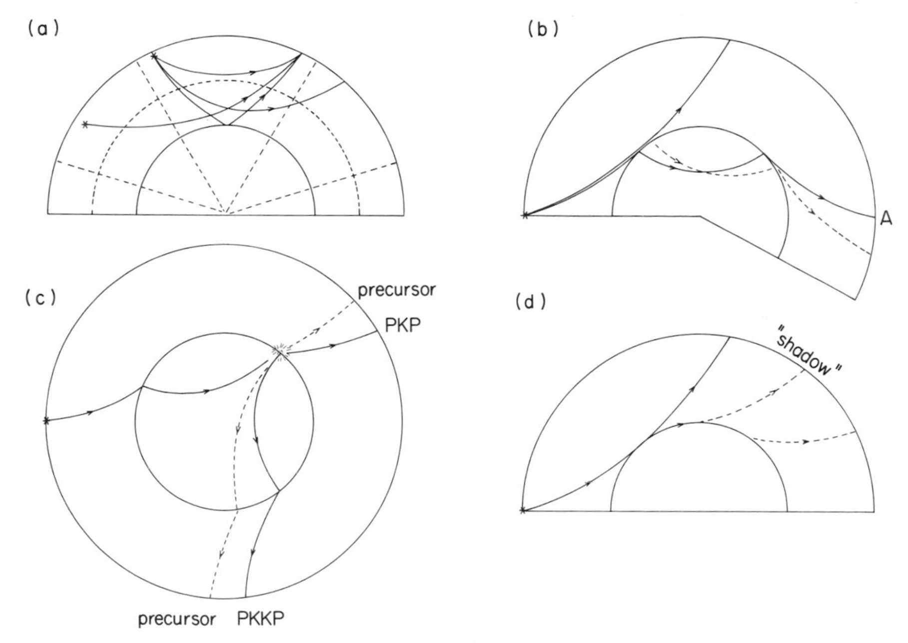

Figure 1. Ray illustration of seismic waves sampling the lower mantle and core–mantle boundary region.

(a) Direct *P*, *S*, *PcP*, *ScS* etc. Travel time data from many sources and receivers have been combined in inversion for velocity structure. Discretization of the model is illustrated by dotted lines. (b) Tunnelling through high velocity region at the base of the mantle. Solid ray: grazing *P* and core wave defining critical point A (of *PKP* in the figure). Dotted ray: tunnelled wave. With decreasing velocity at the base of the mantle, point A shifts to greater distances. (c) Scattering of core phases at core–mantle boundary. Solid rays: ordinary *PKP* and *PKKP*. Dotted rays: forward scattering into the core shadow zone gives precursors to *PKP*; backscattering gives precursors to *PKKP*. (d) Diffraction around the core (dotted rays).

rays. In the presence of lateral variations the eigenfrequencies are split, and linearized inversion procedures for splitting data have been given (Dahlen, 1974; Jordan, 1978), together with some preliminary results (Dahlen, 1976) which again suggest lateral variations in the lower mantle. A major practical problem here is that splitting is not easily observed or measured, and other effects may corrupt the splitting data; averaging procedures have sometimes been proposed to overcome some of these problems (e.g. Dahlen, 1974).

As pointed out in the previous section, the core–mantle boundary (CMB) region plays an important role in testing against dynamic models. The so called D″ region at the base of the mantle has been suspected for a long time to be anomalous, although the evidence has been controversial until recently (Cleary, 1974). The CMB region is also the site of a number of interesting seismic wave phenomena (diffraction, tunnelling, scattering), whose observation and analysis appears to be very fruitful in making inferences on the structure. In Fig. 1(d), diffraction around the core is illustrated. Geometric ray theory predicts a shadow zone at epicentral distances greater than about 100°. In reality there are diffracted waves decaying with increasing epicentral distance in the shadow. In seismology, this phenomenon was first given a theoretical basis by Scholte (1956). Alexander and Phinney (1966) and Phinney and Alexander (1966) first made use of it to develop models for the base of the mantle consistent with the frequency dependent decay of *P*, but theoretical limitations at that time did not yet allow realistic models to be included. Many of these limitations have now been removed, and results of a recent inversion of *P* and *S* diffraction data will be presented in the next section.

Observations of short-period *PnKP* (core phases multiply reflected inside the core) beyond the critical point A in standard Earth models, have been quoted as evidence for a low-velocity zone at the base of the mantle (e.g. Buchbinder, 1971). However, Richards (1973) has shown that tunnelling through a high-velocity zone may produce a similar effect. In the latter case the wave has a turning point above CMB, the wave field below the turning point decays exponentially with depth but is non-zero at the interface, and this accounts for transmission into the core. Tunnelling back into the mantle proceeds similarly (Fig. 1(b)). Tunnelling is more efficient at longer periods and this property has been invoked in modelling the upper mantle (Fuchs and Schultz, 1976), but for the base of the mantle, spectral analysis of the relevant *PnKP* phases has not yet been attempted.

Diffraction and tunnelling as described above are very sensitive to velocity structure above CMB, but not much to detailed shape of the boundary (Doornbos and Mondt, 1979a). On the other hand, elastic wave interaction with topography on the boundary leads to scattered waves, and this effect may be observable as precursors to certain short-period core phases. Figure 1(c) illustrates two possibilities. The most widely studied case are the precursors to *PKIKP* in the approximate distance range 125°–140°. Their interpretation as scattered waves follows a suggestion by Haddon (1972). They can be identified because the distance range shorter than 140° is a shadow zone for ordinary short-period outer core phases, and the inner core phase *PKIKP* arrives later. It has been shown that both a slightly rough CMB and a slightly heterogeneous lower mantle above CMB can explain most of the observed amplitudes (Doornbos, 1978), although in some cases a scattering source can be located on the basis of arrival time and direction of approach (Doornbos, 1976). The other possibility illustrated in Fig. 1(c) gives precursors to *PKKP* due to backscattering at CMB. This leads to precursors because *PKKP* (BC and DF branches) is a maximum time phase. This possibility was first mentioned by Chang and Cleary (1978) in connection with some remarkable *PKKP* observations at the LASA array, and the phenomenon has now been used to estimate the topography on CMB. The results obtained so far will be presented in the last section.

The base of the mantle

"Standard models" of the Earth, based on body wave travel times and free oscillation eigenfrequencies or a combination of the two, do not usually exhibit anomalous structure at the base of the mantle, and they are close to a state of adiabatic compression. Recent examples of resolution analysis applied to travel times (Sengupta and Julian, 1978) and eigenfrequencies (Masters, 1979) give for the base of the mantle resolving lengths of the order of hundred km (travel times) or more (eigenfrequencies). On the other hand, recent dynamic models suggest that the (globally averaged) anomalous zone may be less than 100 km thick, and so these models cannot be adequately tested with the "standard data".

Early evidence of a velocity reversal at the base of the mantle (assembled by Cleary, 1974) was obtained mostly on the basis of a ray geometrical interpretation. Since diffraction effects can be significant, this evidence has been criticized. More recently, diffraction theory has been used to interpret the data. An extensive data set of diffracted P and S waves has been given by Mondt (1977a), together with an interpretation indicating a velocity reversal for S waves at the base of the mantle (Mondt, 1977a, b). Based on the concept of "turning point level", Doornbos and Mondt (1979a, b) have shown that diffraction data do have the required resolution to test against the dynamic models. Several different features of the model have an effect on the diffraction data and in particular on the (frequency dependent) decay, but the effects are usually different for P, SH and SV and can be isolated by a joint inversion of these data. SV is usually not observed in the shadow (as expected theoretically) and the data set consists of P and SH decay spectra and $dT/d\Delta$. Figure 2 shows an average of Mondt's data together with computational results for a standard model (PEM-C from Dziewonski, Hales and Lapwood, 1975), and for the revised model, PEMC-L01, which has for P and S a comparable velocity decrease with depth in a rather thin (~ 75 km) transition zone at the base of the mantle (Doornbos and Mondt, 1979b). More generally, it was concluded that the data set allows a transition zone of thickness 50–100 km and the P and S velocities both decrease with depth. There is a suggestion of an additional low-Q zone; it would improve the fit to the decay data at the higher frequencies of SH, but it is not absolutely required due to the large standard errors associated with these data. Model PEMC-L01 is shown in Fig. 3 together with two different interpretations, as a density effect or as a temperature effect. The procedure has been described by Doornbos and Mondt (1979b) in some detail and it will suffice here to summarize. Common to both interpretations is the combining of P and S velocity gradients to form a so called "index of inhomogeneity" η (Bullen, 1975). If the medium is adiabatic but chemically inhomogeneous, the same η appears in a generalization of the Adams–Williamson equation from which the density gradient can then be obtained. (η

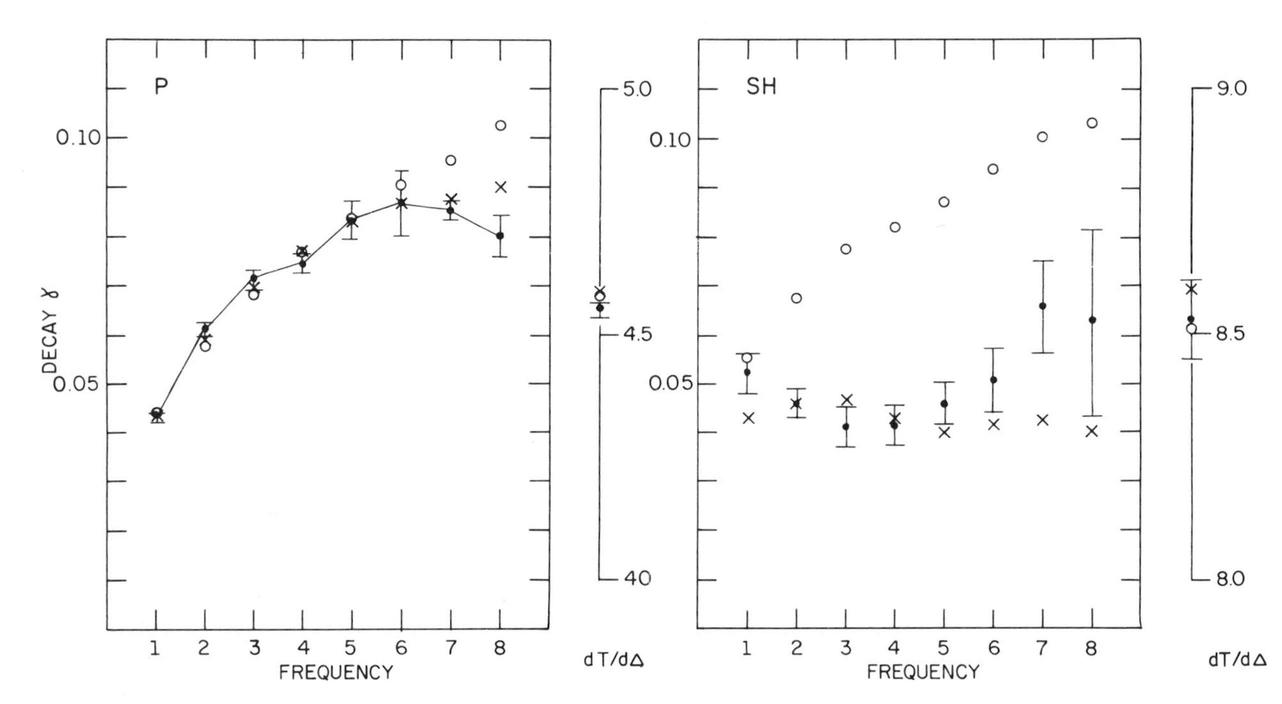

Figure 2. Decay spectra and $dT/d\Delta$ of *P* and *SH*. The decay γ measures logarithmic amplitude decay per degree of epicentral distance and corrected for geometrical spreading. Frequency points 1,...,8 correspond to 0·015625,...,0·125 Hz in steps of 0·015625. ●: observational means with standard deviations; ○: theoretical data for PEM-C model; ×: theoretical data for PEMC-L01 model (after Doornbos and Mondt, 1979b).

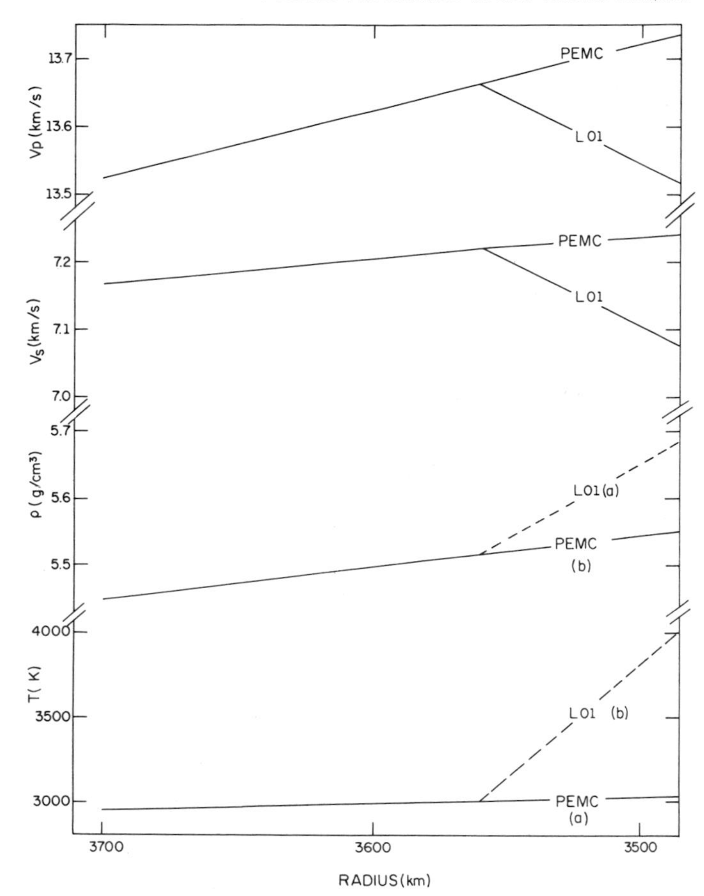

Figure 3. Models (PEM-C and PEMC-L01) for P and S velocities v_p, v_s, density ρ, temperature T in the lower mantle.

(a) The PEMC-L01 density curve assumes adiabatic conditions ($\equiv$ PEM-C temperature curve). (b) The PEMC-L01 temperature curve assumes chemical homogeneity ($\equiv$ PEM-C density curve). Assumed values for Grüneisen ratio $\gamma = 1{\cdot}5$, thermal expansion coefficient $\alpha = 1{\cdot}5 \times 10^{-5}$. Absolute temperature at 75 km above CMB has been arbitrarily taken as 3000 K.

has sometimes been defined by this generalized equation, e.g. Masters, 1979.) If the medium is homogeneous but non-adiabatic, η is used to calculate a superadiabatic temperature gradient (Birch, 1952). For "reasonable" values of thermodynamic quantities, the resulting temperature profile agrees quite well with temperatures predicted in recent dynamic boundary layer models (cf. Fig. 2 of Elsasser, Olson and Marsh, 1979). The result must yet be treated with caution, since not only are the thermodynamic quantities known within wide limits, but also the value for the "index of inhomogeneity" η is rather sensitive to apparently small changes in the velocity model.

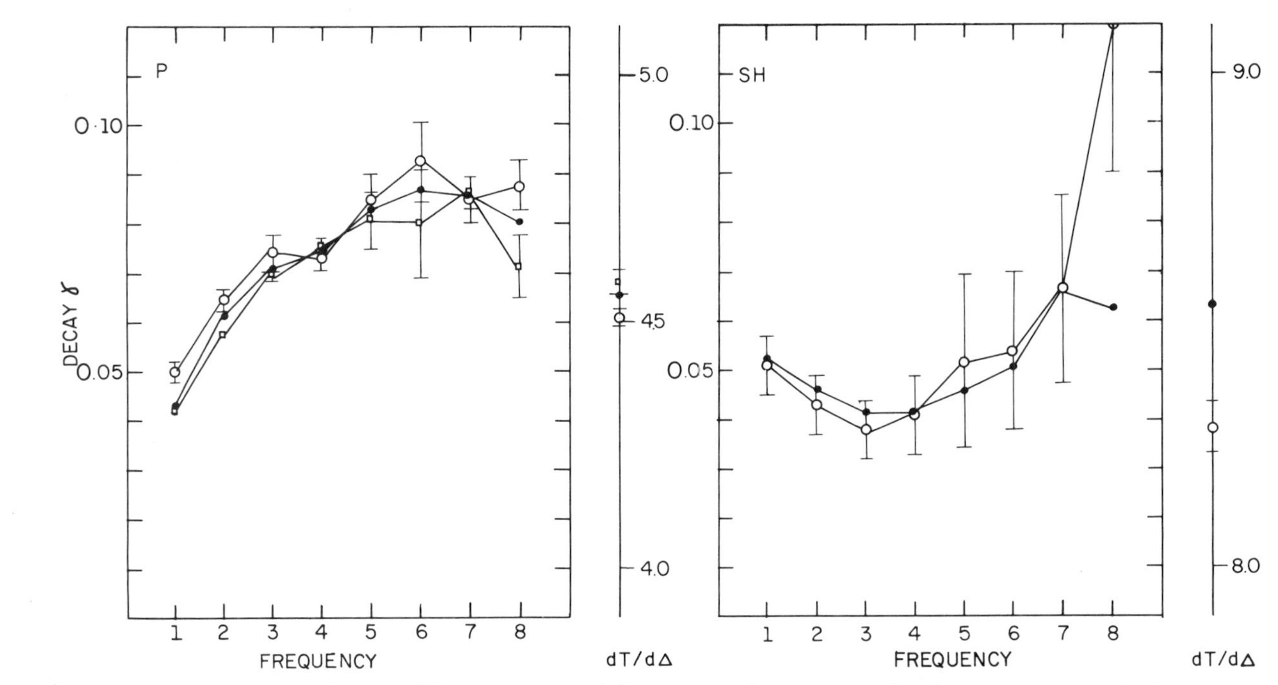

Figure 4. Decay spectra and $dT/d\Delta$ for *P* and *SH*. Decay γ and frequency points 1,...,8 as in Fig. 2. ●: average all data; ○: average data sampling core–mantle boundary beneath Pacific, with standard deviation (for *SH* this is no average as there was only one path beneath the Pacific). □: average data sampling other regions.

It would be of interest to test for the presence of lateral variations at the base of the mantle, and if possible to map these variations. Large-scale variations between different lines of stations most likely have their origin in the deep mantle near CMB. There are indications that these variations exist, both from $dT/d\Delta$ data (e.g. Espinosa, 1967) and from amplitude decay data (Phinney and Alexander, 1969) of diffracted *P* waves. In a statistical sense, Doornbos and Mondt (1979b) have tested that the effect on the data of the large-scale

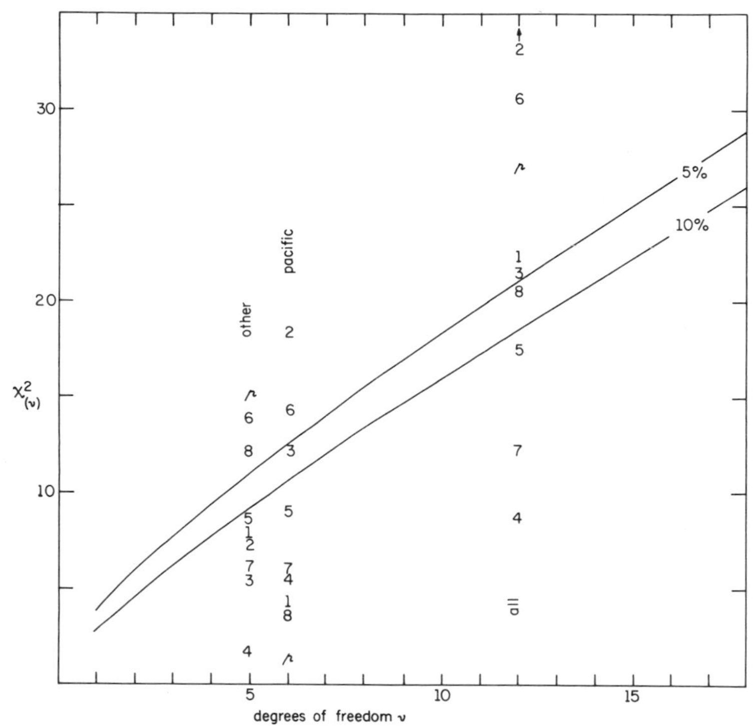

Figure 5. χ^2 test for significance of large-scale lateral variations on *P* diffraction data. χ^2 of decay at frequency points 1,..., 8, and $dT/d\Delta$ indicated "*p*", 5 and 10% significance levels are also given. Null hypothesis assumes no effect of large-scale lateral variations between different lines of stations. The division of all data ("all") in data sampling Pacific region ("pacific") and other regions ("other") corresponds to that in Fig. 4.

variations is significant. With the presently available data, the application of a formal mapping procedure does not yet seem to be warranted. For the *P* data however, a meaningful division in two subsets can be made; of the 13 great circle paths sampling the CMB region for *P*, 7 are beneath the Pacific. Figure 4 shows the result of averaging the data on these 7 paths, as compared to an average over the other 6 and to the global average (all data). The differences thus obtained are systematic, and they are significant. Doornbos and Mondt (1979b) applied a χ^2 test for the significance of large-scale lateral variations. Figure 5 shows the same test also applied to the two subsets of *P*. Clearly, the effect of large-scale variations within the subsets is considerably reduced; note in particular the consistency of $dT/d\Delta$ for the Pacific paths. Tentatively, the *P* data are indicative of somewhat higher than average velocities near CMB beneath the Pacific, although in this case other possibilities like anisotropy cannot be ruled out.

Topography of the core–mantle boundary

Decade fluctuations of several milliseconds in the length of the day (Munk and MacDonald, 1960) have been interpreted as being due to coupling between the mantle and the core (e.g. Rochester, 1970). Of the possible coupling mechanisms, topographic coupling seems to be the most efficient (Hide, 1977). The details of this mechanism and the critical parameters are still poorly

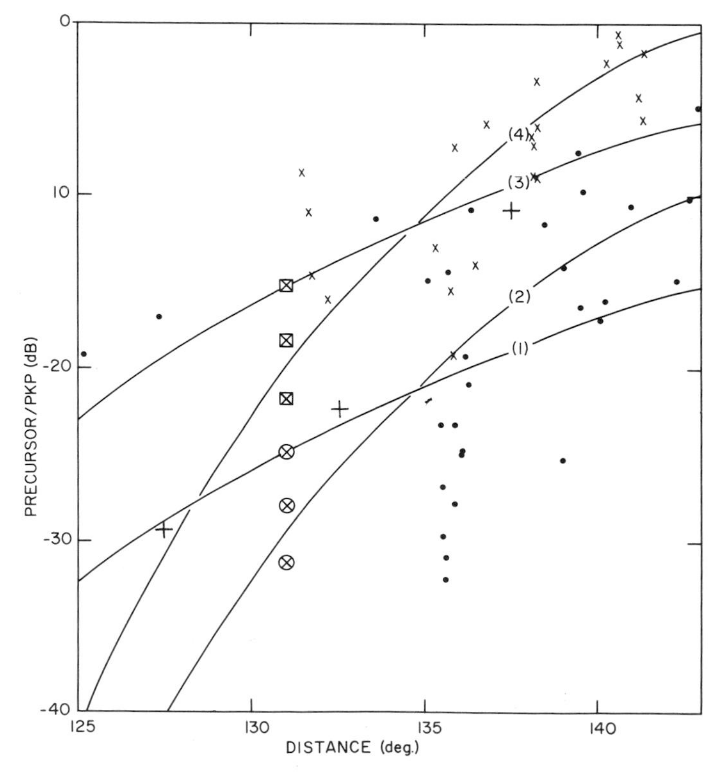

Figure 6. Precursor energy at 1 Hz, relative to *PKP*. ×: NORSAR array data; ●: UKAEA array data; +single station records, averaged over 5 degree intervals.

Curves (1)–(4) represent precursors generated by rough CMB. (1): $\xi = 100$ m, $\sigma = 10$ km; (2): $\xi = 100$ km, $\sigma = 20$ km; (3): $\xi = 300$ m, $\sigma = 10$ km; (4): $\xi = 300$ m, $\sigma = 20$ km. ξ = average amplitude of CMB irregularity; σ = scale length of irregularity. Symbols ⊠ and ⊗ indicate energy reduction due to restricting slowness window of incoming wavefield as in beamforming (after Doornbos, 1978).

known, but rough order of magnitude estimates suggest that undulations less than 1 km would be sufficient.

In principle, a straightforward application of reflection seismology would be to map the boundary by *PcP* and/or *ScS* travel times. In practice, the scatter in these data due to anomalies higher in the mantle and in particular the crust, is so large that meaningful conclusions cannot be reached (see, for example, the *PcP* data of Taggert and Engdahl (1968) and shown by Johnson (1969)). From differential (*PcP–P*) travel times, Engdahl and Johnson (1974) infer an upper bound of 5–10 km on CMB undulations, if all the scatter in the data were to be attributed to this effect. Indeed, there is other evidence, e.g. travel times of *PnKP* (Bolt and Qamar, 1972), suggesting a much smoother boundary, at least in the regions sampled by these data.

In recent years, a different approach to studying the CMB region has been followed. Precursors to certain short-period core phases must be interpreted as scattered waves on or above CMB as mentioned earlier, and the analysis of these waves provides clues as to the scattering structures. We are particularly interested here in characterizing topography of the boundary. It is convenient, and sufficient for the present purpose, to characterize it by an average amplitude and lateral scale length. The procedure in an application to scattering by a rough CMB has been described by Doornbos (1978) in some detail. One result is shown in Fig. 6, which gives relative amplitudes of precursors to *PKP* at frequency 1 Hz, for several assumed amplitudes (100 and 300 m) and scale lengths (10 and 20 km) of topography. Note that the diameter of individual obstacles on the interface would be about twice the scale length used in this figure. Also shown is a compilation of observational data, both from single stations and from arrays. Clearly, most observed amplitudes are explained quite well by a slightly rough CMB, although there are exceptions in particular for some data observed at the NORSAR array. More detailed analysis has revealed that these data are also anomalous in other respects (Doornbos, 1976).

It must be mentioned at this point that, although scattering by topography on the boundary is an attractive interpretation, it is not an unambiguous one since scattering by volume heterogeneity in the mantle above CMB may produce a similar effect. This is illustrated in Fig. 7, where the same data are compared to scattering by density fluctuations above CMB. A more definite conclusion with regard to topography of CMB now appears possible however, since a rough CMB has an observable effect on *PKKP* as well. The effect is due to backscattering $PK \wedge KP$ at CMB, and it may be observed as precursive arrivals because *PKKP* (BC and DF branches) is a maximum time phase. Observations of this type would not be explained as backscattering by volume heterogeneity since this is generally insignificant, whereas backscattering by a rough interface may be equally efficient as forward scattering. Observations of "*PKKP*" and precursors were reported by Chang and Cleary (1978); the data were from Novaya Zemlya explosions observed at the LASA array, at an epicentral distance of about 60°. It was then shown by Doornbos (1980) that precursors to *PKKP* at the NORSAR array are explained by a slightly rough CMB (undulations 100–200 m). The results were summarized in a figure which is reproduced here as Fig. 8. It also shows that the "*PKKP*" and precursors from the Novaya Zemlya explosions are anomalously large, and cannot be

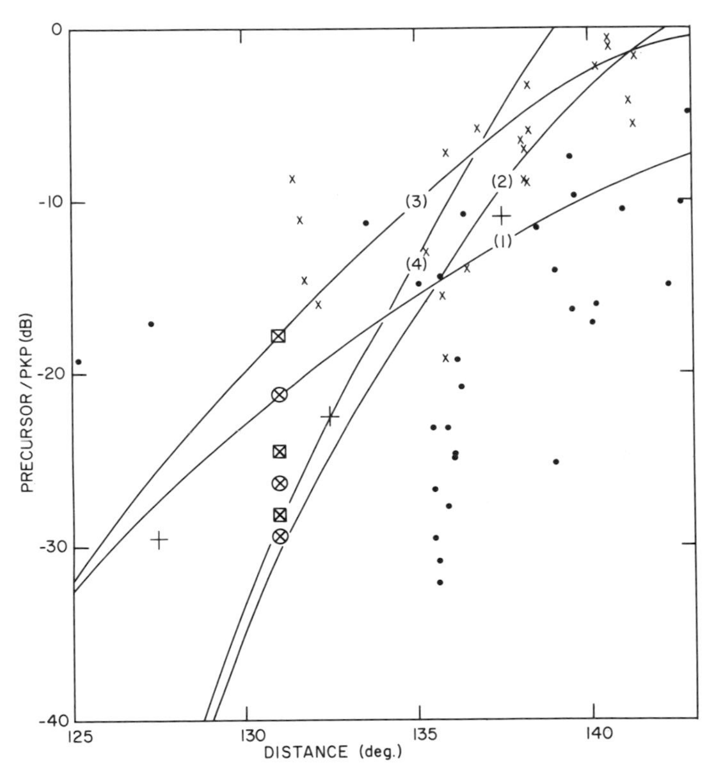

Figure 7. Same data as in Fig. 6, but curves (1)–(4) represent precursor energy generated by lower mantle heterogeneity characterized by an average relative density perturbation of 1% (1): $d = 200$ km, $\sigma = 10$ km; (2): $d = 200$ km, $\sigma = 20$ km; (3): $d = 1000$ km, $\sigma = 10$ km; (4): $d = 1000$ km, $\sigma = 20$ km. d = thickness of scattering layer on top of the core; σ = scale length of heterogeneity. Other details as in Fig. 6.

explained by small undulations. Strong scattering or reflection from a dipping boundary may be considered (results according to the latter possibility are given in the figure); in either case, large topographic features seem to be required in the region sampled by these data, in contrast to small relief in the regions sampled by all other data.

Finally, the results give rise to a few comments: (1) It may appear remarkable that undulations as small as a few hundred metres cause observable scattering. Besides the favourable observational conditions, this is due to the large contrasts in both rigidity (solid–liquid) and density across CMB. Indeed, undulations in solid–liquid interface not accompanied by a large density jump (cf. the inner core boundary) would probably not be observable. (2) These "small" undulations may be significant not only in a seismological, but also in a dynamical context. In case of finite viscosity of the

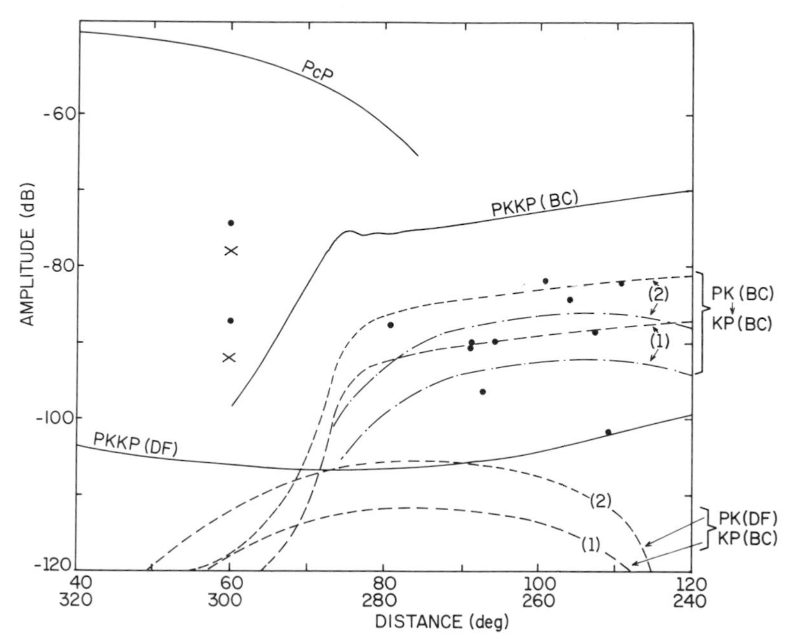

Figure 8. Relative amplitudes of *PKKP* and *PcP* (solid lines, absolute amplitude level arbitrary), and scattering by a rough CMB (dashed lines); the frequency is 1 Hz.

(1) CMB irregularity with average amplitude $\xi = 100$ m. (2) $\xi = 200$ m; ---- maximum precursor energy (~ 1 s before *PKKP*); —·—·— precursor energy 5 s before *PKKP*.

Data at $\Delta > 75°$ are from NORSAR and are taken relative to *PKKP*. Data at $\Delta = 60°$ represent Novaya Zemlya event of 27 October 1973 observed at LASA and are taken relative to *PcP*. $\times$: theoretical amplitudes for reflection at dipping CMB; reflection from BC branch to BC branch (upper $\times$) and from DF branch to BC branch (lower $\times$) (after Doornbos, 1979).

mantle the undulations would tend to smooth out, and for the often quoted viscosity of 10^{22} poise (Cathles, 1975), a very rough estimate of the exponential decay time is 2×10^5 yr (Doornbos, 1980). Since this time is short compared to the time of existence of the core, the conclusion is that these undulations must be dynamically produced, if current estimates of viscosity are correct. In this regard, several possibilities (convection, erosion, freezing) have been considered; they were summarized previously by Jacobs (1975) and also by Hide (1977).

Acknowledgement

The work on diffraction reported in the third section of this paper, was done in collaboration with Dr J. C. Mondt, with whom I have benefited from many discussions on this subject.

References

Aki, K., Christoffersson, A. and Husebye, E. S., 1977. Determination of the three-dimensional seismic structure of the lithosphere. *J. Geophys. Res.* **83**, 277–296.

Alexander, S. S. and Phinney, R. A., 1966. A study of the core–mantle boundary using *P* waves diffracted by the Earth's core. *J. Geophys. Res.* **71**, 5943–5958.

Birch, F., 1952. Elasticity and constitution of the Earth's interior. *J. Geophys. Res.* **57**, 227–286.

Bolt, B. A. and Qamar, A., 1972. Observations of pseudo-aftershocks from underground nuclear explosions. *Phys. Earth Planet. Int.* **5**, 400–402.

Buchbinder, G. G. R., 1971. A velocity structure of the Earth's core. *Bull. Seism. Soc. Am.* **61**, 429–456.

Bullen, K. E., 1975. *The Earth's Density*, Chapman and Hall, London.

Cathles, L. M., 1975. *The Viscosity of the Earth's Mantle*, Princeton University Press, Princeton, NJ.

Chang, A. C. and Cleary, J. R., 1978. Precursors to *PKKP*. *Bull. Seism. Soc. Am.* **68**, 1059–1079.

Cleary, J. R., 1974. The D″ region. *Phys. Earth Planet. Int.* **9**, 13–27.

Dahlen, F. A., 1974. Inference of the lateral heterogeneity of the Earth from the eigenfrequency spectrum: a linear inverse problem. *Geophys. J. R. Astr. Soc.* **38**, 143–167.

Dahlen, F. A., 1976. Models of the lateral heterogeneity of the Earth consistent with eigenfrequency splitting data. *Geophys. J. R. Astr. Soc.* **44**, 77–105.

Davies, G. F., 1977. Whole-mantle convection and plate tectonics. *Geophys. J. R. Astr. Soc.* **49**, 459–486.

Doornbos, D. J., 1976. Characteristics of lower mantle inhomogeneities from scattered waves. *Geophys. J. R. Astr. Soc.* **44**, 447–470.

Doornbos, D. J., 1978. On seismic wave scattering by a rough core–mantle boundary. *Geophys. J. R. Astr. Soc.* **53**, 643–662.

Doornbos, D. J., 1980. The effect of a rough core–mantle boundary on *PKKP*. *Phys. Earth Planet. Int.* **21**, 351–358.

Doornbos, D. J. and Mondt, J. C., 1979a. Attenuation of *P* and *S* waves diffracted around the core. *Geophys. J. R. Astr. Soc.* **57**, 353–379.

Doornbos, D. J. and Mondt, J. C., 1979b. *P* and *S* waves diffracted around the core and the velocity structure at the base of the mantle. *Geophys. J. R. Astr. Soc.* **57**, 381–395.

Dziewonski, A. M., Hales, A. L. and Lapwood, E. R., 1975. Parametrically simple Earth models consistent with geophysical data. *Phys. Earth Planet. Int.* **10**, 12–48.

Dziewonski, A. M., Hager, B. H. and O'Connell, R. J., 1977. Large-scale heterogeneities in the lower mantle. *J. Geophys. Res.* **82**, 239–255.

Elsasser, W. M., Olson, P. and Marsh, B. D., 1979. The depth of mantle convection. *J. Geophys. Res.* **84**, 147–155.

Engdahl, E. R. and Johnson, L. E., 1974. Differential *PcP* travel times and the radius of the core. *Geophys. J. R. Astr. Soc.* **39**, 435–456.

Espinosa, A. F., 1967. *P* in the shadow of the Earth's core. I. *Pure Appl. Geophys.* **67**, 5–14.

Fuchs, K. and Schulz, K., 1976. Tunnelling of low-frequency waves through the subcrustal lithosphere. *J. Geophys.* **42**, 175–190.

Gilbert, F. and Dziewonski, A. M., 1975. An application of normal model theory to the retrieval of structural parameters and source mechanisms from seismic spectra. *Phil. Trans. R. Soc.* **A287**, 187–269.

Haddon, R. A. W., 1972. Corrugations on the core–mantle boundary, or transition layers between inner and outer cores? *Trans. Am. Geophys. Union* **53**, 600.

Heiskanen, W. A. and Vening Meinesz, F. A., 1958. *The Earth and its Gravity Field*, McGraw-Hill, New York.

Hide, R., 1977. Towards a theory of irregular variations in the length of the day and core–mantle coupling. *Phil. Trans. R. Soc.* **A284**, 547–554.

Hide, R. and Malin, S. R. C., 1970. Novel correlations between global features of the Earth's gravitational and magnetic fields. *Nature* **225**, 605–609.

Jacobs, J. A., 1975. *The Earth's Core*, Academic Press, New York.

Jeffreys, H., 1970. *The Earth*, Cambridge University Press.

Johnson, L. R., 1969. Array measurements of *P* velocities in the lower mantle. *Bull. Seism. Soc. Am.* **59**, 973–1008.

Jones, G. M., 1977. Thermal interaction of the core and the mantle and long-term behaviour of the geomagnetic field. *J. Geophys. Res.* **82**, 1703–1709.

Jordan, T. H., 1978. A procedure for estimating lateral variations from low-frequency eigenspectra data. *Geophys. J. R. Astr. Soc.* **52**, 441–445.

Khan, M. A., 1971. Some geophyscial implications of the satellite-determined geogravity field. *Geophys. J. R. Astr. Soc.* **23**, 15–43.

Masters, G., 1979. Observational constraints on the chemical and thermal structure of the Earth's deep interior. *Geophys. J. R. Astr. Soc.* **57**, 507–534.

Mondt, J. C., 1977a. Full wave theory and the structure of the lower mantle. Ph.D. Thesis, Univ. of Utrecht.

Mondt, J. C., 1977b. *SH* waves: theory and observations for epicentral distances greater than 90 degrees. *Phys. Earth Planet. Int.* **15**, 46–59.

Munk, W. H. and MacDonald, G. J. F., 1960. *The Rotation of the Earth*, Cambridge University Press.

O'Connell, R. J., 1977. On the scale of mantle convection. *Tectonophysics* **38**, 119–136.

Oxburgh, E. R. and Turcotte, D. L., 1978. Mechanisms of continental drift. *Rep. Progr. Phys.* **41**, 1249–1312.

Phinney, R. A. and Alexander, S. S., 1966. *P* wave diffraction theory and the structure of the core–mantle boundary. *J. Geophys. Res.* **71**, 5959–5975.

Phinney, R. A. and Alexander, S. S., 1969. The effect of a velocity gradient at the base of the mantle on diffracted *P* waves in the shadow. *J. Geophys. Res.* **74**, 4967–4971.

Richards, P. G., 1973. Calculation of body waves for caustics and tunnelling in core phases. *Geophys. J. R. Astr. Soc.* **35**, 243–264.

Richter, F. and McKenzie, D. P., 1978. Simple plate models of mantle convection. *J. Geophys.* **44**, 441–471.

Rochester, M. G., 1970. Core–mantle interactions: geophysical and astronomical consequences. In *Earthquake Displacement Fields and the Rotation of the Earth* (Eds. Mansinha, L., Smylie, D. E. and Beck, A. E.), 136–148, Reidel, Dordrecht, Holland.

Runcorn, S. K., 1972. Dynamical processes in the deeper mantle. *Tectonophysics* **13**, 623–637.

Sammis, C. G., Smith, J. C., Schubert, G. and Yuen, D. A., 1977. Viscosity–depth profile of the Earth's mantle: effects of polymorphic phase transitions. *J. Geophys. Res.* **82**, 3747–3761.

Scholte, J. G. J., 1956. On seismic waves in a spherical Earth. *Kon. Ned. Meteorol. Inst. Publ.* **65**, 1–55.

Sengupta, M. K., 1975. The structure of the Earth's mantle from body wave observations. Ph.D. Thesis, Massachusetts Institute of Technology, Cambridge, MA.

Sengupta, M. K. and Julian, B. R., 1978. Radial variation of compressional and shear velocities in the Earth's lower mantle. *Geophys. J. R. Astr. Soc.* **54**, 185–219.

Taggert, J. and Engdahl, E. R., 1968. Estimation of *PcP* travel times and the depth to the core. *Bull. Seism. Soc. Am.* **58**, 1293–1303.

Tozer, D. C., 1972. The present thermal state of the terrestrial planets. *Phys. Earth Planet. Int.* **6**, 182–197.

Outgassing and the Power Source for Plate Tectonics

T. GOLD

Center for Radiophysics and Space Research, Cornell University, Ithaca, New York, USA

Summary

The plate movements of the Earth are not likely to be powered simply by a heat engine deriving its energy from the heat flow. Earthquakes alone at the present time are estimated to represent in acoustic energy 0·001 of the total terrestrial heat flow energy. If one now considers that much of the free energy required to drive plates will not appear as acoustic energy of earthquakes, but be merely dissipated in friction, then surely the free energy requirement would go up by a substantial factor. If then one considers that large areas of the Earth are merely losing heat, without generating any motion, one would suppose that all the heat flow in those regions contributes nothing to the heat engine and the free energy supply. Taking these factors into consideration, the heat engine efficiency in the areas where it is operative would have to be in the region of at least several per cent. A heat engine that uses rock as its working fluid is quite unlikely to achieve such efficiencies.

The gravitational sorting of materials represents a source of free energy, and for all that is known could be large enough. Outgassing of the Earth (Gold, 1979; Gold and Soter, 1980) and migration of volatile substances towards the surface represents itself also the direct supply of free energy, but it also provides the possibility of a vapour being the working fluid of a heat engine and thus the possibility of much higher efficiencies than could occur with rocky materials only.

The total outgassing that seems to have taken place over geologic time is quite significant for these considerations. Three hundred atmospheres of water are in the oceans, and a hundred atmospheres of CO_2 in the carbonate rocks. The total outgassing has thus amounted to a few per cent by mass of the entire lithosphere. Both the gravitational sorting and the conversion of heat flow to free energy in these volatiles may well represent a very significant contribution to the powering of the plate movements.

The entire outgassing process of the Earth has received far too little attention in the past. A better understanding of the physics of the process may be relevant, not only to the general powering of plate tectonics, but to the details of the earthquake process. The chemical evolution of the Earth's atmosphere is of course a critical matter in much of the geologic discussion. The origin of fuels, and in particular of methane, may be found to be related to the outgassing process.

References

Gold, T., 1979. Terrestrial sources of carbon and earthquake outgassing. *J. Petrol. Geol.* **1**, 3.
Gold, T. and Soter, S., 1980. The deep Earth gas hypothesis. *Sci. Am.* (June).

Astronomical Evidence of Relationships between Polar Motion, Earth Rotation and Continental Drift

A. POMA, E. PROVERBIO

Istituto di Astronomia dell'Università di Cagliari, Italy

Earth rotation: theoretical approach and astronomical evidence

On the basis of Euler's rigid body motion theory, the Earth should rotate, assuming external forces to be zero, at a constant rate about an axis which, in turn, should undergo a free nutation about a principal axis of inertia (Eulerian polar motion). As a consequence of this, astronomers should observe that the sidereal day remains constant and that a ten-month variation in latitude occurs. However, this is not so. What has been shown through astronomical evidence is that the Earth's rotation is a more complex phenomenon than that suggested by Euler. Both polar motion and length of day exhibit secular, periodic and irregular variations because of several geophysical and meteorological causes, some at present still unexplained (Table I). From a theoretical point of view a departure from rigidity has been necessary and more "realistic" Earth models have been developed. Assuming the Earth to be an axially symmetrical body, the perturbations in the angular velocity vector of the Earth induced by any excitation function **L** is well expressed by the Liouville equation

$$\mathbf{L} = \mathbf{H} + \boldsymbol{\omega} \times \mathbf{H}, \tag{1}$$

where $\mathbf{H} = C\omega_3\,\mathbf{k} + A\omega_2\,\mathbf{j} + A\omega_1\,\mathbf{i}$ is the angular momentum vector relative to the Earth, ω is the angular speed of the terrestrial reference system, and A and C are respectively the equatorial and along-the-axis-of-symmetry moments of inertia, referred to the terrestrial conventional reference system, the origin of which is the so called Conventional International Origin (CIO). The CIO is approximated by the position assumed by the Earth's axis of figure in 1903.

TABLE I *Principal periods in the rotation vector*

Period (approximate) (yr)	Direction (m_1, m_2)	References	Rate (m_3) (ms/day)	References	Explanations
0·5			0·32	(BIH)	Tidal period
1·0	0″·137	(Poma and Proverbio, 1979)	0·43	(BIH)	Annual period
1·2	0″·183	(Poma and Proverbio, 1979)			Chandler period
2·0	0″·006	(Proverbio, 1974)	0·08	(Iijima and Okasaki, 1972)	Atmospheric circulation
9·0	0″·010	(Proverbio, 1974	0·18	(Luo Shi-fang *et al.*, 1977)	Sunspot activity
12·0	0″·006	(Proverbio, 1974)	0·14	(Luo Shi-fang *et al.*, 1977)	Sunspot activity
18·0	0″·007	(Proverbio, 1974)	0·52	(Luo Shi-fang *et al.*, 1977)	Lunar node period
22·0			0·43	(Luo Shi-fang *et al.*, 1977)	Sunspot polarities
30·0	0″·026	(Proverbio, 1974)	0·52	(Luo Shi-fang *et al.*, 1977)	
60·0			1·24	(Luo Shi-fang *et al.*, 1977)	Sunspot activity

Using the notation introduced by Munk and MacDonald (1960)

$$\omega_1 = m_1 \Omega,$$

$$\omega_2 = m_2 \Omega,$$

$$\omega_3 = (1 + m_3)\Omega,$$

where Ω is the mean angular velocity of the Earth, the Liouville Eqn (1) then reduces to the subsidiary equations

$$L_1 = A\dot{m}_1 \Omega - m_2 \Omega^2 (C - A),$$

$$L_2 = A\dot{m}_2 \Omega + m_1 \Omega^2 (C - A),$$

$$L_3 = C\dot{m}_3 \Omega,$$

where for small nutation the products and squares of m_i have been neglected. These equations show that, in general, any excitation source **L** must cause at the same time an irregular change in all the m_i quantities involved; that is, in the length of the day, the Chandler wobble, and the secular drift of the pole. Several theoretical models have been proposed in order to furnish a plausible explanation of the observed fluctuation in the Earth's rotation and polar motion. Among recently published papers, two kinds of solutions involving external meteorological excitation sources (Lambeck and Casenave 1973, 1974; Wilson and Haubrich 1976) or internal forces to the Earth are proposed. Some authors suggest that sudden displacements of the pole of figure relative to the Earth are important in explaining polar wobble (Mansinha and Smylie, 1967), whilst other investigations argue that displacements of the angular momentum vector relative to the Earth are responsible for the Earth's observed fluctuations (Runcorn, 1970). However, the conclusions of these investigations show that neither the contribution of meteorological excitation nor that due to mass displacement of the crust during earthquakes appears to be less than about 25% of the Chandler wobble variance and therefore not able to explain completely the mean and long term fluctuations observed.

It may appear puzzling that in the theoretical and experimental work carried out so far insufficient attention has been given to the fact that both fluctuations in the Earth's rotation and those observed in polar motion may have a common geophysical cause. It is interesting to observe that Runcorn (1968) has pointed out the peculiarity of such a relationship, arguing that irregular changes in the length of the day may arise from the same physical mechanism as polar displacement.

Evidence of a relationship between the amplitude of the free nutation of the Earth and irregular variations in the Earth's rotation has been noticed previously (Stoyko, 1966). An additional relationship between the amplitude of a 0·33 year wave in the Earth's rotation and the angular distance of the instantaneous pole with respect to the mean pole of the epoch was emphasized by Fleer (1973).

During the past twenty years considerable attention has been devoted to another important phenomenon which may in some way simulate an apparent displacement of the Earth's pole as a consequence of station drifting due to absolute plate motion. Although until now no definitive explanation has been made concerning the cause of the plate movement, the widely-accepted theory of plate tectonics suggests the possibility that crustal sheets, or plates, are moving. Theoretical approaches also suggest the existence of driving forces such as that first noticed by Eötvös (1912) or those arising from sudden or secular decelerations of the Earth's rotation (Knopoff and Leeds, 1972) which could cause a whole crust to slide over the upper mantle.

The possible apparent simultaneous displacements of lithospheric plates with an actual or apparent secular motion of the pole have recently been emphasized by Proverbio and Quesada (1973, 1974a, b) starting from latitude and longitude long series data. In our opinion, this result demonstrates that quite probably the irregular and long-term variations observed in polar motion and in the Earth's rotation are caused both by real displacements of the Earth's crust and by change in the angular momentum relative to the Earth. Probably also, to a lesser extent, by sudden displacement of the Earth's pole of figure.

The interesting question now arises as to whether and how it will be possible to separate all these different geophysical effects, and whether it is possible to correlate the latter using a theoretical mechanism by which stress builds up on the rigid crust or on the Earth as a whole.

On this subject it is interesting to note that recent investigations have shown that differential mass displacements from lithospheric plate motion give no significant contribution to the displacement of the Earth's axis of rotation and to the resulting polar motion observed (Han-Shon Liu, Carpenter and Agreen, 1974; Soler and Mueller, 1978). Another important consideration is that the same changes in mass distribution at the surface of the Earth cause only second-order effects on the change of the length of day (De Sitter, 1927). Bearing in mind all these considerations, we are led to think that if an actual polar displacement exists (caused by impulsive excitation displacement of the angular momentum vector or by changes of the pole of figure relative to the Earth) a relationship should be discovered between changes in the *direction* of the polar axis and variations in the *rate* of the Earth's rotation.

These real variations in the rotation vector are probably superimposed on tectonic movements of the crust on continental and global scales. The existence of these movements can be detected by the study of relative motions between different plates carried out by analysing the components m_1 and m_2 of the rotation vector and correlating these variations with possible variations of the m_3 component of the same rotation vector. As we have already mentioned, recent work carried out in this direction seems to lead to fairly convincing conclusions as to the existence of tectonic movements on a global scale (Proverbio and Poma, 1975; Soler and Mueller, 1978). In our opinion the most exciting problem is now that of trying to prove or disprove the existence of a connection at the theoretical and observational levels between the change observed in the direction of the axis and variations in the length of day.

The classical astronomical data

Polar motion and Earth rotation are determined by classical astronomical measurements of latitude and time and, more recently, also by techniques based on satellite observations (Laser and Doppler methods) and radio interferometric methods (Very Long Base-line Interferometry).

The direction of the instantaneous rotation vector axis given by m_1 and m_2, is usually referred to a right-handed reference system fixed on the Earth: the x_1 axis along the Greenwich meridian, the x_2 axis along 90° East. The x_3 axis of the system is nearly along the axis of figure pointing to some reference point (CIO).

From an operative point of view, this Earth-fixed frame is defined by the conventional values of the terrestrial coordinates (namely latitude φ_{oi} and longitude λ_{oi}) of a set of observatories. Since latitude is the angle that the direction of the observatory's zenith makes with the direction of the rotation axis, if the latter changes so will latitude and one can observe the variation

$$\Delta\varphi_i = \varphi_i - \varphi_{oi} = m_1 \cos \lambda_{oi} - m_2 \sin \lambda_{oi},$$

from which m_1 and m_2 can be derived by the least-squares method. Just 80 years ago, a chain of observatories situated on the 39° 08′ parallel, the International Latitude Service (ILS) was organized for the purpose of determining how latitudes vary. (At present five stations are operating; in Carloforte (Italy), Kitab (USSR), Mizusawa (Japan), Gaithesburg and Ukiah (USA).

The series of coordinates $x = m_1$ and $y = -m_2$ obtained from ILS results since 1899 until now, is indicated here as the ILS series. For its length and homogeneity it represents the principal source for research on polar motion, especially concerning secular and long period terms. In spite of these advantages, however, the ILS series appears to be affected by various kinds of errors. In 1962, therefore, following the XIth General Assembly of the International Astronomical Union, a new service, the International Polar Motion Service (IPMS), was established.

The IPMS has continued the series of polar coordinates not only from ILS results but also utilizing an independent set of observatories and instruments (at present more than 75). Also, the Bureau International de l'Heure (BIH), starting from 1956, publish a series of polar coordinates derived from both time and latitude observations carried out by a large number of instruments operating in the collaborating stations. In 1968 all the results were referred to a homogeneous system, the 1968 BIH System (Guinot, Feissel and Laclare, 1970).

By using the same observations the BIH also provides the values of the astronomical scale of Universal Time (UT) referred to the international atomic time scale denoted by IAT. Since the IAT can be supposed uniform with good approximation, the differences between UT and IAT are ascribed to variations in the rotation of the Earth. The values published are UT1 and UT2 (Universal Time corrected for changes in direction of the Earth's axis and corrections to UT1 arising from seasonal (annual and semiannual) variations respectively).

TABLE II *International Services monitoring polar motion*

	ILS	IPMS	BIH
No. observing stations	5	47[a]	54[b]
No. instruments	5	70[a]	77[b]
Type of instrument	VZT	VZT, PZT, astrolabe transit instrument	VZT, PZT, astrolabe transit instrument
Period	1900–1979	1962–1979	1956–1979

[a] IPMS Annual Report for 1975.
[b] BIH Annual Report for 1977.

The principal characteristics of ILS, IPMS and BIH series are summarized in Table II. Some remarks, however, are needed. First, the three series are independent. Both IPMS and BIH also use the results of the five ILS stations, but their weight on the global results is small. Secondly, the reference point for the x_3 axis is defined by the coordinates, and the weights attributed to the observing stations; nominally ILS, IPMS and BIH refer to the Conventional International Origin but, for example, the origin of the BIH system differs from the CIO (Poma and Proverbio, 1976). Finally, secular, periodic and irregular differences between the various reference systems may exist; they could arise from instrumental and observational errors, but also from possible zenith displacements of the observatories lying on different plates.

Relationship between polar motion and Earth rotation

In order to detect the existence of any relationship between changes in the speed of rotation and secular and irregular variations in the polar motion, we compare the observed quantity $m_3(t)$ (i.e. the deviation of the spin from uniformity) with variations in the polar motion components $m_1(t)$ and $m_2(t)$ after removing the Chandler and annual wobbles.

Figure 1 shows the quarterly values of m_3, expressed in ms/day for the years 1955·75–1977·50. Values are obtained from BIH data for UT2 – IAT by means of the first difference

$$m_3 = \Delta(\mathrm{UT2} - \mathrm{IAT})/\Delta t.$$

Since the quarterly mean of published data were considered, the nominal value of Δt is 0·25 years. The diagram shows some irregular long-term variations superimposed upon which are accelerations of varying magnitude and frequency. Some have been partially explained, while others remain a puzzle.

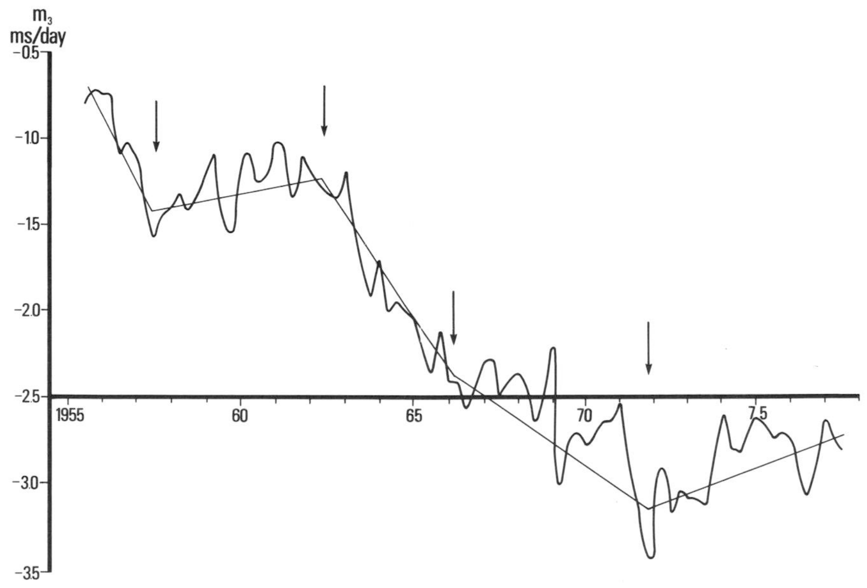

Figure 1. Variations in the rate of the Earth's rotation (from BIH data for UT2-TAI).

Sudden irregular changes do not appear to take place. Following Brouwer's suggestion (1952), Markowitz (1970) pointed out that such sudden changes do not occur in velocity but in acceleration. That can easily be seen in Fig. 1, and Fig. 2 where quarterly values of acceleration given by $dm_3/dt = \Delta^2(\mathrm{UT2} - \mathrm{IAT})/\Delta t^2$ are plotted. Because, as emphasized by Markowitz (1970), a span of three months or longer in present astronomical data can detect the speed of rotation of the Earth with the required accuracy, we can presume the changes observed to be real.

The values of observed long-term rotational acceleration related to different periods are given in Table III. We can ascribe all these variations (irregular or periodic) in speed of rotation to core–mantle coupling or to change in moment of inertia.

We now investigate the three series of polar coordinates $x = m_1$ and $y = -m_2$ supplied for every 1/20th year by ILS and BIH from 1956 to 1977 and by IPMS (1962–1977). To derive mean pole positions it is customary to use numerical filtering methods (for instance six-year running means) but,

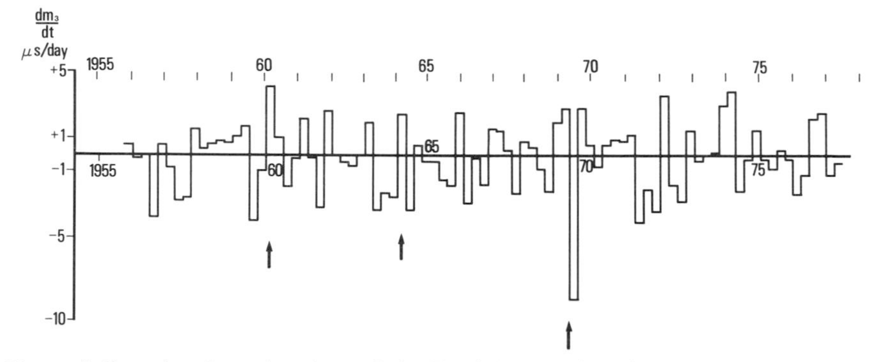

Figure 2. Rotational accelerations of the Earth (quarterly values).

TABLE III *Irregular long term variations in the Earth's acceleration*

Interval	$\dot{m}_3$ (10^{-10}/yr)
1955·5–1957·2	−39·7
1957·2–1962·4	+3·6
1962·4–1966·2	−34·8
1966·2–1971·9	−15·5
1971·9–1977·5	+8·1

because the available interval is too short we prefer to subtract a Chandlerian mean and a mean annual wobble.

By least squares spectral analyses we found (Poma and Proverbio, 1979) for the Chandler frequency f_c the value 0·845 (432 days) and the amplitudes

$$x_c = 0{\cdot}126 \sin(2\pi f_c t + 346^\circ),$$

$$y_c = 0{\cdot}133 \sin(2\pi f_c t + 76^\circ),$$

and an approximate annual frequency $f_a = 0{\cdot}993$ (368 days) with amplitudes

$$x_a = 0{\cdot}103 \sin(2\pi f_a t + 209^\circ),$$

$$y_a = 0{\cdot}091 \sin(2\pi f_a t + 302^\circ),$$

both phases corresponding to 1962·0.

By removing $(x, y)_c$ and $(x, y)_a$ from the above data, we computed a new series from which evidence should be obtainable regarding:

(1) progressive motion;

(2) residuals arising from changes in the amplitude of the Chandler motion;

(3) irregular fluctuations.

The computed values are grouped in Fig. 3. At first sight there is a similar trend for all three series, but it is also interesting to note that apparently no random differences occur. We shall discuss this point later.

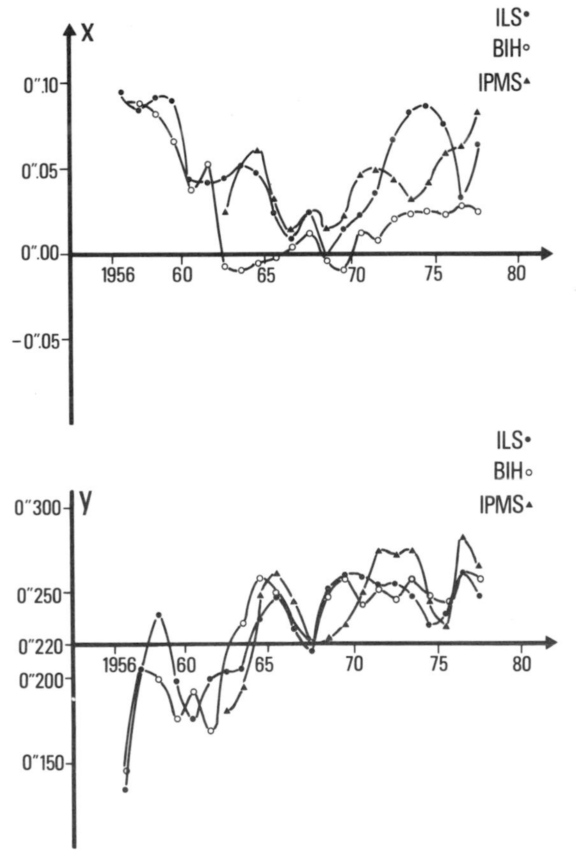

Figure 3. Coordinates of the pole derived from ILS, IPMS and BIH data after removal of a mean annual and Chandler term.

Finally, the above computed values m_1, m_2, m_3 are compared. For each components m_i the BIH data have been used, and the annual means are computed and plotted in Fig. 4. The comparison clearly indicated that changes in polar motion are linked to variations in the rate of the Earth's rotation. It is interesting to note that, disregarding irregular fluctuations, sharp variations in the trend of quantities m_1 and m_2 evidently correspond to the epoch in which a sudden change in acceleration occurs.

Taking into account that also the variations in m_1 and m_2 are in good agreement, a fairly plausible explanation is that the changes we emphasized in polar motion are due to changes in the amplitude of the Chandler wobble. Guinot (1972) has discussed these variations and found that sudden changes occur in amplitude without accompanying sudden changes in the mean pole. Moreover, the curve of his results for the years 1956–1970 shows very interesting agreement with that shown in Fig. 4 for m_1 and m_2 (similar rapid changes nearly up to 1960, 1962 and 1967).

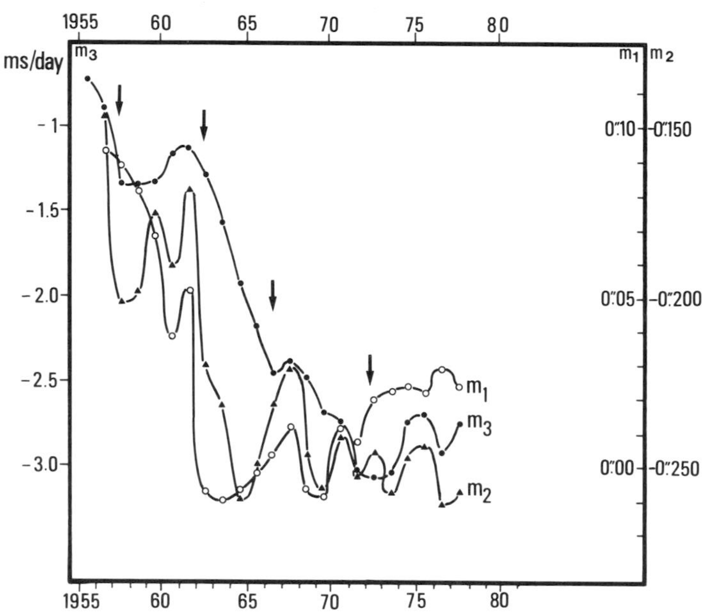

Figure 4. Comparison between fluctuations in the direction (m_1 and m_2) and in rate (m_3) of the rotation vector.

From these considerations it follows that the already-mentioned hypothesis (Runcorn, 1968) that the existence of torques asymmetrical with respect to the equatorial plane and thus capable of both influencing the Earth's rotation and also causing the axis of angular momentum (and as a consequence also Chandlerian motion) to vary, may be a possible and correct explanation of our results.

In our opinion, however, assuming that the barycentre of the different sets of stations of BIH, IPMS and ILS will lie on different plates, even the hypothesis of displacements of large plates of the Earth's surface indirectly caused by the same force responsible for the variations in the Earth's rotation cannot be disregarded. As already mentioned, and as shown in Figs 5 and 6, the differences between the coordinates of polar motion obtained from different sets of observatories appear to be responsible for long period terms which thus far have received insufficient attention. (In particular a period of about 8 years seems to be present.)

By spectral analysis we found that there is the same period in the acceleration of the Earth. Similar results have been found by O'Hora (1975). Moreover, there seems to be some connection between changes in acceleration and the relative variations between different plates shown in Figs 5 and 6. In addition to these periodical relative movements it is interesting to observe that changes in relative velocity of movement seems to correspond with the epoch in which abrupt variations in the acceleration of the Earth's motion are observed (see Figs 2, 5 and 6).

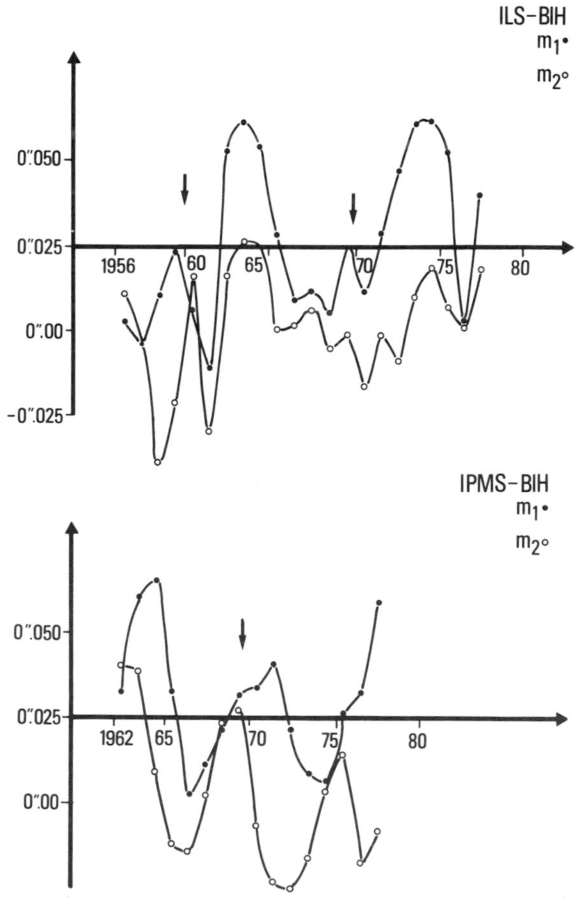

Figure 5. Differences between ILS and BIH coordinates (top); differences between IPMS and BIH coordinates (bottom)

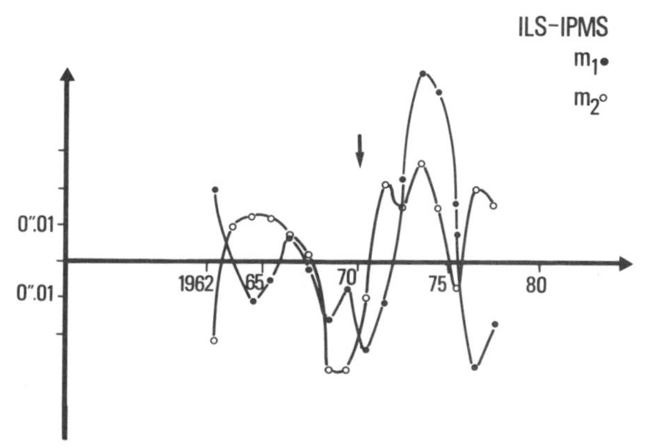

Figure 6. Differences between IPMS and ILS coordinates.

Although our data are preliminary and the period examined is perhaps insufficient, investigations on this subject appear to give interesting results. Further research is in progress.

Conclusions

It has been found that a significant relationship exists between the change in the rate of the Earth's rotation and the variation in polar motion. To a large extent these common fluctuations can be considered as real and attributed to the same physical cause originating non-axially symmetrical forces. However, the phenomenon of the Earth's rotation is so complex that phenomena which may contribute to real or apparent fluctuations in all the components of the rotation vector cannot be excluded.

The same results suggest that relative displacements between different lithospheric plates are evinced in connection with sudden change in the acceleration of the Earth. Torques applied over large regions by Earth's change in velocity or acceleration could explain in part the lithospheric motions observed.

References

Brouwer, D., 1952. *Astron. J.* **57**, 125–146.
De Sitter, 1927. *Bull. Astron. Inst. Neth.* **4**, 21–38.
Eötvös, R., 1912. *Verh. 17 Konf. Int. Eerdmessung*, 111.
Fleer, A. G., 1973. *Soviet Astron.* **16**, 1046–1047.
Guinot, B., 1972. In *Rotation of the Earth* IAU Symp. 48 (Eds Melchior, P. and Yumi, S.), Reidel, Dordrecht, Holland.
Guinot, B., Feissel, M. and Laclare, F., 1970, Rapport Annual pour 1969, BIH, Paris.
Han-Shon Liu, Carpenter, L. and Agreen, L. W., 1974. *J. Geophys. Res.* **79**, 4379–4382.
Iijima, S. and Okasaki, S., 1972. *Publ. Astr. Soc. Japan* **24**(1), 109–125.
Knopoff, L. and Leeds, A., 1972. *Nature* **237**, 93–94.
Lambeck, K. and Casenave, A., 1973. *Geophys. J. R. Astr. Soc.* **32**, 79–93.
Lambeck, K. and Casenave, A., 1974. *Geophys. J. R. Astr. Soc.* **38**, 49–61.
Luo Shi-fang, Lian Shi-guang, Ye Shu-hua, Yan Shao-zhang and Li Yuan-xi, 1977. *Chinese Astronomy* **1**, 221–227.
Mansinha, L. and Smylie, D. E., 1967. *J. Geophys. Res.* **72**, 4731–4743.
Markowitz, W. M., 1970. In *Earthquake Displacement Fields and the Rotation of the Earth*, (Eds Mansinha, L. *et al.*) Reidel, Dordrech, Holland.
Munk, W. H., and MacDonald, G. J. F., 1960. *The Rotation of the Earth*, Cambridge University Press.
O'Hora, N. P. J., 1975. In *Growth Rhythms and the History of the Earth's Rotation* (Eds Rosenberg, G. D. and Runcorn, S. K.), Wiley, New York.
Poma, A. and Proverbio, E., 1976. *Astron. Astrophys.* **47**, 105–111.
Poma, A. and Proverbio, E., 1979. In *Time and Earth's Rotation*, IAU Symp. 84, (Eds McCarthy, D. D. and Pilkington), Reidel, Dordrecht, Holland.
Proverbio, E. and Poma, A., 1975. In *Growth Rhythms and the History of the Earth's Rotation* (Eds Rosenberg, G. D. and Runcorn, S. K.), Wiley, New York.
Proverbio, E. and Quesada, V., 1973. *Bull. Geod.* **109**, 281–291.

Proverbio, E. and Quesada, V., 1974a. *Bull. Geod.* **112**, 187–212.
Proverbio, E. and Quesada, V., 1974b. *J. Geophys. Res.* **79**, 4941–4943.
Proverbio, E. and Uras, S., 1974. In Proceedings Symp. Space Astrometry, ESRO SP-108, 83.
Runcorn, S. K., 1968. In *Continental Drift, Secular Motion of the Pole and Rotation of the Earth*, (Eds Markowitz, W. M. and Guinot, G.), *Reidel*, Dordrecht, Holland.
Runcorn, S. K., 1970. In *Earthquake Displacement and the Rotation of the Earth* (Eds Mansinha, L. *et al.*) Reidel, Dordrecht, Holland.
Soler, T. and Mueller, I. I., 1978. *Bull. Geod.* **59**, 39–57.
Stoyko, A., 1966. *C. r. Acad. Sci. Paris* **262**, 1098–1100.
Wilson, C. R. and Haubrich, R. A., 1976. *Geophys. J. R. Astr. Soc.* **46**, 707–744.

Subject Index